Recent Trends in Surface and Colloid Science

Statistical Science and Interdisciplinary Research

Series Editor: Sankar K. Pal *(Indian Statistical Institute)*

Description:
In conjunction with the Platinum Jubilee celebrations of the Indian Statistical Institute, a series of books will be produced to cover various topics, such as Statistics and Mathematics, Computer Science, Machine Intelligence, Econometrics, other Physical Sciences, and Social and Natural Sciences. This series of edited volumes in the mentioned disciplines culminate mostly out of significant events — conferences, workshops and lectures — held at the ten branches and centers of ISI to commemorate the long history of the institute.

Vol. 4 Advances in Multivariate Statistical Methods
 edited by A. SenGupta *(Indian Statistical Institute, India)*

Vol. 5 New and Enduring Themes in Development Economics
 edited by B. Dutta, T. Ray & E. Somanathan
 (Indian Statistical Institute, India)

Vol. 6 Modeling, Computation and Optimization
 edited by S. K. Neogy, A. K. Das and R. B. Bapat
 (Indian Statistical Institute, India)

Vol. 7 Perspectives in Mathematical Sciences I: Probability and Statistics
 edited by N. S. N. Sastry, T. S. S. R. K. Rao, M. Delampady and
 B. Rajeev *(Indian Statistical Institute, India)*

Vol. 8 Perspectives in Mathematical Sciences II: Pure Mathematics
 edited by N. S. N. Sastry, T. S. S. R. K. Rao, M. Delampady and
 B. Rajeev *(Indian Statistical Institute, India)*

Vol. 9 Recent Developments in Theoretical Physics
 edited by S. Ghosh and G. Kar *(Indian Statistical Institute, India)*

Vol. 10 Multimedia Information Extraction and Digital Heritage Preservation
 edited by U. M. Munshi *(Indian Institute of Public Administration,
 India)* & B. B. Chaudhuri *(Indian Statistical Institute, India)*

Vol. 11 Machine Interpretation of Patterns: Image Analysis and Data Mining
 edited by R. K. De, D. P. Mandal and A. Ghosh
 (Indian Statistical Institute, India)

Vol. 12 Recent Trends in Surface and Colloid Science
 edited by Bidyut K. Paul *(Indian Statistical Institute, India)*

Platinum Jubilee Series

Statistical Science and
Interdisciplinary Research — Vol. 12

Recent Trends in Surface and Colloid Science

Editor

Bidyut K. Paul
Indian Statistical Institute, India

Guest Editor

Satya P. Moulik
Centre for Surface Science, India

Series Editor: **Sankar K. Pal**

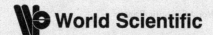

NEW JERSEY · LONDON · SINGAPORE · BEIJING · SHANGHAI · HONG KONG · TAIPEI · CHENNAI

Published by

World Scientific Publishing Co. Pte. Ltd.
5 Toh Tuck Link, Singapore 596224
USA office: 27 Warren Street, Suite 401-402, Hackensack, NJ 07601
UK office: 57 Shelton Street, Covent Garden, London WC2H 9HE

British Library Cataloguing-in-Publication Data
A catalogue record for this book is available from the British Library.

RECENT TRENDS IN SURFACE AND COLLOID SCIENCE
Statistical Science and Interdisciplinary Research — Vol. 12

Copyright © 2012 by World Scientific Publishing Co. Pte. Ltd.

All rights reserved. This book, or parts thereof, may not be reproduced in any form or by any means, electronic or mechanical, including photocopying, recording or any information storage and retrieval system now known or to be invented, without written permission from the Publisher.

For photocopying of material in this volume, please pay a copying fee through the Copyright Clearance Center, Inc., 222 Rosewood Drive, Danvers, MA 01923, USA. In this case permission to photocopy is not required from the publisher.

ISBN-13 978-981-4299-41-1
ISBN-10 981-4299-41-3

Printed in Singapore by World Scientific Printers.

Foreword

The Indian Statistical Institute (ISI) was established on 17th December, 1931 by a great visionary Professor Prasanta Chandra Mahalanobis to promote research in the theory and applications of statistics as a new scientific discipline in India. In 1959, Pandit Jawaharlal Nehru, the then Prime Minister of India introduced the ISI Act in parliament and designated it as an *Institution of National Importance* because of its remarkable achievements in statistical work as well as its contribution to economic planning.

Today, the Indian Statistical Institute occupies a prestigious position in the academic firmament. It has been a haven for bright and talented academics working in a number of disciplines. Its research faculty has done India proud in the arenas of Statistics, Mathematics, Economics, Computer Science, among others. Over eighty years, it has grown into a massive banyan tree, like the institute emblem. The Institute now serves the nation as a unified and monolithic organization from different places, namely Kolkata, the headquarters, Delhi, Bangalore, Chennai and Tezpur four centers, a network of five SQC-OR Units located at Mumbai, Pune, Baroda, Hyderabad and Coimbatore, and a branch (field station) at Giridih.

The platinum jubilee celebrations of ISI have been launched by Honorable Prime Minister Dr. Manmohan Singh on December 24, 2006, and the Government of India has declared 29th June as the "Statistics Day" to commemorate the birthday of Professor Mahalanobis nationally.

Professor Mahalanobis, was a great believer in interdisciplinary research, because he thought that this will promote the development of not only Statistics, but also the other natural and social sciences. To promote interdisciplinary research, major strides were made in the areas of computer science, statistical quality control, economics, biological and social sciences, physical and earth sciences.

The Institute's motto of 'unity in diversity' has been the guiding principle of all its activities since its inception. It highlights the unifying role of statistics in relation to various scientific activities.

In tune with this hallowed tradition, a comprehensive academic programme, involving Nobel Laureates, Fellows of the Royal Society, and other dignitaries, has been implemented throughout the Platinum Jubilee year, highlighting the emerging areas of ongoing frontline research in its various scientific divisions, centres, and outlying units. It includes internatinal and national-level seminars, symposia, conferences and workshops, as well as series of special lectures. As an outcome of these events, the Institute is bringing out a series of comprehensive volumes in different subjects under the title *Statistical Science and Interdisciplinary Research*, published by World Scientific Publishing, Singapore.

The present volume titled *Recent Trends in Surface and Colloid Science* is the twelfth one in the series. It has twenty two chapters, written by eminent scintists from different parts of the world, dealing with various aspects of the subjects, for example, modern concept on colloidal stability, mixed micellar and mixed protein-surfactant systems, microemulsion and related systems with special reference to fundamental phenomena and wide applications in nanoscience and nanotechnology, drug delivery, reaction medium, biocatalysis and enzymatic reactions, photochemistry. The volume provides a state-of-the-art in significant developments and advances that have been made in the field of surface and colloid science over the past decades in a comprehensive manner.

The volume will serve as a fountain-head for further exciting developments in this field, and will be useful to students, researchers and practitioners.

Thanks to the contributors for their excellent research contributions and to volume editors, Dr. Bidyut K. Paul and Professor S.P. Moulik (guest editor) for their sincere effort in bringing out the volume nicely. Thanks are also due to World Scientific for their initiative in publishing the series and being a part of the Platinum Jubilee endeavor of the Institute.

February, 2012
Kolkata

Sankar K. Pal
Series Editor and
Former Director, ISI

Preface

Colloid and surface science is a fascinating interdisciplinary field in science, where modern development and knowledge of physics, chemistry, biology, material science, pharmacy, engineering etc. have been extensively exploited and adopted, with ample scope of fundamental research and extensive potentials for applications. The progress of research in this important field has been remarkable throughout the world during the last four decades, and it has developed for the benefit of the society and people all over the world. This volume contains extended version of some articles presented at *International Symposium on Recent Trends in Surface and Colloid Science* (ISSCS), in addition to the contributions made by some of the members of the International Advisory Committee of this International Symposium. With a summary of recent advances in this multifaceted field, the chapters of this volume provide critical information to enrich the literature by presenting the basic concepts of organized systems in relation to their practical significance. It would be a potential addition to the literature and also should be of interest to beginners as well as experts working in the field of colloid and surface science.

In the first chapter, Kunz and Boström have proposed an efficient but physically well-based modified DLVO theory considering specific ion effects around water profiles near colloidal surfaces and molecular dynamics simulation. The new appraoch is expected to be useful to explain specific ion adsorption on electrode surfaces. In the second chapter, electrokinetic theories in suspensions of soft particles (particle surfaces containing ion-penetrable polyelectrolyte layers) have been critically reviewed by Ohshima. In the reviewing process, electrokinetic phenomena viz., electrophoresis, electroosmosis, streaming potential and sedimentation potential have been summarized. In the subsequent chapter, phase diagram analysis of an evaporating emulsion comprising water, linalool and a commercial surfactant, Laureuth 4 has been presented by Friberg and Aikens. Based on an equilibrium between the vapor and emulsion, variations in the fractions of compounds and phases have been determined. The process has been found to be influenced by the relative humidity. In Chapter 4, Thaker, Sengupta and Sengupta discussed on the stability of an emulsion system containing NaOH with aging. The presence of NaOH has been found to influence both viscosity and salinity of the system. Increased internal phase volume has affected the viscosity but not the emulsion salinity. Aging has influenced the droplet size distribution. In Chapter 5, the work of Moulik and Mitra entails enthalpy of micellization determined by van't Hoff procedure and by direct calorimetry on a number of ionic and nonionic surfactant systems. The results do not agree, the discripencies are more for ionic systems than nonionics. This anomaly arises because of the basic differences between the two evaluation methods; while the van't Hoff procedure is a differential approach, the calorimetry is an integral approach. The first method evaluates only the enthalpy of micellization whereas the second measures the total heat produced in the system, micellization or otherwise. In the next chapter, the salt free binary cataionic system of hexadecyltrimethylammonium octylsulfonate has evidenced a miscibility gap for the formal lamellar phase, as presented by Silva, Marques and Olsson. Coexistance of a swollen and a collapsed lamellar phase has been observed in a two-phase region with linear swelling in each phase. Increase of temperature has caused a vesicle-to-micelle transition of the system. A theoretical cell model has been proposed to provide insight to the observed miscibility gap in the system. In Chapter 7, Miller and coworkers have proposed models for quantitative analysis of the adsorption layer formed from mixed solutions comprising proteins and surfactants for understanding the hydrodynamics, adsorption kinetics and dilational rheology. The procedure used was drop shape analysis using a capillary pressure tensiometer. The findings are encouraging and needs analysis by way of improved theories and better experimentations. Factors that affect mixed aggregation of surfactants have been summarized by Schulz in Chapter 8. The interaction parameter, β in regular solution theory has been separated into β_{ph} (interaction between polar head groups) and β_{core} (interaction of amphiphile tails in the micellar core); thus $\beta = \beta_{ph} + \beta_{core}$. The role of various factors like differences in chain lengths, steric constraints, presence of double bond on the β has been also discussed. In Chapter 9 Chanda, Singh and Ismail have focused on the micellization of the surfactant AOT in the presence of both sodium chloride (NaCl) and sodium salicylate (NaSl). The counter-ion binding in 0.02 mol dm^{-3} NaCl has been found to increase two fold which did not occur in the presence of NaSl. It is considered that at 0.02 mol dm^{-3} NaCl both the aggregation number and micellar polarity of AOT suddenly change. The shape of the micelle has also been expected to change. In the subsequent chapter, the results on clouding and solubilization of the amphiphilic drug, promazine hydrochloride in presence of additives, viz., alkanols, sugars and amino acids have been presented by Kabir-Ud-Din and coworkers. The results have been analyzed in the light of drug aggregation and modification of the physico-chemistry of the medium by the additive effects. The influence of the amide urea on surfactant self-assembly in solution has been reviewed by Souza, Alvarez and Politi in Chapter 11. An attempt has been made to rationalize this effect in the light of electrostatic interaction as well as hydrophobic repulsion wherein the increase in the dielectric permittivity by urea has a role to play. In Chapter 12, Romsted presented a model for the role of ion-pair formation at the micellar interface by way of counterion binding in the sphere-to-rod transition of ionic micelles and potentially other structural transitions of different types of association colloids. The merits and demerits of the model has been presented in view of its future applications. The review article of Xenakis (Chapter 13) describes uses of microemulsions and related systems to immobilize lipase and causing favourable catalytic esterification of fatty acids. The catalytic activity of lipase remains effective when the external organic solvent is replaced by supercritical CO_2. The stability of lipase and its reusability have been assessed and presented. The review article of Gupta comprises an overview of different types of nanoscale colloidal drug delivery systems in Chapter 14. The advantages and disadvantages of these systems have been highlighted with relevant examples. Nanoscale self-assembly of gelatin molecules and clusters spread on hydrophilic surface has been discussed by Bohidar and Gupta using AFM technique in Chapter 15. The nano-structured assemblies formed compact fractal objects resulting from a diffusion limited aggregation process. Koetz and coworkers have reported on water soluble polymers which can modify w/o microemulsions and used as templates for the synthesis of nanoparticles. The system poly (diallydimethyl ammonium chloride, PDADMAC) incorporated into inverse microemulsion (water / toluene / heptanol / sulfobetaine) droplets can be successfully used as template for the formation of spherical $BaSO_4$ nanoparticles in Chapter 16. Addition of nonionic polymer, polythylene glycol in water / toluene / SDS / polyampholytes, can produce nanorods of $BaSO_4$ in the template. Uptake of nickel oxide in water / AOT / isoctane w/o microemulsion has been discussed by Nassar and Husein in Chapter 17. Nanoparticle uptake increased with increasing AOT and precursor salt concentration. Maximum uptake occured at water to AOT mole ratio, $R = 3.0$. The particle size increased with increasing [AOT] and [precursor salt] due to intermicellar nucleation and growth. Synthesis and size dependent photocatalytic behavior of luminescent semiconductor quantum dots, viz., CdS and CdTe have been presented and discussed in Chapter 18 by Saha and coworkers. The methods of preparation and catalytic efficacies of the materials have been presented with physicochemical justifications. Possible development of fluorescence-based nonoparticle sensors has been discussed. The article of Das, Sarkar and Chattopadhyay deals with a fluorometric strategy to enhance the efficiency of a quencnching-based cationic fluorosensor for Cu^{2+} by several order of magnitude in Chapter 19. This strategy is considered to be applicable for physiochemical fluids for detecting trace metal ions like Cu^{2+}. According to Mejuto and coworkers, the nanodroplets in microemulsion can act as scalable microreactors in which synthesis of materials and dynamics of reactions, can be studied (Chapter 20). This chapter deals with pseudophase approach to study the transnitrosation reactions in microemulsion media. The role of interface between water and oil has been proposed to be vital. In Chapter 21, Sinha and Tarafdar have focused on the introduction of electric field for the formation of gel of synthetic clay, laponite. Rapid gel formation at the anode has been observed, which is slow in the absence of the electric field. The structure and properties of these gels are reported to be different in the two conditions. The final chapter by (Nandi) Ganguly and coworkers describes molecular association (dimerization etc.) of cinnamic acid molecules that exists in the crystalline form by intermolecular hydrogen bonding. Positron annihilation spectroscopic and DSC studies with varying temperature, have been conducted to understand the changes in the molecular structure in the crystalline phase.

At the end, our pleasant task of thanking those who helped in many and different ways to bring this volume to fruition. First of all we express our heart felt gratitude towards the contributors, who are internationally renowned researchers, of this volume. In total, 62 individuals from 11 countries contributed to this volume. We are also grateful to members of the organizing committee for their constant encouragement. Our special thanks are due to the reviewers for their valuable comments, as peer-review is a requirement to preserve the highest standard of publication. We also express our gratitude to the Indian Statistical Institute Platinum Jubilee Core Committee and the series editor for giving us opportunity to edit this volume. Finally, the help of Mr. Dibyendu Bose to prepare the camera-ready version is sincerely acknowledged with thanks and appreciation. Our appreciation goes to both Rhaimie B. Wahap and Hwee Yun Tan for their excellent job done on the production of this edited volume.

Bidyut Kumar Paul
Editor

Satya Priya Moulik
Guest Editor

Contents

Foreword   v

Preface   vii

1. Specific Ion Effects in Colloid and Surface Science: A Modified DLVO Approach   1
   Werner Kunz and Mathias Boström

2. Electrokinetics in a Suspension of Soft Particles   11
   Hiroyuki Ohshima

3. Relative Humidity and Evaporation of a Simple Fragrance Emulsion   25
   Stig E. Friberg and Patricia A. Aikens

4. Aging and Stability of W/O Emulsions with NaOH in Aqueous Phase   39
   Rujuta Thaker, Bina Sengupta and Ranjan Sengupta

5. Energetics of Micelle Formation: Non Agreement between the Enthalpy Change Measured by the Direct Method of Calorimetry and the Indirect Method of van't Hoff   51
   Satya P. Moulik and Debolina Mitra

6. Unusual Phase Behavior in a Two-Component System   69
   Bruno F.B. Silva, Eduardo F. Marques and Ulf Olsson

7. Mixed Proteins/Surfactants Interfacial Layers as Studied by
   Drop Shape Analysis and Capillary Pressure Tensiometry   85
   *V.S. Alahverdjieva, D.O. Grigoriev, A. Javadi, Cs. Kotsmar,
   J. Krägel, R. Miller, V. Pradines and A.V. Makievski*

8. Factors Affecting Mixed Aggregation   105
   Pablo C. Schulz

9. Micellization Characteristics of Sodium Dioctylsulfosuccinate:
   An Overview   131
   S. Chanda, O.G. Singh and K. Ismail

10. Phase Separation Study of Surface-Active Drug Promazine
    Hydrochloride in Absence and Presence of Organic Additives 143
    *Kabir-ud-Din, Mohammed D.A. Al Ahmadi,
    Andleeb Z. Naqvi and Mohd. Akram*

11. Effect of Urea on Surfactant Aggregates: A Comprehensive
    Review   155
    Silvia M.B. Souza, E.B. Alvarez and Mario J. Politi

12. Specific Ion-Pair/Hydration Model for the Sphere-To-Rod
    Transitions of Aqueous Cationic Micelles. The Evidence from
    Chemical Trapping   171
    Laurence S. Romsted

13. Biocatalytic Studies in Microemulsions and Related Systems 199
    Aristotelis Xenakis

14. Colloidal Dispersions for Drug Delivery   207
    Syamasri Gupta

15. Nanoscale Self-Organization of Polyampholytes   231
    H.B. Bohidar and Amarnath Gupta

16. Polymer-Modified Microemulsions as a New Type of Template for the Nanoparticle Formation — 243

    Joachim Koetz, Carine Note, Jennifa Baier and Stefanie Lutter

17. Maximizing the Uptake of Nickel Oxide Nanoparticles by AOT (W/O) Microemulsions — 257

    Nashaat N. Nassar and Maen M. Husein

18. A Brief Overview on Synthesis and Size Dependent Photocatalytic Behaviour of Luminescent Semiconductor Quantum Dots — 271

    A. Priyam, S. Ghosh, A. Datta, A. Chatterjee and A. Saha

19. Dramatic Enhancement in the Cation Sensing Efficiency in Anionic Micelles: A Simple and Efficient Approach Towards Improving the Sensor Efficiency — 299

    Paramita Das, Deboleena Sarkar and Nitin Chattopadhyay

20. Organic Reactivity in AOT-Based Microemulsions: Pseudophase Approach to Transnitrosation Reactions — 309

    G. Astray, A. Cid, J.C. Mejuto and L. García-Río

21. Electric Field Induced Gel Formation and Fracture in Layers of Laponite — 337

    Suparna Sinha and Sujata Tarafdar

22. Temperature Dependent Structural Insignia of Cinnamic Acid — 345

    B. Nandi Ganguly, Nagendra Nath Mondal, S.K. Bandopadhyay and Pintu Sen

Chapter 1

Specific Ion Effects in Colloid and Surface Science: A Modified DLVO Approach

Werner Kunz
*Institute of Physical and Theoretical Chemistry,
University of Regensburg, D-93040 Regensburg,
Germany*

Mathias Boström
*Division of Theory and Modeling,
Department of Physics, Chemistry and Biology,
Linköping University,
SE-581 83 Linköping, Sweden*

In this contribution a short history is given about the introduction of non-electrostatic interactions in the Poisson Boltzmann equation, starting from insufficient first-order dispersion forces up to very recent approaches, in which both water-surface and ion-surface profiles inferred from molecular dynamics simulations are utilized. This new approach can also be used to explain specific ion adsorption on electrodes.

Contents

1.1 Introduction 1
1.2 Modified DLVO Theory 3
1.3 Conclusions 8
References 9

1.1. Introduction

Since many decades DLVO theory is used to describe interactions between colloidal systems. And indeed it is a very successful theory, because it can reproduce numerous phenomena.[1,2] However, specific ion effects are difficult to be taken into account in this simplified theory. This has different reasons: DLVO is made of a sum of electrostatic and van der Waals interactions. As

Ninham pointed out,[3] this combination is not made in a rigorous way; it is in fact rather an arbitrary sum of these two terms.

However, even when this basic theoretical problem is resolved, it is still a so-called primitive model, in which the solvent structure is not explicitly taken into account. Only the dielectric constant describes it. However, the solvation of ions is a crucial factor that discriminates one ion from another one. And the description of solvent structure, especially around ions, is not in the focus of DLVO.

Usually ion hydration is taken into account by different ion sizes. But such a simplified correction via "decorated" ions is by no means sufficient to explain drastic differences in colloidal systems, when for example replacing hydroxide ions by bromide ions.[4,5] Ternary phase diagrams of water, ionic surfactants and oil can be completely different, depending on the type of the counterion that is used with the surfactant.[6] Further, vesicular structures composed of surfactants may considerable swell or shrink, when counterions are exchanged, even when the charge of the ions is always the same.[7]

So what to do ? Probably the simplest way of testing specific ion effects in colloid science is to consider the air-solution interface of simple aqueous salt solutions. In a pioneering work Jungwirth and Tobias[8] could show that it is probably crucial to take into account the polarizability of the ions, such as Ninham and Yaminsky[9] proposed in their landmark paper several years earlier. But in contrast to Ninham and Yaminsky, Jungwirth and Tobias used molecular dynamics (MD) simulations in order to predict surface tensions. They found out that it is necessary to consider the influence of the ion polarizability on the solvent structure in order to get the right ordering of the surface tension. And they also could demonstrate that the surface tension of a salt solution can increase, despite the fact that in the very first surface layer ions are adsorbed. Simply, because, according to the Gibbs equation, the surface tension is an integral over the ion profiles and does not give information about local adsorption or desorption processes on the surface. This paper by Jungwirth and Tobias and several others initiated a Renaissance in the research of specific ion effects and Hofmeister series.[10]

However, some problems remained. Although MD simulations very roughly predicted the right Hofmeister series of surface tensions, such simulation techniques are heavy to carry out. Further, the simulation box is small, leading to possible artifacts in the calculation of the Gibbs integral. Due to the statistical noise, thermodynamic properties cannot be calculated with high precision. It is even very difficult to approximate the surface tension of pure water. In any case, MD simulations are not appropriate for rapid calculations of surface properties.

In this respect the approach by Ninham and Yaminsky is much easier to use. In principle the influence of solvent structure can be taken into account within the DLVO model by using a convenient Lifshitz-like ansatz.[9] There, all non-electrostatic interactions are taken into account via frequency summations overall electromagnetic interactions that take place in the solutions. If done rigorously, the result should be more or less exact. As a proof of principle, Boström and Ninham made a first attempt in this direction. The classical DLVO ansatz was replaced by a modified Poisson Boltzmann (PB) equation, in which a simplified so-called dispersion term was added to the electrostatic interaction. In this way ion specificity came in quite naturally via the polarizability and the ionization potential of the ions. However, it turned out that this first-order approximation of the non-electrostatic interactions was not sufficient to predict the Hofmeister series of surface tension.[11] Heavier ions, such as iodide had to be supposed to have smaller polarizabilites as smaller ions such as chloride. Although the exact polarizabilities of ions in water are still under debate, this is not physical.

Further on, Kunz et al.[12] could show that this failure of the simplified dispersion model is not a consequence of the weakness of the Poisson-Boltzmann equation. More elaborate statistical mechanics, using the so-called hypernetted chain equation (HNC), yielded basically the same result. Obviously the problem comes from the neglect of ion-water interactions and their changes near the surface. To introduce such interactions in primitive model calculations Boström et al.[13] used Jungwirth's water profile perpendicular to the surface as a basis to model a distance dependent electrostatic function, instead of a static dielectric constant. Such ideas were used several times over the years, for instance to model activity coefficients of electrolyte solutions.[14]

With such a modification the description of ion profiles on surfaces became more physical within this modified DLVO approach. However, some of the profiles were still not satisfactory, probably because of the still too crude approximation of the so-called dispersion forces. In the following section we will present the newest DLVO theory and discuss some exciting results.

1.2. Modified DLVO Theory

Recently, Horinek and Netz[15] could successfully infer from MD simulations potentials of mean force (pmf) of ions near the interface between a self-assembled hydrophobic monolayer (SAM) and aqueous salt solutions. Interestingly it was by no means possible to interpret these pmf by simple analytical physical models. They seem to be a complex balance of different interactions between ion and water and surface. Anyway, they can be used in numerical form to replace the dispersion term in the Poisson Boltzmann equation previously used.

In total this means that the initial Poisson Boltzmann equation is modified in two respects: the solvent structure is taken into account via the dielectric function and the total interactions between ions and the surfaces, modified by the presence of water, by the pmf, which can be directly implemented in the interaction term of the PB equation. With these two essential modifications, a new and truly much improved alternative to the old DLVO theory is available now. It should be noted that no adjustable parameters are used here.

The modified PB equation is as follows:

$$\epsilon_0 \frac{d}{dx}\left(\epsilon(x)\frac{d\psi}{dx}\right) = -e\sum_i C_{0,i} z_i \exp\left[-(z_i e\psi + U_i(x))/k_B T\right], \quad (1)$$

$$\left.\frac{d\psi}{dx}\right|_{x=L/2} = 0, \quad \left.\left(\epsilon(x)\frac{d\psi}{dx}\right)\right|_{surface} = -\frac{\sigma}{\epsilon_0}. \quad (2)$$

where k_B is the Boltzmann constant, z_i is the charge number of the ions, e is the elementary charge, T is the temperature of the system, ψ is the self consistent electrostatic potential, x_{min} is the minimum cutoff distance, σ is the surface charge, here taken to be zero, U_i is the ionic non-electrostatic potential inferred from MD simulations (pmf) and acting between the surface and the ion. The expression used for the inhomogeneous dielectric constant $\epsilon(x)$ for the SAM-water system is given by Lima et al.,[16] also according to MD simulation data. Eq. (1) can be solved using finite volume method[16] to determine the ion distributions between two SAM.

In the following we show some results of this very new approach. The obtained electrostatic potential profiles and ion distributions can in principle be used to calculate surface or interfacial tensions. However, up to now only few pmf for ion-water surface interactions are available from MD simulations and no reliable experimental data of interfacial tensions for SAM-solution interfaces. Therefore it is not yet possible to check if the correct Hofmeister series can be obtained with this new approach.

But another property can be inferred, such as the pressure between surfaces. The MD data from Horinek et al.[15] allowed us to calculate the

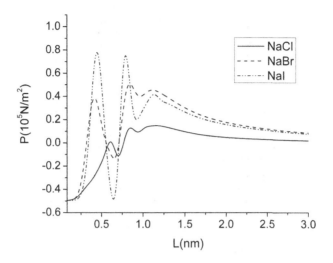

Fig. 1.1 Double layer pressure between two SAM surfaces interacting at different salt solutions: NaCl(solid line), NaBr(dashed line) and NaI(dash-double dotted line) at a temperature of 298.15 K and a concentration of 0.01M (from Ref. 16 with permission).

double layer pressure between two planar surfaces at a distance L simply by the differentiation of the free energy of the system:

$$P = -\frac{\partial}{\partial L}\left(\frac{A}{area}\right), \qquad (3)$$

and the free energy per unit of area is expressed by

$$\frac{A}{area} = \frac{e}{2}\int_{x_{min}}^{L}\psi\sum_i c_i z_i dx + \int_{x_{min}}^{L}\sum_i c_i U_i dx \\ + k_B T \int_{x_{min}}^{L}\sum_c c_i \left[ln\left(\frac{c_i}{c_{0,i}}\right) - 1\right] dx. \qquad (4)$$

The first two terms in the right hand side of the Eq.(4) are the energy contributions (electrostatic and the ionic potential of mean force contribution, respectively) to the free energy of the system and the third term is the entropic contribution. The derivative of the free energy can be solved numerically or developed analytically.[16]

Figure 1.1 shows the contribution of the ions to the interactions of two self-assembled monolayers (SAM) as a function of their distance. Of course,

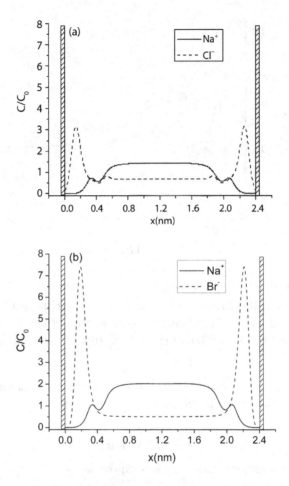

Fig. 1.2 Concentration profile of ions between two uncharged SAM surfaces at a bulk concentration $C_0 = 0.01$ mol/L and effective distance $L = 2.4$ nm. We consider two salt solutions: (a) NaCl and (b) NaBr. The batched vertical bars represent the SAM surfaces (from Ref. 16 with permission).

this is just a calculation and no comparison to experiment is made in the moment. However, several interesting features can be seen.

We observe that for distances greater than roughly 1 nm, the pressure between two SAM surfaces becomes more repulsive in the order $NaCl > NaBr >$ NaI, following an inverse Hofmeister series, where the more po-

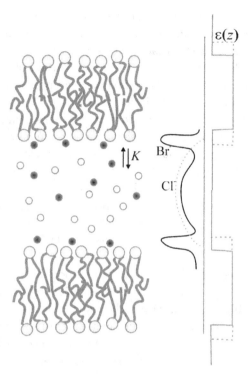

Fig. 1.3 Conceptual drawing showing the interfacial accumulation of Br ions (solid line) compared with nonbinding Cl (dashed line) (from Ref. 17 with permission).

larizable the anion, the more repulsive the pressure. But there can also be more surprising effects such as a reversal of the Hofmeister series with plate separation (at about 0.7 nm). Using surface force measurements between surfaces coated with self-assembled monolayers one should be able to observe this transition from a direct to a reversed Hofmeister sequence. Even more, not only the series is reversed, but the pressure can go from a positive (repulsive) one to a negative (attractive) one. If this model is true, then it is not surprising that ion effects are not really understood, given the complexity of these results.

The question is if there is any experimental evidence in favor of these results. The answer is yes. To see this we consider the ion profiles near the surfaces, at a fixed distance of the layers, see Fig. 1.2. We observe that the ion specific pmf interactions give rise to a negatively charged adsorption layer near to each surface, so charging the surface and resulting in a repulsive double layer pressure. The concentration of bromide near the surfaces is around eight times the bulk concentration. Because of the higher concentration of anions near the surfaces, the pressure is more repulsive for bromide than for chloride. One can observe that the concentration of Na$^+$ between the plates increases when the anion is more polarizable. However, it is not sufficient to screen out the electrostatic repulsion between the two negative adsorption layers. Although the surface is considered to be rigid and perfectly smooth in our model, this result is in reasonable agreement with very careful experiments on the preferential adsorption of bromide ions to neutral lipid membranes, see Fig. 1.3.[17] This is a first independent check of the validity of our new and modified DLVO approach based on MD results.

If it is true that bromide, and even more iodide, can strongly adsorb on hydrophobic surfaces due to non-electrostatic forces, then this can have also a considerable impact in electrochemistry. There, since more than 80 years Otto Stern's idea of an obscure non-electrostatically adsorbed layer is an ad-hoc assumption that is not really explained as little as specific interactions of ions with equally charged electrodes.

1.3. Conclusions

We have presented a plausible way to treat specific salt effects, but one should stress that it is not the only way. One can modify the Poisson-Boltzmann equation itself (as is done here), or one can modify the boundary conditions. Petrache et al.[17] for example used a charge regulated boundary condition to model specific ion effects. In general both charge regulation due to ion binding and physisorption due to ion-surface potentials can influence forces between surfaces.[18]

We have shown that the asymmetry of ion-surface potentials of mean force can give rise to strong Hofmeister effects for the double layer force between uncharged surfaces. Our results have shown that it is possible to obtain a change from direct to reversed Hofmeister sequence by changing the distance between uncharged surfaces. Using surface force measurements between gold surfaces coated with self-assembled monolayers one should be able to observe this transition from a direct to a reversed Hofmeister sequence. It is however important to point out that at short distances below 1 nm we expect hydrophobic interactions between the surfaces that within the treatment in the paper are neglected, it could well modify the forces and possibly change the ionic ordering. In any case, the attraction of the ions to the surface is a consequence of both their sizes and the polarizabilities on one hand, and of the repulsion of water from the surface on the other hand.

It seems that we are not far from a better understanding of specific ion effects in colloid and surface chemistry. The key appears to be the use of ion and water profiles near surfaces inferred from molecular dynamics simulation and their appropriate use in solvent-averaged models in order to derive an efficient, but physically well-based alternative to DLVO.

Acknowledgments

We thank the Swedish Research Council the German Arbeitsgemeinschaft industrieller Forschungvereinigungen Otto von Guericke e.V. (AiF). This work would not have been possible without numerous discussions with many colleagues, especially with Barry W. Ninham, whose significant contribution to modern colloidal chemistry and specific ion effects is gratefully acknowledged. We also thank Dr. R. Neueder for his help in editing the manuscript.

References

1. S. Marcelja, *Curr. Opin. Colloid Interface Sci.*, 9, 165 (2004).
2. R. M. Pashley, P. M. McGuiggan, B. W. Ninham, J. Brady and D. F. Evans, *J. Phys. Chem.*, 90, 1637 (1986).
3. B. W. Ninham, *Adv. Colloid Interface Sci.*, 83, 1 (1999).
4. B. W. Ninham, D. F. Evans and G. J. Wie, *J. Phys. Chem.*, 87, 5020 (1983).
5. B.W. Ninham, S. Hashimoto and J.K. Thomas, *J. Colloid Interface Sci.*, 95, 594 (1983).
6. M. Nyden, O. Soderman, P. Hansson, *Langmuir*, 17, 6794 (2001).
7. M. E. Karaman, B. W. Ninham and R. M. Pashley, *J. Phys. Chem.*, 98, 11512 (1994).
8. P. Jungwirth and D. J. Tobias, *J. Phys. Chem. B*, 105, 10468 (2001).
9. B. W. Ninham and V. Yaminsky, *Langmuir*, 13, 2097 (1997).
10. E. K. Wilson, A Renaissance for Hofmeister, *Chem. & Eng. News.*, 85, 47 (2007).
11. M. Boström, D. R. M. Williams, and B. W. Ninham, *Langmuir*, 17, 4475 (2001).
12. W. Kunz, L. Belloni, O. Bernard and B. W. Ninham, *J. Phys. Chem. B*, 108, 2398 (2004).
13. M. Boström, W. Kunz, and B. W. Ninham, *Langmuir*, 21, 2619 (2005).

14. W. Kunz, J. M'Halla and S. Ferchiou, *J. Phys. Cond. Matt.*, 3, 7907 (1991).
15. D. Horinek and R. R. Netz, *Phys. Rev. Lett.*, 99, 226104/1 (2007).
16. E. R. A Lima, D. Horinek, R. R. Netz, E. C. Biscaia, F. W. Tavares, W. Kunz and M. Boström, *J. Phys. Chem. B*, 112, 1580 (2008).
17. H. I. Petrache, T. Zemb, L. Belloni, V. A. Parsegian, *Proc. Nat. Acad. Sci. USA*, 103, 7982 (2006).
18. M. Boström, E. R. A. Lima, F. W. Tavares, B. W. Ninham, *J. Chem. Phys.*, 128, 135104/1-135104/4 (2008).

Chapter 2

Electrokinetics in a Suspension of Soft Particles

Hiroyuki Ohshima

Faculty of Pharmaceutical Sciences,
Tokyo University of Science
2641 Yamazaki, Noda Chiba 278-8510, Japan

Theories of electrokinetics in suspension of soft particles, i.e., particles covered with an ion-penetrable surface layer of polyelectrolytes are reviewed. Various electrokinetic phenomena such as electrophoresis, sedimentation potential, electro-osmosis, and streaming potential are discussed.

Contents

2.1 Introduction 11
2.2 Electrophoresis of Soft Particles 12
2.3 Sedimentation Potential 18
2.4 Electrokinetic Flow between Two Parallel Soft Plates 19
2.5 The Physical Meaning of the Softness Parameter 23
2.6 Conclusions and Outlook 24
References 24

2.1. Introduction

Electrokinetic phenomena in a suspension of colloidal particles covered with an ion-penetrable surface layer of polyelectrolytes, which are called soft particles are quite different from those of bare particles.[1-6] For such particles one must consider the potential distribution and the liquid flow distribution not only outside but also inside the surface charge layer. These effects can be taken into account with the help of the model of Debye and Bueche[7] that the polymer segments are regarded as resistance centers distributed in the polyelectrolyte layer, exerting frictional forces on the liquid flowing in the polyelectrolyte layer. In this article, we review theories of electrokinetics of soft particles and present simple analytic equations describing various electrokinetic phenomena.

2.2. Electrophoresis of Soft Particles

Consider a spherical soft particle moving with a velocity $\mathbf{U}$ in an electrolyte solution in an applied electric field $\mathbf{E}$ (Fig. 2.1). We assume that the uncharged particle core of radius a is coated with an ion-penetrable layer of polyelectrolytes of thickness d. The polymer-coated particle has thus an inner radius a and an outer radius $b \equiv a + d$ (Fig. 2.1). The origin of the spherical polar coordinate system (r, θ, ϕ) is held fixed at the center of the particle core and the polar axis ($\theta = 0$) is set parallel to $\mathbf{E}$. Let the electrolyte be symmetrical with valence z and bulk concentration (number density) n. We denote the drag coefficient of cations by λ_+ and that of anions by λ_-. We adopt the model of Debye-Bueche[7], in which the polymer segments are regarded as resistance centers distributed in the polyelectrolyte layer, exerting frictional forces $-\gamma \mathbf{u}$ on the liquid flowing in the polymer layer, where $\mathbf{u}$ is the liquid velocity and γ is the frictional coefficient. We also assume that ionized groups of valence Z are distributed with a constant number density of N. The density of fixed charges in the surface layer thus becomes ZeN, where e is the elementary electric charge.

The main assumptions in our analysis are as follows. (i) The Reynolds numbers of the liquid flows outside and inside the polyelectrolyte layer are small enough to ignore inertial terms in the Navier-Stokes equations and the liquid can be regarded as incompressible. (ii) The applied field $\mathbf{E}$ is weak so that the particle velocity $\mathbf{U}$ is proportional to $\mathbf{E}$ and terms of higher order in $\mathbf{E}$ may be neglected. (iii) The slipping plane is located on the particle core. (iv) No electrolyte ions can penetrate the particle core. (v) The polyelectrolyte layer is permeable to electrolyte ions. (vi) The relative permittivity ε_r takes the same value both inside and outside the polyelectrolyte layer.

The fundamental electrokinetic equations are given by

$$\eta \nabla \times \nabla \times \mathbf{u} + \gamma \mathbf{u} + \nabla p + \rho_{el} \nabla \psi = \mathbf{0}, \quad a < r < b \tag{1}$$

$$\eta \nabla \times \nabla \times \mathbf{u} + \nabla p + \rho_{el} \nabla \psi = \mathbf{0}, \quad r > b \tag{2}$$

$$\nabla \cdot \mathbf{u} = 0 \tag{3}$$

$$\mathbf{v}_\pm = \mathbf{u} - \frac{1}{\lambda_\pm} \nabla \mu_\pm \tag{4}$$

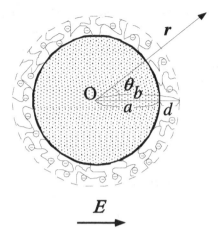

Fig. 2.1 A soft particle in an applied electric field **E**. The soft particle consists of the unchanged particle core of radius a coated with an ion-penetrable layer of polyelectrolytes of thickness d. $b = a + d$.

$$\nabla \cdot (\eta_\pm \mathbf{v}_\pm) = 0 \tag{5}$$

$$\rho_{el}(\mathbf{r}) = ze\{n_+(\mathbf{r}) - n_-(\mathbf{r})\} \tag{6}$$

$$\mu_\pm(\mathbf{r}) = \mu_\pm^\infty \pm ze\, \psi(\mathbf{r}) + kT\, ln n_\pm(\mathbf{r}) \tag{7}$$

$$\nabla \psi(\mathbf{r}) = -\frac{\rho(\mathbf{r})}{\varepsilon_r \varepsilon_0} \tag{8}$$

where ε_0 is the permittivity of a vacuum, k is Boltzmann's constant, T is the absolute temperarure, $\mathbf{u}(\mathbf{r})$ is the liquid velocity at position $\mathbf{r}$ relative to the particle ($\mathbf{u}(\mathbf{r}) \to -\mathbf{U}$ as $r \equiv |\mathbf{r}| \to \infty$), $\mathbf{v}_\pm(\mathbf{r})$ is the velocity of cations and anions, $p(\mathbf{r})$ is the pressure, $\rho_{el}(\mathbf{r})$ is the charge density resulting from electrolyte ions given by Eq. (6), $\psi(\mathbf{r})$ is the electric potential, $\mu_\pm(\mathbf{r})$ and $n_\pm(\mathbf{r})$ are, respectively, the electrochemical potential and the concentration (the number density) of cations and anions, and $\mu_\pm^0$ is a constant term in $\mu_\pm(\mathbf{r})$. Equations (1) - (3) are the Navier-Stokes equation and the equation of continuity for an incompressible flow (assumption (i)). Equation (4) expresses that the flow $\mathbf{v}_\pm(\mathbf{r})$ of cations and anions is caused by the liquid flow $\mathbf{u}(\mathbf{r})$ and the gradient of the electrochemical potential $\mu_\pm(r)$, given by Eq. (7). Equation (5) is the continuity equation for cations and anions, and Eq. (8) is Poisson's equation.

Under assumption (ii), we may write

$$n_i(\mathbf{r}) = n_i^{(0)}(r) + \delta n_i(\mathbf{r}) \tag{9}$$

$$\psi(\mathbf{r}) = \psi^{(0)}(r) + \delta\psi(\mathbf{r}) \tag{10}$$

$$\mu_i(\mathbf{r}) = \mu_i^{(0)} + \delta\mu_i(\mathbf{r}) \tag{11}$$

$$\rho(\mathbf{r}) = \rho^{(0)}(r) + \delta\rho(\mathbf{r}) \tag{12}$$

where the quantities with superscript (0) refer to those at equilibrium, i.e., in the absence of $\mathbf{E}$, and $\mu_i^{(0)}$ is a constant independent of $\mathbf{r}$. By substituting Eqs. (9)-(12) into Eqs. (1)-(5), we obtain

$$\eta \nabla \times \nabla \times \nabla \times \mathbf{u} + \gamma \nabla \times \mathbf{u} = \nabla \delta n_+ \times \nabla n_+^{(0)} + \nabla \delta n_- \times \nabla n_-^{(0)}, \quad a < r < b \tag{13}$$

$$\eta \nabla \times \nabla \times \nabla \times \mathbf{u} = \nabla \delta n_+ \times \nabla n_+^{(0)} + \nabla \delta n_- \times \nabla n_-^{(0)}, \quad r > b \tag{14}$$

and

$$\nabla \cdot \left(n_\pm^{(0)} \mathbf{u} - \frac{1}{\lambda_\pm} n_\pm^{(0)} \nabla \delta \mu_\pm \right) = 0 \tag{15}$$

where $n_\pm^{(0)}(r)$ is the equilibrium concentration (number density) of cations and anions ($n_\pm^{(0)} \to n$ as $r \to \infty$), and $\delta n_\pm^{(0)}(\mathbf{r})$ is the deviation of $n_\pm^{(0)}(r)$ due to the applied field $\mathbf{E}$.

Symmetry considerations permit us to write

$$\mathbf{u}(\mathbf{r}) = \left(-\frac{2}{r} h(r) E \cos\theta, \frac{1}{r}\frac{d}{dr}(rh(r)) E \sin\theta, 0 \right) \tag{16}$$

$$\delta\mu_\pm(\mathbf{r}) = \pm z e \phi_\pm(r) E \cos\theta \tag{17}$$

where $E = |\mathbf{E}|$. The fundamental electrokinetic equations (13)-(15) can be transformed into equations for $h(r)$ and $\phi_\pm(r)$ and the electrophoretic mobility $\mu = U/E$ (where $U = |\mathbf{U}|$) can be calculated from

$$\mu = 2 \lim_{r \to \infty} \frac{h(r)}{r} \tag{18}$$

The result is

$$\mu = \frac{b^2}{9}\int_b^\infty \left[3\left(1-\frac{r^2}{b^2}\right) - \frac{2L_2}{L_1}\left(1-\frac{r^3}{b^3}\right)\right]G(r)dr$$
$$+\frac{2L_3}{3\lambda^2 L_1}\int_a^\infty \left(1+\frac{r^3}{2b^3}\right)G(r)dr - \frac{2}{3\lambda^2}\int_a^b \left[1-\frac{3a}{2\lambda^2 b^3 L_1}\right]$$

$$\{(L_3 + L_4\lambda r)\cosh[\lambda(r-a)] - (L_4 + L_3\lambda r)\sinh[\lambda(r-a)]\}G(r)dr \quad (19)$$

with

$$L_1 = \left(1+\frac{a^3}{2b^3}+\frac{3a}{2\lambda^2 b^3}-\frac{3a^2}{2\lambda^2 b^4}\right)\cosh[\lambda(b-a)]$$
$$-\left(1-\frac{3a^2}{2b^2}+\frac{a^3}{2b^3}+\frac{3a}{2\lambda^2 b^3}\right)\frac{\sinh[\lambda(b-a)]}{\lambda b} \quad (20)$$

$$L_2 = \left(1+\frac{a^3}{2b^3}+\frac{3a}{2\lambda^2 b^3}\right)\cosh[\lambda(b-a)] + \frac{3a^2}{2b^2}\frac{\sinh[\lambda(b-a)]}{\lambda b} - \frac{3a}{2\lambda^2 b^3} \quad (21)$$

$$L_3 = \cosh[\lambda(b-a)] - \frac{\sinh[\lambda(b-a)]}{\lambda b} - \frac{a}{b} \quad (22)$$

$$L_4 = \sinh[\lambda(b-a)] - \frac{\cosh[\lambda(b-a)]}{\lambda b} + \frac{\lambda a^2}{3b} + \frac{2\lambda b^2}{3a} + \frac{1}{\lambda b} \quad (23)$$

where

$$\lambda = (\gamma/\eta)^{1/2} \quad (24)$$

$$G(r) = -\frac{zen}{\eta r}\frac{dy}{dr}(e^{-y}\phi_+ + e^{+y}\phi_-) \quad (25)$$

$$y(r) = \frac{ze\psi^{(0)}(r)}{kT} \quad (26)$$

is the scaled equilibrium potential and $1/\lambda$ is called the electrophoretic softness.

For the special case of $a \to \infty$, we obtain from Eq. (19)

$$\mu = \frac{\varepsilon_r\varepsilon_0}{\eta}\frac{1}{\cosh(\lambda d)}\left[\psi(-d) + \lambda\int_{-d}^0 \psi(x)\sinh(\lambda(x+d))dx\right]$$
$$+\frac{ZeN}{\eta\lambda^2}\left[1-\frac{1}{\cosh(\lambda d)}\right] \quad (27)$$

which is the electrophoretic mobility for a plate-like soft particle.

Now consider the case where $\lambda a \gg 1$, $\kappa a \gg 1$, $\lambda d \gg 1$ and $\kappa d \gg 1$. In this case, as shown in Fig. 2.2, the potential distribution inside the surface charge layer can be approximated by

$$\psi(x) = \psi_{DON} + (\psi_0 - \psi_{DON})\exp(\kappa_m x) \qquad (28)$$

with

$$\kappa_m = \kappa\left[1 + \left(\frac{ZN}{2zn}\right)^2\right]^{1/4} \qquad (29)$$

$$\kappa = \left(\frac{2z^2 e^2 n}{\varepsilon_r \varepsilon_0 kT}\right)^{1/2}$$

$$\psi_{DON} = \frac{kT}{ze}\ln\left[\frac{ZN}{2zn} + \left\{\left(\frac{ZN}{2zn}\right)^2 + 1\right\}^{1/2}\right] \qquad (31)$$

$$\psi_0 = \left(\frac{kT}{ze}\ln\left[\frac{ZN}{2zn} + \left\{\left(\frac{ZN}{2zn}\right)^2 + 1\right\}^{1/2}\right]\right.$$

$$\left. + \frac{2zn}{ZN}\left[1 - \left\{\left(\frac{ZN}{2zn}\right)^2 + 1\right\}^{1/2}\right]\right) \qquad (32)$$

where $x = r - b$, ψ_{DON} is the Donnan potential in the polyelectrolyte layer, ψ_0 is the potential at the boundary $x = 0$ (or, $r = b$) between the polyelectrolyte layer and the surrounding solution, which we call the surface potential of the soft particle, and κ_m is the Debye-Hückel parameter of the polyelectrolyte layer that involves the contribution of the fixed-charges ZN. Note that when κd is sufficiently large, the contribution of the surface charge on the particle core (i.e., at $r = a$) becomes negligible, unless it is very large, and the mobility becomes insensitive to the position of the slipping plane. This implies that the concept of the zeta potential, i.e., the potential at the slipping plane, $r = a$ ($x = -d$), loses its physical meaning.

By substituting Eq. (28) into Eq. (19) yields, after some algebra, we obtain the following approximate expression for the electrophoretic mobility of a soft particle:

$$\mu = \frac{\varepsilon_r \varepsilon_0}{\eta}\frac{\psi_0/\kappa_m + \psi_{DON}/\lambda}{1/\kappa_m + 1/\lambda} + \frac{ZeN}{\eta\lambda^2} \qquad (33)$$

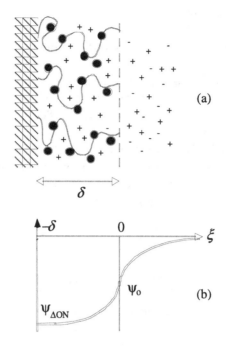

Fig. 2.2 Ion distribution (a) and potential distribution (b) across a soft surface.

Note that Eq. (33) dose not involve the thickness of the surface charge layer or the position of the slipping plane. This implies that the zeta potential and the slipping plane both lose their meaning. In the limit of $\lambda \to \infty$, Eq. (33) becomes the following Smoluchowski equation, which is most widely used for the electrophoretic mobility of a hard particle:

$$\mu = \frac{\varepsilon_r \varepsilon_0}{\eta} \psi_0 \qquad (34)$$

On the other hand, in the limit of high electrolyte concentrations, Eq. (33) becomes a non-zero value, viz.,

$$\mu \to \mu^\infty \equiv \frac{ZeN}{\eta \lambda^2} \qquad (35)$$

This is the most remarkable difference between Eq. (33) and Smoluchowski's equation (Eq. (34)).

2.3. Sedimentation Potential

When charged spherical paricles are falling steadily under gravity, the electrical double layer around each particle loses its spherical symmtery because of the fluid motion (the relaxation effect). A microscopic electric field arising from the distortion of the double layer reduces the falling velocity (the sedimentation velocity) of the particle and the fields from the individual particles are then superimposed to give rise to a macroscopic field $\mathbf{E}_{SED}$ (the sedimentation field), which is uniform for a homogeneous dispersed suspension. Experimantally, the sedimentation field is observed as a sedimentation potential. For the sedimentation problem, the Navier-Stokes equations (1) and (2) for the flow velocity $\mathbf{u}(\mathbf{r})$ are replaced by

$$\eta \nabla \times \nabla \times \mathbf{u} + \gamma \mathbf{u} + \nabla p + \rho_{el} \nabla \psi - \rho_0 \mathbf{g} = 0, \quad a < r < b \quad (36)$$

$$\eta \nabla \times \nabla \times \mathbf{u} + \nabla p + \rho_{el} \nabla \psi - \rho_0 \mathbf{g} = 0, \quad r > b \quad (37)$$

On the basis of the same approximation method as used in the electrophoresis problem, we obtain the following expression for the sedimentation potential $\mathbf{E}_{SED}$[8]:

$$\mathbf{E}_{SED} = -\frac{(\phi_c \Delta \rho_c + \phi_s \Delta \rho_s)}{K} \mu \mathbf{g} \quad (38)$$

With

$$\Delta \rho_c = \rho_c - \rho_0 \quad (39)$$

$$\Delta \rho_s = \rho_s - \rho_0 \quad (40)$$

$$K = nz^2 e^2 \left(\frac{1}{\lambda_+} + \frac{1}{\lambda_-} \right) \quad (41)$$

where ρ_c and ρ_s are the mass densities of the particle core and the polymer segments, ϕ_c is the volume fraction of the particle core and ϕ_s is the volume fraction of the polymer segments, μ is the electrophoretic mobility of a soft particle, $\mathbf{g}$ is the gravity, K is the electrical conductivity of the electrolyte solution. Equation (38) is an Onsager relation between electrophoretic mobility, μ and sedimentation potential, $\mathbf{E}_{SED}$.

2.4. Electrokinetic Flow between Two Parallel Soft Plates

Consider the steady flow of an incompressible aqueous solution of a symmetrical electrolyte through a pore between two parallel fixed similar plates, 1 and 2, at separation h, under the influence of a uniform electric field E and a uniform pressure gradient P, both applied parallel to the plates.[9]

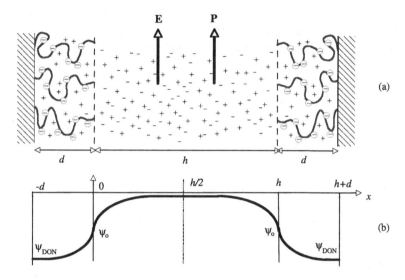

Fig. 2.3 Schematic representation of a channel of two parallel similar plates 1 and 2 at separation h covered by ion-penetrable surface charge layers of thickness d under the influence of an electric field E and a pressure gradient $P(a)$ and potential distribution (b).

The plates are covered by ion-penetrable surface charge layers of thickness d. The plate areas are assumed large enough for edge effects to be neglected. We take the x-axis to be perpendicular to the plates with its origin at the surface of plate 1 and the y-axis to be parallel to the plates so that E and P are both in the y direction (Fig. 2.3) and are given by $E = -\partial\Psi/\partial y$ and $P = -\partial p/\partial y$, where Ψ is the electric potential and p is the pressure. In this case the Navier-Stokes equations (1) and (2) become

$$\eta\frac{d^2u}{dx^2} + \rho_{el}(x)E + P = 0, \quad 0 < x \leq \frac{h}{2} \tag{42}$$

$$\eta\frac{d^2u}{dx^2} - \gamma u(x) + \rho_{el}(x)E + P = 0, \quad -d < x < 0 \tag{43}$$

We can write the electric potential $\Psi(x,y)$ as the sum of the equilibrium electric potential $\psi(x)$ and the potential of the applied field $-Ey$, viz.,

$$\Psi(x,y) = \psi(x) - Ey \tag{44}$$

We assume that charged groups of valence Z are distributed in the surface charge layers at a uniform density N. Since end effects are neglected, the fluid velocity is in the y direction, depending only on x, and the present system can be considered to be at thermodynamic equilibrium with respect to the x direction. We define $u(x)$ as the fluid velocity in the y direction and $\rho_{el}(x)$ as the volume charge density of electrolyte ions, both independent of y.

The Navier-Stokes equations (42) and (43) can be solved to give

$$u(x) = u(0) + \frac{\varepsilon_r \varepsilon_0 E}{\eta}\{\psi(x) - \psi(0)\} - \frac{P}{2\eta}x^2 + \frac{Ph}{2\eta}x, \ 0 \le x \le \frac{h}{2} \tag{45}$$

$$u(x) = \frac{\varepsilon_r \varepsilon_0 E}{\eta}\left[\psi(x) - \frac{\cosh(\lambda x)}{\cosh(\lambda d)}\psi(-d) + \lambda \int_{-d}^{x}\psi(x')\sinh(\lambda(x-x'))dx'\right.$$
$$\left. - \frac{\lambda}{\cosh(\lambda d)}\int_{-d}^{0}\psi(x')\cosh(\lambda x)dx \cdot \sinh(\lambda(x+d))\right]$$
$$-\frac{ZeNE - P}{\eta\lambda^2}\left[1 - \frac{\cosh(\lambda x)}{\cosh(\lambda d)}\right] + \frac{Ph}{2\eta\lambda}\frac{\sinh(\lambda(x+d))}{\cosh(\lambda d)}, \ -d \le x \le 0 \tag{46}$$

Let us consider the situation where $P = 0$ and h is infinitely large, and obtain the liquid velocity at a large distance from plate 1. This velocity is called the electro-osmotic velocity, which we denote by U_{eo}. By setting $P = 0$ and $h \to \infty$ in Eqs. (45) and (46) and noting that $\psi(x) \to 0$ far from the plate, we find that

$$U_{eo} = u(0) - \frac{\varepsilon_r \varepsilon_0}{\eta}\psi(0)E$$
$$= -\frac{\varepsilon_r \varepsilon_0 E}{\eta}\frac{1}{\cosh(\lambda d)}\left[\psi(-d) + \lambda \int_{-d}^{0}\psi(x)\sinh(\lambda(x+d))dx\right] \tag{47}$$
$$-\frac{ZeNE}{\eta\lambda^2}\left[1 - \frac{1}{\cosh(\lambda d)}\right]$$

Note that the ratio $-U_{eo}/E$ agrees exactly with the expression for the electrophoretic mobility μ given by Eq. (27), viz.,

$$\mu = -\frac{U_{eo}}{E} \tag{48}$$

The volume flow per unit time, V, transported by electro-osmosis is obtained by integrating $u(x)$, viz.,

$$V = 2 \int_{-d}^{h/2} u(x)|_{P=0} dx \tag{49}$$

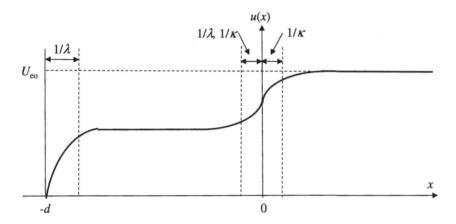

Fig. 2.4 Liquid velocity distribution across a soft surface.

The liquid flow accompanies the flow of electrolyte ions, that is, an electric current. The density of the electric current $i(x)$, which is in the y direction, is given by the flow of the positive charges minus that of the negative charges, viz.,

$$i(x) = ze[n_+(x)v_+(x) - n_-(x)v_-(x)] \tag{50}$$

where $v_+(x)$ and $v_-(x)$ are, respectively, the velocity of cations and that of anions, both independent of y. The flow of electrolyte ions is caused by the liquid flow (convection) and the gradient of the electrochemical potential of the ions (conduction). Let μ_+ and μ_- be, the electrochemical potentials of cations and anions respectively, which are given by

$$\mu_\pm(x,y) = \mu_\pm^0 \pm ze\Psi(x) + kT \ln n_\pm(x) \tag{51}$$

where $\mu_\pm^0$ are constant. By substituting Eq. (44) into Eq. (51), we obtain

$$\mu_\pm(x,y) = \mu_\pm^0 \pm ze[\psi(x) - Ey] + kT \ln n_\pm(x) \tag{52}$$

Then the ionic velocities $v_+(x)$ and $v_-(x)$ can be expressed as

$$v_\pm(x) = u(x) - \frac{1}{\lambda_\pm}\frac{\partial \mu_\pm}{\partial y} \tag{53}$$

By substituting Eq. (48) into Eq. (49), we obtain

$$v_\pm(x) = u(x) \pm \frac{ZeE}{\lambda_\pm} \tag{54}$$

Substituting Eq. (54) into Eq. (50), we obtain

$$i(x) = \rho_{el}(x)u(x) + z^2 e^2 \left[\frac{n_+(x)}{\lambda_+} + \frac{n_-(x)}{\lambda_-}\right] E \tag{55}$$

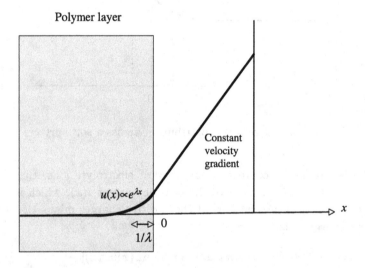

Fig. 2.5 Liquid velocity distribution across a surface layer under constant velocity gradient field.

The first term corresponds to the convection current and the second to the conduction current. The total electric current I flowing in the y-direction is then given by

$$I = 2\left[\int_{-d}^{0} i(x)dx + \int_{0}^{h/2} i(x)dx\right] \tag{56}$$

The streaming potential E_{st} is equal to the value of E such that $I = 0$.

We use the approximate Eq. (28) to calculate the electro-osmotic velocity U_{eo}, the volume flow V, the electric current I, and the streaming potential E_{st}. The results are

$$U_{eo} = -\mu E \qquad (57)$$

$$V = U_{eo}\, h = -\mu E h \qquad (58)$$

$$I = (KE - \mu P)h \qquad (59)$$

$$E_{st} = \frac{\mu}{K} P \qquad (60)$$

Equations (57)-(59) are all related to the approximate expression (33) for the electrophoretic mobility, u. It can easily be shown that the following Onsager's relation is satisfied between electro-osmosis and streaming potential.

$$\left.\frac{E}{P}\right|_{I=0} = \left.\frac{V}{I}\right|_{P=0} = \frac{\mu}{K} \qquad (61)$$

Makino et. al.[10] found that Eq. (57) holds between the electro-osmotic velocity U_{eo} on a poly(N-isopropylacrylamide) hydrogel-coated solid surface and the electrophoretic mobility μ of a poly(N-isopropylacrylamide) hydrogel-coated latex particle.

2.5. The Physical Meaning of the Softness Parameter

In the limit of $1/\lambda \to 0$ ($\lambda \to \infty$), a soft particle becomes a hard particle, we call $1/\lambda$ the softness parameter. On the other hand, Debye and Buehe,[7] showed that the parameter $1/\lambda$ is a shielding length, because beyond this distance the liquid flow $u(x)$ vanishes. Figure 2.4, however, exhibits that under the applied fields E and P the liquid flow $u(x)$ remains a non-zero value beyond the distance of order $1/\lambda$. Indeed, it follows form Eq. (46) that this non-zero value is given by

$$-\frac{ZeNE - P}{\eta \lambda^2} \qquad (62)$$

The reason for the existence of the non-zero value of the electrophoretic mobility μ is the same as that for the electro-osmotic velocity. Note that only when $E = 0$, $P = 0$, and the third and fourth terms on the left-hand side of Eq. (43) cancel with each other, Eq. (43) becomes

$$\eta \frac{d^2 u}{dx^2} - \gamma u(x) = 0, \quad -d < x < 0 \qquad (63)$$

whose solution is

$$u(x) \propto \exp(\lambda x), \quad -d < x < 0 \qquad (64)$$

This implies the liquid velocity $u(x)$ vanishes beyond the distance $1/\lambda$ and the parameter $1/\lambda$ has the meaning of the shielding length. Such a situation is possible when, for example, a constant velocity gradient field is applied in the y-direction outside the surface layer (Fig. 2.5).

2.6. Conclusions and Outlook

We have reviewed theories of electrokinetics of soft particles and present simple analytical a equations describing various electrokinetic phenomena. Electrokinetic phenomena in a suspension of soft particles are characterized by two parameters, that is, the density of fixed charges ZN distributed in a polyelectrolyte layer covering the core of a soft particle and the electrophoretic softness parameter $1/\lambda$. In the case of hard particles without surface structures, on the other hand, their electrokinetic behavior is characterized by only one parameter, i.e., zeta potential. This is the most remarkable difference between soft and hard particles.

References

1. H. Ohshima, *J. Colloid Interface Sci.*, 163, 474 (1994).
2. H. Ohshima, *Adv. Colloid Interface Sci.*, 62, 189 (1995).
3. H. Ohshima, *J. Colloid Interface Sci.*, 233, 142 (2001).
4. J.F.L. Duval and H. Ohshima, *Langmuir*, 22, 3533 (2006).
5. H. Ohshima, *Electrophoresis*, 27, 526 (2006).
6. H. Ohshima, *Theory of Colloid and Interfacial Electric Phenomena*, Academic Press, (2006).
7. P. Debye and A. Bueche, *J. Chem. Phys.*, 16, 573 (1948).
8. H. Ohshima, *J. Colloid Interface Sci.*, 229, 140 (2000).
9. H. Ohshima and T. Kondo, *J. Colloid Interface Sci.*, 135, 443 (1990).
10. K. Makino, K. Suzuki, Y. Sakurai, T. Okano and H. Ohshima, *J. Colloid Interface Sci.*, 174, 400 (1995).

Chapter 3

Relative Humidity and Evaporation of a Simple Fragrance Emulsion

Stig E. Friberg

Chemistry Department
University of Virginia
Charlottesville, VA USA

Patricia A. Aikens

BASF Corporation
Stony Brook, NY USA

The variation in the fractions of compounds and phases were estimated from the phase diagram for the evaporation of an emulsion of water, linalool and a commercial surfactant Laureth 4, assuming equilibrium between the vapor and the emulsion. In the investigated part of the system, the vapor pressure of water was presumed to be approximately constant, which allowed the time dependence of the evaporation changes to be assessed. The influence of the relative humidity on the process was not, as may first be surmised, limited to the time of exhaustion of the water, but actually influenced the direction of the evaporation path in a decisive manner.

3.1. Introduction

Emulsions are of decisive importance in several fields, such as pharmaceutics, cosmetics, preparation of nanoparticles, and others. Hence, general treatments of their properties are readily available.[1,2] Evaporation is a vital part of the over all process in many emulsion applications; for example dermatology, food technology, cosmetics, coatings, preparation of nanoparticles, cleaning of computer chips to mention as few. As a consequence, the fundamentals of the evaporation progression have attracted the attention of outstanding research groups; in particular, the Chemistry Department at Hull University in the United Kingdom has been a leading institution during the last two decades. Their research focused on the evaporation rate during the initial process while the emulsion still consists of the two original liquid phases and clarified the path of the evaporating compounds from the dispersed phase.[3–8]

The authors argued that the evaporation from the drops takes place via diffusion of the compound through the continuous phase; the drops per se not interacting with the surface of the emulsion.

In contrast to these major investigations on the initial part of the evaporation process, the later stages have not attracted a corresponding concerted effort on systems in which the main body of the vapor comes from the continuous phase or the evaporation from a bi-continuous structure, in which the diffusion rates of the compounds are of a similar magnitude as in the continuous phase.[9–12]

The most important feature of the phase diagram approach is the fact that the composition of the departing vapor is obtained without having to analyze the gas phase. Among the more specific aspects of this phenomenon to be addressed by the approach is the fact that the direction of sufficiently slow evaporation from a simple emulsion of a volatile organic compound and water critically depends on both the O/W ratio in the original emulsion and surprisingly, on the relative humidity in the atmosphere into which the evaporation takes place. In fact, a variation in the relative humidity well within what may be expected under normal climatic conditions may cause the most pronounced changes in the number and structure of the phases encountered during the evaporation.[12] We found the later relationship, which is intuitively not evident and has not been treated, of sufficient originality and interest to justify an analysis. In the present contribution, an emulsion of water, linalool, and Laureth 4 (a commercial tetraethylene glycol dodecyl ether) was used to illustrate the decisive importance of the relative humidity for the evaporation direction and the potential disparity in the phases traversed.

3.2. The Concept

The general features of the phase diagram to be utilized for the analysis are shown in Figure 3.1, adapted from Al-Bawab.[13] The emulsion system is a traditional two-phase emulsion with all the surfactant in the oil phase at surfactant/hydrocarbon ratios less than one and a three-phase emulsion of an aqueous phase (Aq), an oil phase, (Oi), and a liquid crystalline phase, (LC), at ratios in excess of this value.

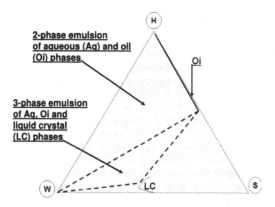

Fig. 3.1 The part of the generic phase diagram used for the analysis. The evaporation is initiated in the 2-phase region and in the process the composition may enter the 3-phase region.

The emulsion is described by its composition expressed as the weights of the three compounds (w_E, l_E, s_E) as well as their weight fractions (W_E, L_E, S_E). The composition in fractions of the compounds in the phases becomes (W_{Aq}, L_{Aq}, S_{Aq}) with the subscripts Oi and LC for the two remaining phases. Since there is only insignificant solubility of the two compounds in the water, the aqueous phase always has the composition in fractions $(1, 0, 0)$. In addition, the fraction of phases is given by A_q (Aqueous), LC (liquid crystal) and Oi (oil).

With this fact in mind, the composition of the oil phase with negligible amounts of water, $(W_{Oi} = 0)$, in the two-phase region immediately follows from the total composition of the emulsion (W_E, L_E, S_E) and

$$L_{Oi} = L_E/(1 - W_E) \qquad (1)$$

In the 3-phase region, the equations for the limiting tie lines are necessary. They are :

$$Aq - Oi;\ W = 1 - L(1 - {}^{Th}W_{Oi})/{}^{Th}L_{Oi} \qquad (2)$$

$$Aq - LC;\ W = 1 - L(1 - {}^{Th}W_{LC})/{}^{Th}L_{LC} \qquad (3)$$

$$LC - Oi;\ W = {}^{Th}W_{LC} + (L - {}^{Th}L_{LC})({}^{Th}W_{Oi} - {}^{Th}W_{LC})({}^{Th}L_{Oi} - {}^{Th}L_{LC}) \quad (4)$$

In the above equations, the pre(super)script "Th" indicates the composition of the equilibrium phases in the three-phase region.

The fraction of the aqueous phase within the three-phase region is obtained by the expression:

$$Aq = (L_J - L_E)/L_J \quad (5)$$

In Eq. (5), L_J is the L fraction at the point where a straight line from the composition of the aqueous phase, $(1,0,0)$ through the emulsion composition, (W_E, L_E, S_E) connects with the tie line from the liquid crystal to the oil phase, equation (4). The line from the $(1,0,0)$ point to the emulsion composition, (W_E, L_E, S_E) is found by rearranging equation (1).

$$W = 1 - L(1 - W_E)/L_E \quad (6)$$

Combination of this equation with equation (4) gives:

$$L_J = [1 - {}^{Th}W_{LC} + {}^{Th}L_{LC}({}^{Th}W_{Oi} - {}^{Th}W_{LC})/({}^{Th}L_{Oi} - {}^{Th}L_{LC})] \\ /[(1 - W_E)/L_E + ({}^{Th}W_{Oi} - {}^{Th}W_{LC})/({}^{Th}L_{Oi} - {}^{Th}L_{LC})] \quad (7)$$

Finally, the fraction of phase 1 is directly obtained by combining Eqs. (5) and (7).

$$A_q = L_E[(1 - {}^{Th}W_{LC})({}^{Th}L_{Oi} - {}^{Th}L_{LC}) + {}^{Th}L_{LC}({}^{Th}W_{Oi} - {}^{Th}W_{LC})/ \\ [(1 - {}^{Th}W_{LC})({}^{Th}L_{Oi} - {}^{Th}L_{LC}) + {}^{Th}L_{LC}({}^{Th}W_{Oi} - {}^{Th}W_{LC})] \quad (8)$$

The fractions of the two remaining phases are calculated analogously. Having established the fundamental algebraic analysis, the importance of the relative humidity is illustrated by a numerical example from the literature.[13]

3.3. The Pertinent Part of the Phase Diagram

The partial phase diagram to be employed has the features of Figure 3.1 and the compositions of the relevant points sufficient for the present treatment. They are: (W, L, S) as (Aq), $(1, 0, 0)$; (Oi), $(0, 0.63, 0.37)$ and (LC), $(0.51, 0.04, 0.45)$. The equations for the tie lines limiting this three-phase region are:

$$(1, 0, 0) \text{ to } (0, 0.63, 0.37): L = 0.63 - 0.63W \quad (9)$$

$$(1, 0, 0) \text{ to } (0.51, 0.04, 0.45): L = 0.08163 - 0.08163W \quad (10)$$

$$(0, 0.37, 0.63) \text{ to } (0.51, 0.04, 0.45): L = 0.63 - 1.157W \quad (11)$$

These calculations assume equilibrium between the vapor and the emulsion and the composition of the vapor is hence, determined entirely by the vapor pressures.

3.4. Vapor Pressure and Composition

The calculations are simplified because for the two volatile compounds, the variation in the vapor pressure of water is insignificant.[13] The solubility of water into the linalool-surfactant solution remains small in the entire range with increasing surfactant content to the point at which the equilibrium phase is changed from water to liquid crystal. Hence, ideal conditions are presupposed with the vapor pressure of water remaining constant. With the vapor pressure of linalool being 0.3874 mm Hg, its molecular weight 154.26, and the molecular weight of the surfactant 346, the vapor pressure of linalool in the linalool- surfactant solution is expressed as:

$$P_L = 0.3874(L_{0i}/154.26)/((L_{0i}/154.26 + S_{0i}/346)) \quad (12)$$

In Eq. (12) L_0 is the weight fraction of linalool in the linalool/surfactant solution with negligible water content assumed. Since $L_{0i} + S_{0i} = 1$, the formula is simplified to:

$$P_L = L_{0i}/(1.1509 + 1.4305 L_{0i}) \quad (13)$$

Furthermore, since

$$L_{0i} = L_E/(1 - W_E) \quad (14)$$

$$P_L = L_E/(1.4305 L_E - 1.1509 W_E + 1.1509) \quad (15)$$

The weight fraction of linalool in the vapor phase, L_V, is

$$L_V = 154.26 P_L/(154.26 P_L + 360(1 - 0.01 RH)) \quad (16)$$

With $RH = 70\%$:

$$L_V = P_L/(P_L + 0.7001) \quad (17)$$

With $RH = 85\%$:

$$L_V = P_L/(P_L + 0.3501) \quad (18)$$

Expressing L_V as a function of L_E is an explicit process, but P_L is *per se* a unit of interest and Eqs. (17) and (18) are used in the numerical calculations. Since the surfactant is absent from the vapor the water weight fraction equals

$$W_V = 1/(1 + 1.428 P_L) \qquad (19)$$

for 70% RH and

$$W_V = 1/(1 + 2.856 P_L) \qquad (20)$$

for 85%. These equations are the basis for the calculation of the evaporation path in an illustrative numerical example.

3.5. Calculation of the Evaporation Path

The initial emulsion composition is (W_E^i, L_E^i, S_E^i) and the composition of the vapor is found from Eqs. (15) to (20). In each step of the evaporation calculation the water weight is changed by an amount Δw in order to make the time scale linear as function of the water weight. The corresponding change of the linalool weight is

$$\Delta l = (l_E/w_E)_S \Delta w \qquad (21)$$

From the changed values of w and l new values of W_E, S_E and L_E are calculated from

$$X_E = x_E / \sum_{x=w}^{x=1} X_E \qquad (22)$$

in which x and X represent water, linalool and surfactant.

3.6. Results

An overview of the evaporation path for the two values of relative humidity is exhibited in Figure 3.2. The results illustrate the main thesis of the investigation. During evaporation under conditions close to equilibrium, the relative humidity has a decisive effect not only on the direction of the evaporation path, but may also lead to a distinct difference in the pattern of phases encountered during the process. In the present case, an increase of the relative humidity from 70% to 85% leads to radically different behavior of the emulsion, in spite of the fact that both humidity values are well within the normal climate range. At 70% relative humidity, the evaporation resulted in a reduction of the number of phases from two to one because the aqueous phase was depleted and the final evaporation stage took place

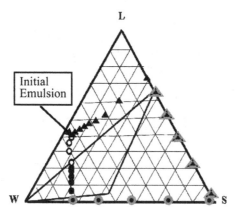

Fig. 3.2 An overview of the evaporation path outlined as the weight fractions of the compounds. Relative humidity (○, ●) 85% and (△) 70%.

from the oil phase with no additional phase changes. A narrative of the phases present in a total evaporation of the emulsion versus time may be schematized as:

$$Aq + Oi \to Oi.$$

Evaporation into an atmosphere of 85% relative humidity, on the other hand, gave a significantly more complex phase pattern with the following steps.

$$Aq + Oi \to Aq + Oi + Lc \to Aq + LC \to LC \to LC + Oi \to Oi$$

A more exact and also more easily comprehended description is to present the fraction of water and oil in a Cartesian coordinate system as depicted in Figure 3.3.

The diagram emphasizes the effect of the relative humidity. At 70% RH, the evaporation of water is far greater than that of the linalool and the weight fraction of water is reduced to zero, while that of linalool is increased; the opposite is true for 85% RH. The difference in behavior is remarkable.

These results accentuate the weight fractions of the compounds, but images presenting the weight of the compounds as function of the water content offer new information because the water weight is a linear function

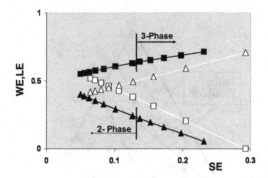

Fig. 3.3 The fraction of water (squares) and oil (triangles) with relative humidities versus the surfactant weight fractions. Squares are water weight fractions; triangles are linalool weight fractions. Open symbols are for 70% RH and filled symbols for 85%. The arrows in the curve for 85% relative humidity mark the change from 2 to 3 phases.

of time, in view of the fact that the vapor pressure of water is constant during the process. It is, hence possible to present both the weights and the weight fractions as a function of time; the latter with an arbitrary but equal scale for the two cases investigated.

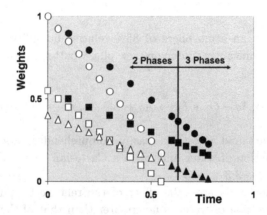

Fig. 3.4 The weights of water (squares), oil (triangles) and emulsion into (circles) with relative humidities 70% (open symbols) and 85% (filled symbols). The arrows in the curve for 75% relative humidity mark the change from 2 to 3 phases.

The curves add essential new information on the path of the process.

First, the time for evaporation is affected by the altered relative humidity. The rate of evaporation is reduced at 85% relative humidity and the time is hence, significantly increased. This reduction of rate is limited to the water, where it is reduced by a factor of 2. Since the rate of evaporation of the water can have no influence on the evaporation of linalool, the latter is not influenced, but due to the reduction in the evaporation rate of water, the evaporation rate of the total emulsion is reduced by a factor 1.3 for the present case, when the relative humidity is increased from 70% to 85%.

The amount of the compounds versus time is an essential part of the evaporation process, but the real amounts of the phases *per se* is probably of even more importance from a practical point of view. In the two-phase region the weight of each phase ie equal to the weight of each of the components, and offers little of additional interest to the results in Figure 3.3. The conditions in the three-phase region are considerably more complex as demonstrated by Eq. (8). However, with the numbers from the phase diagram available, the calculations are made significantly less arduous.

The equation for the line between the aqueous phase and the emulsion composition is:

$$L = L_E(1 - W)/(1 - W_E) \qquad (23)$$

The junction between this line and the one between the liquid crystalline phase and the oil phase, Eq. (11), is

$$W_{Aq-Em/LC-Oi} = [(L_E/(1 - W_E) - 0.63/[(L_E/(1 - W_E)1.157] \qquad (24)$$

The weight fraction of the different phases can now be determined:

$$Aq = (W_E - W_{Aq-Em/LC-Oi})/(1 - W_{Aq-Em/LC-Oi}) \qquad (25)$$

$$LC = (1 - Aq)W_{Aq-Em/LC-Oi}/{}^{Th}W_{LC} \qquad (26)$$

$$Oi = (1 - Aq)({}^{Th}W_{LC} - W_{Aq-Em/LC-Oi})/{}^{Th}W_{LC} \qquad (27)$$

Although, the variation versus evaporation time is essential to obtain a realistic view of the process, the total weight of the emulsion is the most easily determined property and it is considered instructive to plot the weights of the phases as function of the total weight of the emulsion to demonstrate the versatility of the phase diagram approach. Multiplying the values for the phases in Eqs. (25) - (27) by the weight of the emulsion results in the weight of the phases.

The weight of the two liquid phases is reduced approximately in parallel; the difference is an upshot of the variation in the vapor pressure of the

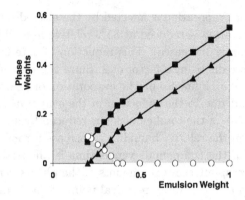

Fig. 3.5 The weight of the aqueous phase (□), the oil phase (△) and the liquid crystalline phase (○) versus the total weight of the emulsion.

linalool and a diagram of this entity is certainly of interest. It is drawn as the pressure in mm Hg in Figure 3.6, revealing two noteworthy features.

At first, the sudden onset of the constant vapor pressure when the liquid crystal is initiated is conspicuous and expected. However, the second feature, the fact that the higher relative humidity results in a lower linalool vapor pressure is not as immediately perceived a consequence. It is the consequence of the difference in the evaporation direction as outlined in the next section.

3.7. Discussion

Prior to discussing the general features of the phase diagram approach to the study of emulsion evaporation it is necessary to understand the relationship between the vapor pressures in Figure 3.6. At a first glance, it may be unexpected to find the vapor pressure of linalool to be less for combinations under greater relative humidity, since the relationship between the linalool and the relative humidity of the surrounding atmosphere appears spurious at best. The reason for this "anomaly" is a consequence of the difference between the vapor pressures, which reflects the equilibrium between the vapor phase and the liquid phase, Eq. (15), and the composition of the escaping vapor, which depends on the relative humidity according to Eq. (16). The initial vapor pressures are consequently equal according to Eq. (15), but after some evaporation, the vapor pressure is reduced under higher relative humidity conditions.

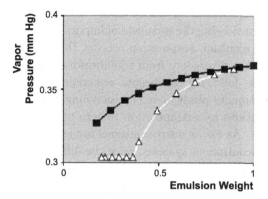

Fig. 3.6 The vapor pressure of linalool versus emulsion weight for emulsions under a relative humidity of 50% (△) and 75% (□).

This relationship is an unequivocal outcome of the dependence of the composition of the oil phase on the total composition of the emulsion. Combining Eqs. (1) and (4), the relationship of the vapor pressure to the emulsion composition is described as:

$$P_L = P_L^0 1/(1 + W_E M_L/L_E M_S) \\ = P_L^0 1/(1 + k W_E/L_E) \tag{28}$$

The linalool vapor pressure is reduced for higher values of W_E/L_E. This Eq. proves that greater values of W_E/L_E lead to reduced vapor pressure of linalool as a general phenomenon. Conversely, it offers no explanation of the relative levels of the two curves in the figure because it is not concerned with relative humidity of the surrounding atmosphere of the evaporation. In other words, it is not inherently related to the evaporation path. The relationship between the evaporation path and relative humidity is established from Eq. (16), for which a selected PL may be written:

$$L_V = 1/(1 - cRH) \tag{29}$$

$$(\partial L_V/\partial RH)_{P(L)} = c/(1 - cRH)^2 \geq 0 \tag{30}$$

A greater value of L_V leads to a greater reduction of lE from Eq. (21) followed by a reduction in P_L according to Eq. (28). As for the more fundamental and general features of the phase diagram approach, the most essential comment is that it presupposes equilibrium between the vapor and the liquid parts of the system. This condition is by no means generally true, but acceptable under controlled conditions as has been demonstrated by the Hull group.[3-8] Considering the wide field of important applications that are influenced by the emulsion evaporation process, the need for a more broad understanding of the divergence from equilibrium appears rather obvious.

At present, the following conclusions are credible. First, the evaporation from the continuous phase in an emulsion may be brought to values for equilibrium conditions by extrapolation to zero of the values for different evaporation rates. As far as microemulsions are concerned the range may be extended to bicontinuous systems, since the diffusion coefficient in such systems is not widely different from those in a solution.[14,15] For the dispersed compound in a microemulsion, the diffusion coefficient may be two orders of magnitude lower than in a solution and more research is obviously needed.

With these qualifications established, it is appropriate to emphasize the specific advantages of the approach. The methodology offers a unique opportunity to obtain information about potential phase changes during the evaporation,[9] the advantage of which may be less apparent in the selected example, but is distinctly evident for emulsions containing solids. These may cause serious processing problems and the prediction of a distillation route to avoid calamities is obviously of great value.

From a fundamental point of view an instructive example is found in the early contribution by Zhou et al.[16] in which specular neutron reflection was used to detect an ordered layering on the surface of a bicontinuous microemulsion. Such a result would actually be expected for an evaporation of the surface layer being faster than the corresponding diffusion to the surface in the system so investigated. A reduction of the fraction of the liquids would inevitably lead to the formation of a lamellar liquid crystal and an ordered structure is expected at the surface.

It should be noted however, that the fact that the phase diagram approach is an efficient tool to estimate the phases encountered during the evaporation leads to more complexity. The information concerning the actual effect on the evaporation rate by the appearance of phases with new structures is available only for isolated cases under less controlled conditions and the need for further information is obvious.

3.8. Conclusion

The evaporation path from an emulsion system was estimated using its phase diagram under the condition of equilibrium between the condensed phases and between them and the vapor. The results showed that under these conditions the relative humidity of the atmosphere into which the vaporization took place had a significant influence on the path.

References

1. J. Sjöblom, Ed. *Handbook of Emulsion Technology*. Ed. J. Sjöblom, Marcel Dekker, Inc. New York, (2001).
2. B.P. Binks, *Modern aspects of Emulsion Science*, Royal Chemical Society, London, (1998).
3. J.H. Clint, P.D.I. Fletcher, and D.I. Todorov, *Phys. Chem. Chem. Phys.*, 1, 5005 (1999).
4. K.J. Beverley, J.H. Clint and P.D.I. Fletcher, *Phys. Chem. Chem. Phys.*, 1, 149 (1999).
5. K.J. Beverley, J.H. Clint, and P.D.I. Fletcher, *Phys. Chem. Chem. Phys.*, 2, 4173 (2000).
6. I. Aranberri, K.J. Beverley, B.P. Binks, J.H. Clint and P.D.I. Fletcher, *Langmuir*, 18, 3471 (2002).
7. I. Aranberri, K.J. Beverley, B.P. Binks, J.H. Clint, and P.D.I. Fletcher, *Chem. Commun.*, 2538 (2003).
8. I. Aranberri, B.P. Binks, J.H. Clint and P.D.I. Fletcher, *Langmuir*, 20, 2069 (2004).
9. S.E. Friberg, T. Huang, and P.A. Aikens, *Colloids and Surfaces*, 121, 1 (1997).
10. S.E. Friberg, *Adv. Colloid Interface Sci.*, 75, 181, (1998).
11. S.E. Friberg, and A. Al-Bawab, *Langmuir*, 21, 9896, (2005).
12. S.E. Friberg, *Can. J. Chem. Engg.*, 85 (5), 602 (2007).
13. A. Al-Bawab, *DIRASAT* (2007).
14. P. Stilbs, M.E. Moseley, and B. Lindman, *J. Magn. Reson.*, 40, 401 (1980).
15. B. Lindman, P. Stilbs, and M.E. Moseley, *J. Colloid Interface Sci.*, 83, 569, (1981).
16. X.L. Zhou, L-T Lee, S-H Chen and R. Strey, *Phys. Rev. A*, 46, 6479, (1992).

Chapter 4

Aging and Stability of W/O Emulsions with NaOH in Aqueous Phase

Rujuta Thaker, Bina Sengupta and Ranjan Sengupta

Chemical Engineering Department
Faculty of Technology and Engineering
The Maharaja Sayajirao University of Baroda, Baroda

Aging and stability behavior of W/O emulsions with NaOH in internal phase was investigated. Emulsifier concentration 3% (wt) of oil phase was found to be optimal for emulsion formation. It was observed that increase in NaOH concentration in the inner phase resulted in reduction of emulsion viscosity. Increase in NaOH concentration also led to a sharp decline in emulsion stability. Increase in internal phase volume fraction increased the viscosity of the emulsion but did not influence the stability significantly. Aging of emulsions resulted in broadening of the drop size distributions. In spite of wide variation of internal phase volume and internal phase reagent concentration it was found that all drop size distributions pivoted around a typical drop size. It was observed that distributions tend to move towards specific drop sizes with aging, showing multiple peaks instead of a flat distribution.

4.1. Introduction

Water-in-oil emulsions containing strong electrolytes in the inner phase are used in the emulsion liquid membrane (ELM) technique to separate solutes such as toxic organics, heavy metals, biomolecules etc. from aqueous streams.[1,2] This technique, invented forty years ago by Li,[3] involves formulation of a water-in-oil emulsion and subsequent dispersion of the emulsion in a continuous phase that has to be treated (Fig. 4.1).

When dispersed the emulsion breaks up into tiny globules and within each globule there exists tiny droplets of the aqueous phase of the emulsion. The emulsion works as an extracting solvent for various ions and molecules. It dissolves the solute at the outer surface of the globule that is in contact with the bulk phase and transports the solute to the inner phase of the emulsion.

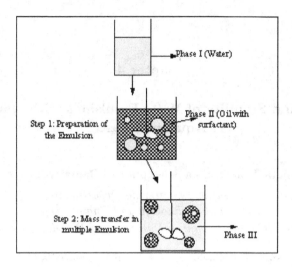

Fig. 4.1. Emulsion liquid membrane separation

After extraction, the emulsion can be separated from the bulk aqueous phase and split to recover the trapped solute and the various emulsion components can be recycled.

Presence of a stripping agent in the inner phase helps strip the solute and encapsulates it within the emulsion, thus phenol can be removed from aqueous streams using emulsion liquid membranes by reacting phenol with NaOH present in the inner phase of the emulsion. Phenol from the feed phase diffuses through the oil phase of the emulsion to reach the inner aqueous phase. An irreversible reaction between phenol and NaOH present within the inner aqueous phase forms sodium phenolate in the internal aqueous phase of the emulsion, being insoluble in the membrane sodium phenolate gets entrapped in the emulsion.[4-7] The oil phase of the emulsion that separates the two aqueous phases works as the liquid membrane.

When the targeted solute has very low solubility in the oil phase such as metals ions in organic solvents, an extractant is added to the oil phase that selectively complexes with the metal ion at the oil external phase boundary. The metal - extractant complex transports to the inner aqueous phase boundary where a reagent induces a stripping reaction to strip the metal in the inner aqueous phase and simultaneously release the extractant in the oil phase that can shuttle back, bind with more ions and transport them to the inner phase of the emulsion. Such transport mechanism is called carrier-mediated transport. One major problem related to ELM separation is the issue of stability of W/O emulsions. Emulsions are inherently unstable hence; suitable surfactants are used to stabilize emulsions. W/O emulsions are usually stabilized by the 'Span' family of emulsifiers particularly Span 80 which is easily available. However, the stability of emulsions prepared using Span 80 with NaOH in the inner aqueous phase has limited stability. This feature is a major hindrance in the commercialization of ELMs for removal of toxic organics from aqueous streams.

The current investigation is an experimental exploration of the aging behavior of W/O emulsions stabilized by Span 80 containing sodium hydroxide in the inner aqueous phase. The objective was to map the parameter space to get an idea of the influence of parameters such as volume fraction of the internal phase of emulsion ϕ, internal reagent phase concentration Cio and surfactant concentration on stability of emulsions. The data can be suitably used to formulate emulsions having greater stability for industrial exploitation of this technique.

4.2. Materials and Reagents

Kerosene supplied by Indian Oil Corporation Limited, having boiling range 152°C - 271°C containing n paraffins(27%), naphthenes(56%), aromatics(16%) and olefins(1%), having ρ_{15} of 821.3 kg/m^{-3} was used as the oil phase. Sorbiton monooleate (Span 80) supplied by S.D. fine chemicals was the emulsifier. All other reagents used were of AR grade.

4.3. Experimental

Kerosene, Span 80 (1–5% wt of oil phase) and sodium hydroxide (internal phase) were blended using a high speed blender at 12,000 rpm to form the W/O emulsion. The internal phase volume fraction (ϕ) was varied from 0.4–0.7 and the internal phase reagent concentration (C_{io}) was varied from 0.3 to 1 $mol\ dm.^{-3}$ The emulsions were characterized by measurement of viscosity and internal drop sizes. Brookfeld cone and plate rheometer model LVDV III + CP was used to determine the viscosity of emulsions at different shear rates.

The internal drop sizes of the emulsions were measured microscopically at regular intervals of time, to study the morphological changes due to aging, using an Olympus microscope, model BH-2 attached with an Olympus photo micrographic system model PM-10AD at a magnification of 600X.

The emulsion samples were diluted with kerosene and more than one thousand internal droplets per sample were counted and the Sauter mean diameter was calculated for each case. The time taken for sedimentation of the emulsion was taken as a measure of emulsion stability.

4.4. Results and Discussion

It is well known that measurable emulsion properties such density, viscosity, drop size distribution *etc.* change with the internal phase volume fraction of the emulsion. Influence of concentration of internal phase reagent i.e. the concentration of NaOH in the inner aqueous phase of the emulsion (C_{io}) on emulsion properties is not well established. Emulsion viscosity is one of the fundamental factors that would influence the dispersion behavior of emulsions. Hence, the influence of ϕ as well as C_{io} on emulsion viscosity were investigated.

4.5. Effect of ϕ and C_{io} on emulsion viscosity

The effect of C_{io} and ϕ on viscosity at shear rate 38.4 sec^{-1} is shown in Fig. 4.2. As expected there is a multifold increase in viscosity with an increase in internal phase volume fraction of the emulsion. The presence of NaOH at low concentration does not influence the viscosities of the emulsion, but on the whole there is a decline in the emulsion's viscosity with an increase in C_{io}. Such decrease in viscosity is not very pronounced at ϕ values up to 0.5, but is significant at higher values of ϕ. Similar trend was also observed at higher shear rates.

Sherman[8] reported that the hydration of emulsifier layer adsorbed around the internal drops influences the rheological properties of an emulsion. Addition of an electrolyte alters the degree of hydration of the emulsifier. An increase in electrolyte concentration results in a decrease of the effective thickness of hydrated adsorbed emulsifier layer. The reduction of emulsion viscosity with increase in sodium hydroxide concentration is attributed to this phenomenon.

It is interesting to observe that at $C_{io} = 0.3$ mol dm^{-3} there is a dip in the viscosity verses Cio curve for all values of ϕ. While this behavior is currently unexplainable, it is worthwhile to note that Gruneisen in 1905 first reported negative curvature at the dilute end of viscosity-concentration curve for dilute electrolyte solutions, the inflection of curve was at 0.3 mol/kg^{-1}. This behavior is known as the Gruneisen effect.[9]

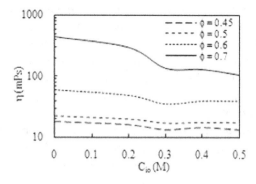

Fig. 4.2. Effect of C_{io} and ϕ on viscosity of emulsion

4.6. Effect of surfactant concentration

W/O emulsions with $\phi = 0.5$ containing 0.3 $mol\ dm^{-3}$ NaOH were prepared with varying surfactant concentrations ranging from 3% to 5% by weight of the oil phase. The internal drop size distribution data is reported in Fig. 4.3. It can be seen that for 3% surfactant concentration more than 50% droplets are of 3.3 μm in size. There is evidence of a second peak with droplet size of around 6.5 μm.

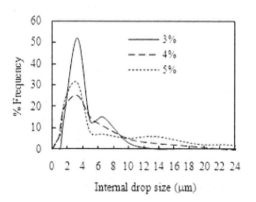

Fig. 4.3. Effect of surfactant concentration on internal drop size distribution ($C_{io} = 0.3 mol\ dm,^{-3}\ \phi = 0.5$)

There is a sharp decline in the peak frequency and a consequent broadening of the internal drop size distribution when surfactant concentration is increased from 3% to 5%. There is an evidence of emergence of new peaks in the distribution when surfactant concentration increases to 5% indicating a polydisperse nature of the emulsion. The Sauter mean diameter d_{32} for these distributions were found to be 5.5 μm for 3%, 14 μm for 4% and 14.6 μm for 5% respectively.

Emulsions were also prepared with surfactant concentration of 1% and 2% (wt of oil phase) but in both these cases the emulsion had very limited stability. The nature of the drop size distribution indicates that 3% wt of surfactant in the oil phase would be optimal for formation of W/O emulsions for ELM separations and hence all emulsions studied were prepared with Span 80 concentration of 3% wt of oil phase.

4.7. Effect of aging

Aging of W/O emulsions was studied by varying the internal phase volume fraction from 0.4−0.7 and internal phase NaOH concentration from 0.3 $mol\ dm^{-3}$ to 0.5 $mol\ dm.^{-3}$ Figs. 4.4, 4.5 and 4.6 show the morphological changes taking place during aging of emulsions at C_{io} 0.3, 0.4 and 0.5 $mol\ dm^{-3}$ respectively. There was considerable decline in emulsion stability with increase in NaOH concentration as shown in Table 4.1. Effect of ϕ did not play an important role in the stability of the emulsions.

Table 4.1 Stablity of Emulsion with Time

C_{io} ($mol\ dm^{-3}$)	ϕ = 0.4	0.5	0.6	0.7
	Time (min.)			
0.2	293	266	253	245
0.3	184	184	180	180
0.4	90	86	84	82
0.5	69	60	60	64

It is seen in Fig. 4.4 that the emulsion undergoes rearrangement of drop sizes as it ages. No significant change in drop distributions was observed for $\phi = 0.4$ and $C_{io} = 0.3$ from 30 min. to 120 min. (Fig. 4.4a) except for the emergence of a second peak at drop size of 9μm. Increasing ϕ from 0.4 to 0.5 gives a drop size distribution with two peaks, the primary peak at 3.3 μm with a frequency of 50% and a secondary peak at 6.6 μm with a frequency of 15%. With passing time the second peak gets consolidated with a corresponding decline in the frequency of the first peak. The drop size distribution remains stable between 60 to 120 min. as seen in Fig. 4.4b.

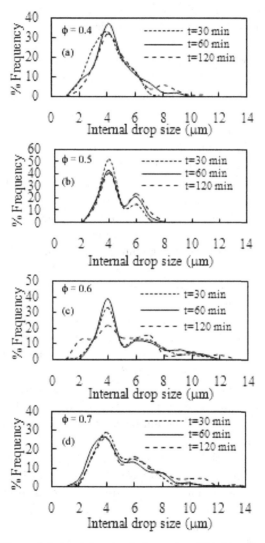

Fig. 4.4. Effect of time on internal drop size distribution, ($C_{io} = 0.3$ $mol\ dm^{-3}$)

Increasing ϕ to 0.6 gives a broader drop size distribution in comparison with that obtained for $\phi = 0.5$. This drop size distribution is quite stable up to 60 min. (Fig. 4.4c) however; further increase in time flattens the distribution significantly indicating existence of large droplets within the emulsion. Such behavior is attributed to internal circulation and coalescence with in the emulsion phase. Further increase in ϕ to 0.7 stabilizes the emulsion and no significant change in the drop size distribution is noticed with time.

Increasing internal phase NaOH concentration to 0.4 $mol\ dm^{-3}$ leads to reduction in emulsion stability. The effect of aging was studied for ϕ 0.4 to 0.7 as shown in Fig. 4.5. In all four cases considerable rearrangement of internal phase occurs with passage of time. In Fig. 4.5a there is moderate decline in the peak frequency with time and a sizeable broadening of the distribution. In Fig. 4.5b the distribution of drop sizes at 15 min. and 60 min. are quite similar with two peaks being prominently visible in the distribution. However, at 30 min. broadening of the distribution is observed with disappearance of the second peak that reappears at 60 min.

In Fig. 4.5c the whole distribution shifts to the right with time indicating an increase in drop sizes with time. Interestingly the frequency of the main peak (3.3 μm) does not change but larger droplets redistribute giving peaks at 6.6 μm and 12 μm. In Fig. 4.5d a distinct second peak emerges with time consequently, there is a decline in the frequency of the first peak with the emergence of the sec.

With $C_{io} = 0.5$ and $\phi = 0.4$ there is moderate broadening of the drop size distribution as seen in Fig. 4.6a but dramatic effects are observed with $\phi = 0.5$ (Fig. 4.6b) there is sharp decline in the peak frequency with time. The distribution is considerably flat indicating large presence of big droplets. Similar effect was observed when $\phi = 0.6$ as well as $\phi = 0.7$ (Fig. 4.6c and 4.6d) the initial drop size distribution shows three peaks that are retained as time passes although peak frequencies decrease resulting in flatter distribution.

It was observed with all drop size distributions that the distribution actually pivots around a typical drop size where the distribution frequency peaks. In this investigation it is 3.3 μm. This peak drop size retains its identity over time as well as changes in ϕ and NaOH concentrations, although changes in frequency with time are observed. It is also interesting to find that distributions tend to move towards some specific drop sizes rather than retaining wide diversity in sizes. Hence the emergence of second peak in the drop size distributions.

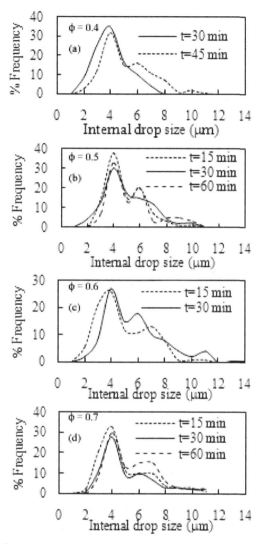

Fig. 4.5. Effect of time on internal drop size distribution ($C_{io} = 0.4$ mol dm^{-3})

It is difficult to attribute a single cause for the aging and unstable behavior of the emulsions; multitude of physical as well as chemical factors could be responsible for the variation in drop size distribution with time. It appears that internal circulation and coalescence of droplets are more

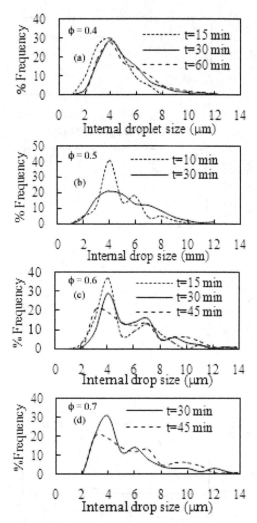

Fig. 4.6. Effect of time on internal drop size distribution ($C_{io} = 0.5$ mol dm^{-3})

likely to change the distribution in the small time scales than Ostwald ripening. However, one cannot rule out the possibility of hydrolysis of the emulsifier (Span 80) by sodium hydroxide that would lead to destruction of the emulsifier film around the internal droplets and assist drop coalescence resulting in emulsion breakage.[10,11]

A more exhaustive investigation is necessary to provide a clear idea on the stability behavior of emulsions and its quantification.

4.8. Conclusions

This experimental investigation indicates that emulsifier Span 80 concentration of 3% (wt of oil phase) is optimal for formulation of W/O emulsions for ELM extractions. Increase in Span 80 concentration leads to an increase in emulsion poly-dispersity and sharp increase in Sauter mean diameter.

Increase in internal phase volume fraction of the emulsion results in sharp increase in the emulsion viscosity but does not significantly influence emulsion stability. Increase in internal phase $NaOH$ concentrations lead to reduction in emulsion viscosity and a sharp decline in emulsion stability.

Drop size distributions broaden with time. It was interesting to note that distributions in all cases pivoted around a typical drop size of 3.3 μm. Although broadening of distribution took place with time but there was a tendency for distributions to move towards specific drop sizes that results in formation of multiple peaks with time instead of a flat distribution profile.

With reference to application of W/O emulsions to ELM separations it could be concluded that $NaOH$ concentration of 0.3 $mol\ dm^{-3}$ in internal aqueous phase would be particularly suitable for formulation of extracting emulsions in view of their reduced viscosity and longer stability that would aid the dispersion behavior as well as provide flexibility of operation in terms of time.

References

1. Z. Gu, W. S. Ho, and N. N. Li, in *Membrane Handbook*, Eds., W. S. Ho and K. K. Sirkar (Chapman and Hall, New York, 1992), 656.
2. W. S. Ho and N. N. Li, in *Chemical Separations with Liquid Membranes*, Eds., R. A. Bartsch and J. Douglas Way, *ACS Symposium Series*, 642 (American Chemical Society, Washington, DC, 1996), 208.
3. N. N. Li. *U.S.Patent 3*, 410, 794 (1968).
4. R. P. Chan and N. N. Li, *Sep Sci.*, 505, (1974).
5. N. N. Li and A. L. Shrier. in *Recent Developments in Separation Science*, (Chemical Rubber Company, Cleveland, 1972), 1, 163.
6. W. S. Ho, T. A. Hatton, E. N. Lightfoot and N.N. Li, *AICHE Journal*, 662, (1982).
7. M. Teramoto, H. Takinaha, M. Shibutani, T. Yuasa and N. Hara, *Sep. Sci. Technol.*, 397, (1983).

8. P. Sherman. in *Encyclopedia of Emulsion Technology*, Ed. P. Becher, (Marcel Dekker, New York, 1983), 1, 405.
9. A. L. Horvath, Ed., *Handbook of Aqueous Electrolyte Solutions* (Ellis-Worwood, New York, 1985).
10. I Abou-Nemah and A. P. Van Peteghem, *Chem. Ing. Tech.*, 420, (1990).
11. I Abou-Nemah and A. P. Van Peteghem, *J Membr Sci.*, 9, (1992).

Chapter 5

Energetics of Micelle Formation: Non Agreement between the Enthalpy Change Measured by the Direct Method of Calorimetry and the Indirect Method of van't Hoff

Satya P. Moulik and Debolina Mitra

Centre for Surface Science
Department of Chemistry
Jadavpur University, Kolkata 700032, India

In this paper, calorimetric and van't Hoff values of change in enthalpy and heat capacity for micellization process in aqueous solution have been compared for different surfactant systems. The measurements taken at different temperatures have shown that both CMCs and enthalpy changes are temperature dependent quantities, and the results obtained by the two procedures significantly differ for ionic surfactants. The van't Hoff process deals only with the self-aggregation of the surfactants whereas calorimetry records contribution of micellization and other associated processes operative in the system. The former although considers the counterion binding effect on the energetic parameters, it normally ignores the micellar aggregation number related component and other factors like non-idealities in the system, inter micellar interaction, *etc.* which inherently contribute to the process enthalpy change. It is worth mentioning that although the enthalpy changes for micellization determined by the two procedures differ, the changes in the heat capacity are mostly similar. The nonionic surfactants produce close agreement of results between the two procedures suggesting appreciable role of the head group charges of the micelle and the electrostatic interaction with the counterions and other related phenomena that deviate the results of ionics from that of nonionics. Nearly constant differences between the van't Hoff and calorimetry derived enthalpies for the herein reported surfactants, and the close agreement between the heat capacity changes obtained from the two procedures envisage necessity for more conceptual inputs in the solution of the problem.

5.1. Introduction

Physical and chemical processes are energetically controlled. Determination of energetic parameters like free energy, enthalpy, entropy and specific heat changes are essential for a complete understanding of the spontaneity or feasibility of such a process. The free energy change can be assessed or evaluated from the knowledge of the equilibrium and the determination of the equilibrium constant. Its temperature dependence helps to evaluate the change in process enthalpy and entropy; the dependence of enthalpy change on temperature in turn provides the specific heat. The equilibrium constant depends on the nature of the process; estimation of the strengths of the free components in the system is required to get the constant. The associated enthalpy variation is either determined from the temperature related values of the equilibrium constant by the use of van't Hoff relation or by direct measurements of heat in a calorimeter. In most studies, the evaluation is made from the van't Hoff rationale because calorimetric measurements are time consuming and requires special measuring arrangements, which are not commonly available in most laboratories. But the two methods have several basic differences whose consideration is essential for the evaluation of the change in enthalpy parameter. Normally it is not critically looked into. The van't Hoff method is a differential process whereas calorimetry is an integral process. Thus, agreement between the two methods is not expected.[1-8] The reasons for the difference can be several but so far not much attention has been given to the issue.

To test the above anomaly, determination of equilibrium constant at different temperatures by a suitable method is required to evaluate the enthalpy change by van't Hoff procedure, and separate determination of enthalpy change in a calorimeter should encounter inaccuracies of the results obtained from the two different procedures. Thus, a true comparison between the two enthalpies (and hence entropies and specific heats) by the above employed procedures is hardly feasible. Hence measurements of physicochemical processes in a calorimeter to get both the free energy and the enthalpy changes should be so far the best proposition. Besides, the method on the whole is easy handling so that minimum errors creep in during measurements and data treatment.

The processes of amphiphile self-association or micellization is a well established equilibrium process,[9-12] where determination of the critical micellization concentration and hence the change in free energy and consequently the enthalpy can be straightforwardly determined using an Isothermal Titration Calorimeter (ITC).[13-23] The method is quick, accurate and helps to determine CMC and related free energy and enthalpy change from the same sets of experiments done at different temperatures.

In the following presentation, results of calorimetric measurements of micellization of a number of ionic (cationic and anionic) and nonionic surfactants will be considered, and the derived thermodynamic (energetic) parameters will be presented and discussed. The probable reasons for agreement and disagreement between the van't Hoff way of determination of enthalpy change, *etc.* with calorimetric results will be discussed, which we believe has not been done so far quantitatively in the past.

5.2. Fundamental Considerations

The following is the basics of the proposition mentioned above. The micelle formation is modeled either as formation of a "Pseudophase" in solution or in terms of the "Mass Action" principle. The pseudophase concept simplifies the process but neglects some important consequences of the self-aggregation process and hence is less rigorous or accurate.[24,25] The "Mass Action" principle does not have such shortcomings and can be treated as reversible thermodynamic chemical equilibrium.[26] In all our discussion we shall consider that the Mass Action principle of micelle formation holds for the systems for consideration.

For nonionic amphiphiles the following equilibrium process holds.

$$nS \rightleftharpoons S_n \text{ or } M \tag{1}$$

where n is the average aggregation number (the number of monomer molecules that constitute a micelle), and S_n or M represents the micelle. The micellization constant K_M (an equilibrium constant) can be written in the form shown below for nonionic surfactants,

$$K_m \rightleftharpoons a_M/(a_S)^n \equiv c_M/(c_S)^n \tag{2}$$

where, a_M and a_S are the activities of the micelle and the free surfactant monomers, respectively; c_M and c_S are their respective concentrations in solution at equilibrium. The activities have been considered equal to concentrations on the ground that (i) solutions are fairly dilute at the CMC points (normally in millimolar range) and (ii) determination of activity of the components in such solution is not an easy proposition.

The standard Gibbs free energy change for micellization, ΔG^o_M can then be obtained from the relation,

$$\Delta G^o_M = -RT \ln X_M + nRT \ln X_S \tag{3}$$

or

$$\Delta G^o_m = (\Delta G^o_M / n) = -(RT/n) \ln X_M + RT \ln X_S \tag{4}$$

to yield
$$\Delta G_m^o = -(RT/n)lnX_M + RTlnCMC, \qquad (4a)$$
where $CMC = X_S$.

In relations 3 and 4, X represents concentration in mole fraction unit, and ΔG_M^o stands for the standard Gibbs free energy change per mole of monomer unit. To get X_M, knowledge on the concentration of monomer in the micellar form as well as the aggregation number, n is required. It is considered that at CMC the concentration of monomers in micelle is within 5% of the total,[27] so that the equivalence is valid.

For ionic surfactants, we can write,
$$nS^\pm + mI^\mp \rightleftharpoons (S_nI_m)^{\pm(n-m)} \text{ or } (M)^{\pm(n-m)}, \qquad (5)$$
where n number of surfactant ions have combined with m number of counterions to form the micelle with resultant charge $\pm(n-m)$, and $n > m$ so that sign of the net micellar charge is always that of the surfactant ion. The micellization constant, K_M is thus,[9,23,28]
$$K_M = [C_{M^{\pm(n-m)}}] / [(C_{S^\pm})^n (C_{I^\mp})^m] \qquad (6)$$
The standard Gibbs free energy change, therefore, becomes
$$\Delta G_m^o = -RTlnX_{M^{\pm(n-m)}} + nRTlnX_{S^\pm} + mRTlnX_{I^\mp} \qquad (7)$$
or
$$\Delta G_m^o = (\Delta G_M^o/n) = -(RT/n)lnX_{M^{\pm(n-m)}} + RT(1+\beta)lnX_{I^\mp} \qquad (8)$$
so that
$$\Delta G_m^o = -(RT/n)lnX_{M^{\pm(n-m)}} + RT(1+\beta)lnCMC, \qquad (9)$$
where $CMC = X_{S^\pm}$ as in Eq. (4a).

In Eqs. (8) and (9), $\beta = m/n$; R and T have their usual significance. The standard state is the hypothetical state of ideal solution of unit mole fraction that applies to both the micelles and the monomers. Normally, for most amphiphiles n is large, where $X_{S^\pm} = X_I^\mp$ so that the first term in Eqs. (4a) and (9) can be neglected because of small magnitude to modify Eqs. (4a) and (9) to Eqs. (10) and (11), respectively. Thus,
$$\Delta G_m^o = RTlnCMC \qquad (10)$$
and
$$\Delta G_m^o = (1+\beta)RTlnCMC \qquad (11)$$

The Eqs. 10 and 11 are used for the micellization of nonionic and ionic amphiphiles respectively. The β involved term is considered as the contribution of the electrical free energy change to the micellization process.[29]

The ΔG_m^o then has two parts, (i) $RTlnCMC$ and (ii) $\beta RTlnCMC$. The first stands for favorable hydrophobic component and unfavorable non-electrostatic component reflecting the repulsion between hydrophilic head groups that functions to limit micellar growth. The second represents the electrical free energy change that arises from electrostatic interaction between opposite charges of micelle and its counterions. There is another contribution that arises from the part of aggregation number (n) to the process of free energy change, which we do not consider for want of reliable data on different systems at different temperatures. It has been found also that, the aggregational contributions to the free energy change and hence to the enthalpy change is small.[3,4,30]

To obtain process enthalpy change for the micellization event, the following relation is used,

$$[\partial(\Delta G_m^o/T)/\partial(1/T)]_p = \Delta H_m^o. \qquad (12)$$

Like the free energy, change in enthalpy has also contributions from hydrophobicity, electrical process and aggregational process. Thus, we may write,

$$\Delta H_m^o = (\Delta H_m^o)_{hy} + (\Delta H_m^o)_{el} + (\Delta H_m^o)_{agg}. \qquad (13)$$

The subscripts 'hy', 'el' and 'agg' represent hydrophobic, electrical and aggregational components, respectively.

From the knowledge of dependence of CMC, β and n on temperature, separate contributions of them on the enthalpy variation of the overall micellization process can be estimated. Alternatively, from the knowledge of ΔG_m^o obtained considering all the above contributing factors, the overall enthalpy change of the process can also be calculated from the temperature dependence of the free energy change. We have followed both, and have found good correlations between them. We reiterate here that the third term in Eq. (13) has been neglected for want of reliable data on many systems and its weak contribution to the process energetics as stated above.

5.3. Calorimetry

The ITC measurements were taken in an OMEGA microcalorimeter of Microcal, Northampton (MA, USA). In the method, a concentrated solution

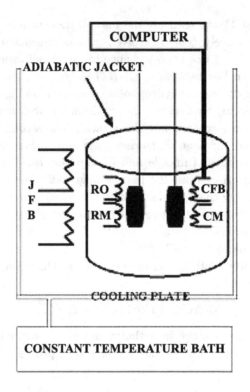

Fig. 5.1. Basics of a cell compartment in an OMEGA ITC (Northampton, MA, USA). R and C are reference and sample cell respectively; CFB and JFB refer to cell and jacket feedback; RM and CM are reference and cell main; RO is reference offset. Operating range of the instrument is from -230 to $+230$ μcal sec.$^{-1}$

of a surfactant ($\sim$ 15–20 times its CMC) was taken into a 350 μL micro syringe fixed in the calorimeter cell having two compartments [one for pure water (reference or control), and the other for addition of surfactant in water]. After equilibration of the system at a constant temperature, 5–10 μL of the surfactant solution was added in 30–35 steps to 1.325 mL water in the calorimeter cell for injection duration of 30 sec. at intervals of 4 min. under constant stirring (300 rpm). The heat absorbed or evolved in the process was monitored in the instrument, and from the integrated heat values for all titration steps, enthalpy change per mole of injectant for all titration steps was calculated. The whole procedure was computer controlled

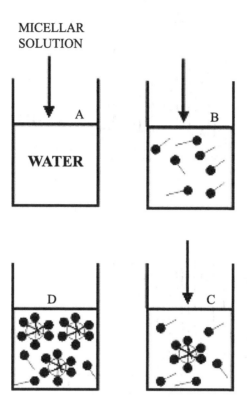

Fig. 5.2. Process of micellar dilution in steps of a concentrated surfactant solution in water in a microcalorimeter. $A - D$ refer to states of the solutions before and after incremental surfactant addition. A, start of addition; B, complete demicellization with monomer dilution on surfactant addition; C, no demicellization, micelles start appearing; D, micellar dilution with enhanced micellar concentration.

including calculations and graphical representations using ITC Microcal Origin 2.9 inbuilt software. Water was circulated within the calorimeter by a NESLAB RTE100 bath at a temperature lower within five degrees of the experimented temperature. The temperature in the cell compartment of the calorimeter was automatically scanned up to the desired temperature of measurement and adjusted with an accuracy of $\pm 0.01K$. The basics of the cell compartment are illustrated in Fig. 5.1 wherein R and C represent reference and sample cell compartments, respectively. Other technical

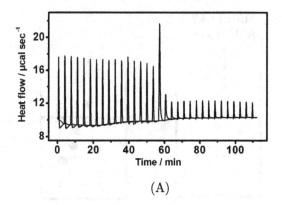

(A)

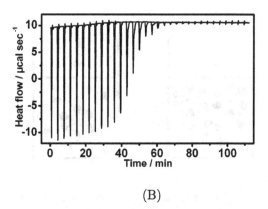

(B)

Fig. 5.3. Raw data for (a) an exothermic and (b) endothermic titration in an ITC at 303K. Injection matrix: 1–16 injections $5\mu L$ each, 17–32 injections $15\mu L$ each. A : CPC (concentration in syringe = 18.063 mM); B : MEGA-10 (concentration in syringe = 101.4679 mM).

part representations require no introduction here. The dilution and micelle formation stages are schematically presented in Fig. 5.4. A represents condition for no addition of surfactant in the cell *i.e.*, the starting point. B illustrates demicellization and monomer dilution on surfactant addition; at C demicellization has stopped: existence of micelles in the solution is indicated. Next, at D, the solution is having increasing presence of micelles with simultaneous micellar dilution. The solution composition is guided by the Mass Action principle. The heat change in the demicellization-micellization

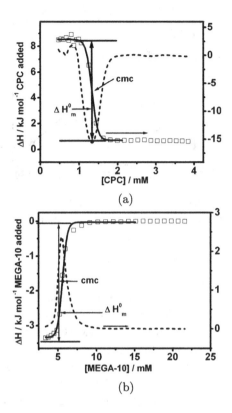

Fig. 5.4. Enthalpy of dilution and its derivative plot at 303K for (a) CPC and (b) MEGA-10. CMC points and ΔH_m^o are indicated in the diagram. For CPC, values of asymptotes are $A_1 = 8.5296 \pm 0.0645$, $A_2 = 0.69802 \pm 0.0856$, and CMC (from SB fitting) = 1.3284 ± 0.0078, CMC (from first derivative plot)=1.33. For MEGA-10, values of asymptotes are $A_1 = -3.3374 \pm 0.0349$, $A_2 = -0.0325 \pm 0.0164$, and CMC (from SB fitting) = 5.5872 ± 0.0183, CMV (from first derivative plot) = 5.5449.

region is significant to yield sigmoidal variation whose analysis is essential for the evaluation of CMC and ΔH_m^o to be discussed subsequently. It should be mentioned that the CMC and ΔH_m^o obtained from ITC measurements hardly depend on the initial concentration of the surfactant taken in the injection syringe.[22] We have used [surfactant]~15−20 times CMC and have also observed very minor differences in CMC upon changing the initial concentration.

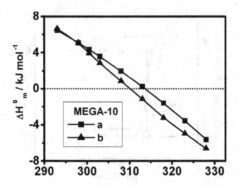

Fig. 5.5. ΔH_m^o vs. T plot for MEGA-10 system to compare (a) van't Hoff and (b) microcalorimetric enthalpy results.

5.4. Results

The titrated raw data (heat evolution and absorption) are documented in Figs. 5.3(a) and 5.3(b) for two amphiphiles, cetylpyridinium chloride (CPC) and decanoyl-N-methylglucamide (MEGA 10), respectively. From these results, the realized enthalpy change of dilution of the said amphiphile (surfactant) solutions at different stages of titration are respectively profiled in Figs. 5.4(a) and 5.4(b). In Fig. 5.4(a), the process was exothermic which was endothermic for the other process documented in Fig. 5.4(b). The inflection points in the profiles were the CMC points, which can be better estimated from their derivative plots (also shown in the diagrams). The differences between the initial and final asymptotes were considered as the enthalpy change of the related micellization process (as indicated in the plots by the perpendicular lines touching the asymptotes). The asymptotes were obtained from the Sigmoidal–Boltzman (SB) statistical relation.[14,31] The ITC derived data were used to get both CMC and the variation in enthalpy through calculation in a computer. Alternatively, the change in enthalpy can be estimated by manually drawing the asymptotes and the CMC from the peaks of the derivative plots shown in Fig. 5.4. The SB procedure yields both CMC and the asymptotes from the fitting of the data, and arbitrariness are thus minimized. A sharp fall of enthalpy values between the two asymptotes would produce better fit and accurate results.

For each system, the measurements were taken at different temperatures, and the data analysis were done as described above. The CMC values and the ΔH_m^o values determined were processed in the van't Hoff

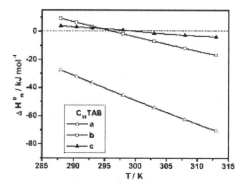

Fig. 5.6. ΔH^o_m vs. T plot for C_{10} TAB system.
(a) van't Hoff derived results using β measured by conductometry, (b) van't Hoff results using $\beta = 1$ (complete counterion condensation) and (c) microcalorimetric enthalpy results.

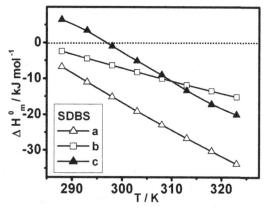

Fig. 5.7. ΔH^o_m vs. T plot for SDBS.
(a) van't Hoff derived results using β measured by conductometry, (b) van't Hoff derived results using $\beta = 1$ (complete counterion condensation) and (c), microcalorimetric enthalpy results.

rationale to get the energetic parameters, and their comparison with the calorimetric results has been made. Since the ITC results are considered accurate, analysis and comparisons as well as related discussions are expected to be more sound compared with data obtained by other methods (in no other method direct evaluation of enthalpy change of the process under investigation is possible). The dependence of van't Hoff and calorimetric enthalpies on temperature for nonionic surfactant (MEGA-10) as well as for ionic (decyltrimethylammonium bromide, C_{10}TAB[3], and sodium dodecylbenzenesulfonate, SDBS[32]) systems are presented in Figs. 5.5-5.7, respectively. The comparative enthalpy variation values for a good number of self-aggregating (micellar) systems are presented in Tables 5.1 and 5.2.

Table 5.1. Temperature dependent enthalpy change of micellization by $(\Delta H^o_m)_{ITC}$ and van't Hoff method $(\Delta H^o_m)^a_{VH}$ for MEGA-10, C_{10} TAB and SDBS.

Temp/K	$(\Delta H^o_m)_{ITC}$ [$(\Delta H^o_m)_{VH}$] / kJ mol^{-1}		
	MEGA-10	C_{10} TAB[3]	SDBS[32]
288	–	3.57±0.11	6.40 ± 0.19
		[-27.5 ± 1.38]	[-6.70 ± 0.34]
290.5	–	2.80 ± 0.08	-
		[-32.2 ± 1.61]	
293	6.63±0.20	2.25±0.07	3.40±0.10
	[7.12±0.32]	[-36.7±1.84]	[-11.0±0.55]
298	5.08±0.15	0.89±0.03	-1.00±0.03
	[5.26±0.25]	[-45.6±2.28]	[-15.1±0.76]
300.5	3.94±0.12	–	–
	[4.36±0.22]		
303	2.86±0.09	-1.44±0.04	-5.10±0.15
	[3.47±0.18]	[-54.2±2.71]	[-19.1±0.96]
308	0.84±0.03	-2.85±0.09	-9.0±0.27
	[1.72±0.10]	[-62.6±3.13]	[-23.0±1.15]
313	-1.14±0.03	-3.97±0.12	-13.4±0.40
	[0.04±0.01]	[-70.6±3.53]	[-26.7±1.33]
318	-3.22±0.10	–	-17.2±0.52
	[-1.59±0.08]		[-30.3±1.52]
323	-4.96±0.15	–	-20.1±0.60
	[-3.17±0.18]		[-33.8±1.69]
328	-6.61±0.20	–	–
	[-4.71±0.28]		

[a] For ionic surfactants, $(\Delta H^o_m)_{VH}$ values were estimated using measured β values by conductometry.

Table 5.2. Temperature dependent enthalpy change of micellization by ITC $(\Delta H^o_m)_{ITC}$ and van't Hoff method $(\Delta H^o_m)^a_{VH}$ for TX-100, CPC, CDMAB and NaC.

Temp/K	$(\Delta H^o_m)_{ITC}$ [$(\Delta H^o_m)_{VH}$] / kJ mol^{-1}			
	TX-100[33a]	CPC[1]	CDMAB[33b]	NaC[16]
285	–	–	–	4.48±0.13
				[2.84±0.14]
286	–	–	–	4.39±0.13
				[2.50±0.13]
288	12.9±0.39	3.65±0.11	-5.90±0.18	–
	[16.1±0.81]	[-43.1±2.16]	[-14.5±0.72]	
293	10.6±0.32	0.50±0.01	-9.40±0.28	–
	[14.1±0.71]	[-43.1±2.16]	[-18.1±0.90]	
298	8.93±0.27	-4.50±0.14	-13.3±0.40	–
	[12.0±0.60]	[-43.1±2.16]	[-21.6±1.08]	
303	6.04±0.18	-7.80±0.23	-16.8±0.50	–
	[9.72±.49]	[-43.1±2.16]	[-24.9±1.25]	
308	3.64±0.11	-10.4±0.31	-20.8±0.62	–
	[7.26±0.36]	[-43.1±2.16]	[-28.2±1.41]	
313	2.40±0.07	-14.4±0.43	-24.4±0.73	-2.43±0.07
	[4.61±0.23]	[-43.1±2.16]	[-31.3±1.57]	[-5.75±0.29]
318	–	-17.9±0.54	-26.3±0.79	–
		[-43.1±2.16]	[-34.4±1.72]	
323	–	-20.6±0.62	-30.2±0.91	-4.40±0.13
		[-43.1±2.16]	[-37.3±1.87]	[-8.46±0.42]
343	–	–	–	-8.70±0.26
				[-13.4±0.67]

[a] For ionic surfactants, $(\Delta H^o_m)_{VH}$ values were estimated using measured β values by conductometry.

5.5. Discussion

The nonionic surfactant MEGA-10 has shown close resemblance between van't Hoff and calorimetric ΔH^o_m values; the agreement was exact below 298K, and nearly followed linear courses with close proximities (Fig. 5.5). Similar were the observations with nonionic surfactants like TX-100[33a] and octyl glucoside.[34] For both C_{10}TAB and SDBS (Figs. 5.6 and 5.7, respectively), the agreements were very poor. When 100% counter-ion binding i.e., $\beta = 1$ was considered, the van't Hoff procedure produced ΔH_m^o values although not in agreement but not too much way off, although it is not expected that all the counter-ions would remain bound to the micelles. The results have clearly indicated that ionic surfactants behaved differently from nonionics: compatibility between the two could not be induced by adjusting the value of β to its maximum i.e., 1. Thus, the reasons for the difference lie elsewhere. Probable explanation has been attempted in our earlier reports,[1,3-5] a further discussion of it will be made with more arguments in the subsequent discussion.

The differences between the two procedures for the surfactant systems herein studied are clearly visible, which has followed the order C_{10}TAB > CPC > CDMAB > SDBS > NaC >TX-100 > MEGA-10. It may be mentioned that β for NaC was low (only 10% of Na$^+$ ions were bound to the maximum).[35,36] Therefore, in the calculation, the weightage of electrical free energy change was also marginal. Further, in comparison, the differences of results between the two procedures for TX-100 were more than that for MEGA-10 (shown in Fig. 5.5). Thus, the chemical nature of the amphiphiles has a say on the observed difference.

It has been already mentioned that the van't Hoff procedure is a differential method (cf Eq. 12) whereas the calorimetry is an integral method. In the latter, along with the process of amphiphile self-aggregation, the contributions of other related or non-related processes are involved. The enthalpies from both procedures are not expected to be identical. In biological field, physicochemical processes (like interaction between macromolecules with small molecules or macromolecular association, denaturation etc.) are always associated with environmental modifications (solvent structure alteration, solvation, desolvation, molecular orientation, component dissociation, molecular organization etc.). All of these should have components in the overall heat or enthalpy change: thus, in a calorimeter the resultant integral heat will be manifested. But the van't Hoff relation holds only for a specific process. Likewise, during the process of micellization, alteration of solvent structure, solvation of the head groups, breaking of icebergs surrounding the amphiphile tails, orientation of amphiphile molecules to end up with a specific geometry, intermicellar interaction, etc. should come into operation; the sum total effect of them can be probed by a sensitive calorimeter, like ITC. The agreement between van't Hoff enthalpy and calorimetric enthalpy changes for the process of micellization is thus, not also expected. Such discrepancies have been mentioned and discussed in our earlier reports.[1,3-5] But knowledge on the types of such effects operative in a system and isolation of the individual effects is a difficult and challenging task.

Table 5.3. Specific heat of micellization by calorimetry and van't Hoff method for different surfactant systems.

System[a]	$\Delta C_{p_m}^o$ / kJ K^{-1} mol^{-1}	
	ITC	VH[c]
AOT[1]	-0.55±0.02	0.00
NaDC[16]	-0.25±0.02	-0.26±0.01
NaC[16]	-0.23±0.01	-0.28±0.01
SDBS[32]	-0.78±0.01	-0.77±0.01
C_{10}TAB[3]	-0.31±0.01	-1.72±0.02
CPC[1]	-0.70±0.02	0.00
CDMAB[33b]	-0.69±0.02	-0.65±0.01
CEDAB[33b]	-0.62±0.02	-1.42±0.02
o-glucoside[34]	-0.38±0.02	-0.36±0.01
MEGA-10[4]	-0.39±0.005	-0.34±0.004
TX-100[33a]	-0.43±0.02	-0.46±0.01

[a] AOT [Sodium bis(2-ethylhexyl-sulfosuccinate)]; NaDC (Sodium deoxycholate); NaC (Sodium cholate); SDBS (Sodium dodecylbenzenesulfonate); C_{10}TAB (Decyltrimethylammonium bromide); CPC (Cetylpyridinium chloride); CDMAB (N-cetyl-N, N-diethanoyl-N -methylammonium bromide); CEDAB (N-cetyl-N-ethanoyl-N,N-dimethylammonium bromide); o-glucoside (octyl glucoside); MEGA-10 (Decanoyl-N-methylglucamide); TX-100 (Triton X-100 i.e., polyethylene glycol tert-octylphenyl ether).
[b] Regression in $(\Delta H_m^o)_{ITC}$ (or $(\Delta H_m^o)_{VH}$) vs. T for all surfactants for estimation of $\Delta C_{p_m}^o$ is 0.99.
[c] ΔH_m^o dependence on T was very nominal for both AOT and CPC. The $\Delta C_{p_m}^o$ values were very small and are shown as zero.

In the Figs. 5.5 – 5.7, it has been seen that although the two types of enthalpies (van't Hoff and calorimetry) differ either largely or marginally, their dependences on temperature were linear. Or, in other words, the emerged specific heat changes of micellization $\left(\Delta C_p^o\right)_m = [d(\Delta H_m^o)/dT]$ are constants. These values for a number of surfactants (both ionic and nonionic) are presented in Table 5.3. Both ITC and VH values are shown for a comparison. It is observed that except $C_{10}TAB$ and CEDAB (N-cetyl-N-ethanoyl-N,N-dimethylammonium bromide), the ITC and VH values fairly agree between them. All values are negative, which is the expected sign for the process of micellization. This is an interesting point of concern since ΔH_m^o are different whereas their temperature dependences observed from two angles (calorimetry and van't Hoff) are identical. Of course, the large differences for the herein reported representatives $C_{10}TAB$ and CEDAB have made the point of concern more thought provoking.

5.6. Conclusion

The non-agreement between the van't Hoff and calorimetric ΔH_m^o is a long known issue but not well documented in literature. It has received mentions in biological phenomena wherein distinction between the two on the basis of differential and intergral heats of interaction/reaction has been stressed without attempts for quantification. It is a challenging issue and calls for accurate measurements with a potential method for its understanding. We have herein considered a simple process, the process of self-aggregation (or micellization) of amphiphiles, and has employed an accurate and versatile method, the method of microcalorimetry to shed more light on this issue. An isothermal titration calorimeter (ITC) has been used to determine CMC (an equilibrium constant related parameter) and the process enthalpy change from direct measurements. The results on analysis has shown that when amphiphile association is treated on the basis of hydrophobic and electrical contributions, the enthalpy variation by van't Hoff and that by calorimetry hardly match with each other for ionic amphiphiles; for nonionic amphiphiles, the corroboration is fair with mild or moderate deviations. In this estimation, the role of micellar aggregation number should have a say which is since minor in most cases, has not been considered in the present investigation for want of accurate temperature dependent data. Besides, non-consideration of monomer non-ideality, electrostatic intermicellar interaction, micellar hydration, water structure modification, *etc.* is expected to have a say on the observed discrepancies between the two evaluation procedures. Although the enthalpy change contributions of these processes are different, their temperature dependences have been found to be corroborative with only minor variations. The agreement between the changes in specific heat of micellization by the van't Hoff and calorimetric measurements has thus produced an apparent puzzle. Based on the possible variations in the standard states particularly for the integral heats measured in a calorimeter contributed by different micelle related and non related processes, the above agreement between the specific heat changes is intriguing. This remains to be sorted out also. However, since the physicochemical processes presented and discussed in this report refer to at or near CMC points (i.e., dilute solutions), the above-mentioned left out phenomena are expected to contribute insignificantly to the measured energetic parameters. But quantitative evaluation of them is warranted at least to frame a near complete picture.

Acknowledgments

S.P.M. thanks the Jadavpur University for an Emeritus Professorship and the Indian National Science Academy for an Honorary Scientist position. D. M. thanks the Council of Scientific and Industrial Research, Govt. of India for financial support.

References

1. A. Chatterjee, S. P. Moulik, S. K. Sanyal, B. K. Mishra and P. M. Puri, *J. Phys. Chem. B*, 105 (51), 12823, (2001).
2. G. C. Kresheck, *J. Phys. Chem. B*, 102 (34), 6596, (1998).
3. I. Chakraborty and S. P. Moulik, *J. Phys. Chem. B*, 102 (34), 3658, (2007).
4. M. Prasad, I. Chakraborty, A. K. Rakshit and S. P. Moulik, *J. Phys. Chem. B*, 110 (20), 9815, (2006).
5. K. Maiti, I. Chakraborty, S. C. Bhattacharya, A. K. Panda and S. P. Moulik, *J. Phys. Chem. B*, 111 (51), 14175, (2007).
6. T. Chakraborty, I. Chakraborty, and S. Ghosh, *Langmuir*, 22 (24), 9905, (2006).
7. D. Matulis, J. Rouzina and V. A. Bloomfield, *J. Am. Chem. Soc.*, 124 (25), 7331, (2002).
8. W-H. Chen, H-M. Huang, C-C. Lin, F−X. Lin and Y-C. Chan, *Langmuir*, 19 (22), 9395, (2003).
9. S. P. Moulik, *Current Sci.*, 71 (5), 368, (1996).
10. Moroi, Y., in *Micelles-Theoretical and Applied Aspects*, Plenum press, New York, Ch. 5, p. 97 (1992).
11. R. Nagarajan and E. Ruckenstein, *Langmuir*, 7 (12), 2934 (1991).
12. Y. Moroi and N. Yoshida, *Langmuir*, 13 (51), 3909, (1997).
13. P. R. Majhi and S. P. Moulik, *Langmuir*, 14 (51), 3986, (1998).
14. S. K. Hait, S. P. Moulik and R. Palepu, *Langmuir*, 18 (7), 2471, (2002).
15. G. C. Kresheck, *J. Am. Chem. Soc.* 120 (42), 10964, (1998).
16. S. Paula, W. Sus, J. Tuchtenhagen and A. Blume, *J. Phys. Chem.*, 99 (30), 11742, (1995).

17. S. Shimizu, P. A. R. Pires and O. A. El Seoud, *Langmuir*, 20 (22), 9551, (2004).
18. J. M. Pestman, J. Kevelam, M. J. Blandamer, H. A. van Doren, R. M. Kellogg and J. B. F. N. Engbert, *Phys. Chem. Chem. Phys.*, 2, 5146, (2002).
19. K. Bijma, M. J. Blandamer and J. B. F. N. Engberts, *Langmuir*, 14 (1), 79, (1998).
20. P. Garidel, A. Hildebrand, R. Neubert and A. Blume, *Langmuir*, 16 (12), 5267, (2000).
21. S. Dai and K. C. Tam, *Colloids and Surfs. A*, 229 (1-3), 157, (2003).
22. (a) M. J. Blandamer, P. M. Cullis and J. B. F. N. Engberts, *J. Chem. Soc. Faraday Trans.*, 94 (61), 2261, (1998). (b) J. Bach, M. J. Blandamer, J. Burgess, P. M. Cullis, L. G. Soldi, K. Bijma, J. B. F. N. Engberts, P. A. Kooreman, A. Kacperska, K. C. Rao and M. C. S. Subha, *J. Chem. Soc. Faraday Trans.*, 91 (8), 1229 (1995). (c) M. J. Blandamer, P. M. Cullis and P. T. Gleeson, *Chem. Soc. Rev.*, 32, 264, (2003).
23. J. Lah, C. Pohan and G. Vesnaver, *J. Phys. Chem. B*, 104 (11), 2522, (2000).
24. M.J. Rosen, in *Surfactants and Interfacial Phenomena* Wiley-Interscience : New York, (1989).
25. K. Shinoda and E. Hutchinson, *J. Phys. Chem.*, 66 (4), 577, (1962).
26. P. Mukerjee, *J. Phys. Chem.*, 76 (4), 565, (1972).
27. N. Funasaki, *Adv. Colloid Interface Sci.*, 43 (1), 87, (1993).
28. K. Matsuoka, Y. Moroi and M. Saito, *J. Phys. Chem.*, 97 (49), 13006, (1993).
29. A. Holtzer and M. F. Holtzer, *J. Phys. Chem.*, 78 (14), 1442, (1974).
30. G. Basu Ray, I. Chakraborty, S. Ghosh, S. P. Moulik, C. Holgate, K. Glenn and R. M. Palepu, *J. Phys. Chem. B*, 111 (33), 9828, (2007).
31. G. Basu Ray, I. Chakraborty, S. Ghosh, S. P. Moulik and R. Palepu, *Langmuir*, 21 (24), 10958, (2005).
32. S. K. Hait, P. R. Majhi, A. Blume and S. P. Moulik, *J. Phys. Chem. B*, 107 (15), 3650, (2003).
33. (a) A. Chatterjee, T. Dey, S. K. Sanyal and S. P. Moulik, *J. Surf. Sci. Technol.*, 17 (1-2), 1, (2001). (b) A. Chatterjee, S. Maiti, S. K. Sanyal and S. P. Moulik, *Langmuir*, 18 (8), 2998, (2002).
34. P. R. Majhi and A. Blume, *Langmuir*, 17 (13), 3844, (2001).
35. A. Bandopadhyay and S. P. Moulik, *Colloid Polym. Sci.*, 266 (5), 455, (1988).
36. H. Sugioka, K. Matsuoka and Y. Moroi, *J. Colloid Interface Sci.*, 259 (1), 156, (2003).

Chapter 6

Unusual Phase Behavior in a Two-Component System Catanionic Surfactant-Water: From Lamellar-Lamellar to Vesicle-Micelle Coexistence

Bruno F.B. Silva, Eduardo F. Marques

Department of Chemistry
University of Porto
Rua do Campo Alegre, 687, 4169-007 Porto, Portugal

and

Ulf Olsson

Physical Chemistry 1
Centre for Chemistry and Chemical Engineering
Lund University, P.O. Box 124, SE-221 00 Lund, Sweden

Salt-free catanionic surfactants form in water binary systems, thus differing from pseudo-ternary equimolar catanionic mixtures, where salt is present. A miscibility gap, an unusual phenomenon in binary systems, is observed for the lamellar phase of the catanionic surfactant hexadecyltrimethylammonium octylsulfonate. Experimental data show the coexistence of a swollen and a collapsed lamellar phase in a wide two-phase region, while linear swelling is observed for each phase. This phase behavior is suggested to stem mainly from a concentration dependence of the charge density of catanionic bilayer, driven by the much higher solubility of the short chain ionic counterpart (octylsulfonate). Thus, a theoretical cell model based on combined DLVO and short range repulsive potentials is presented in order to provide physical insight into the miscibility gap. Furthermore, the surfactant forms at high dilution a solution phase and exhibits a very low critical micelle concentration (0.0035 wt%). The dilute lamellar phase is in equilibrium with the isotropic solution, and small vesicles can also be observed, apparently as a dispersion of the swollen lamellae in the solution. Upon temperature increase, a vesicle-to-micelle transition occurs. These unusual equilibria can also be qualitatively rationalized by the short chain solubility model.

6.1. Introduction

Surfactants can be broadly classified according to their charge and molecular structure. This classification is natural, since from these parameters one can predict many features of surfactant phase behavior in solution.[1] For instance, it is expected that ionic surfactants bearing two alkyl chains self-assemble into bilayers, while ionic surfactants with only one alkyl chain, generally form micelles as the first aggregate in aqueous solution. In turn, nonionic surfactants with polyoxyethylene headgroups can form bigger aggregates, such as elongated micelles, due to weaker electrostatic repulsions between the polar heads. With temperature increase, the headgroups become less hydrophilic, and as a result, bilayer aggregates form.

Catanionic surfactants are a relatively recent class of surfactants which can be prepared by the equimolar mixing of two oppositely charged surfactants, followed by inorganic counterion removal.[2-5] A catanionic surfactant-water system differs from a so-called catanionic mixture, since in the latter counterions are also present and the cationic-anionic surfactant mixing ratio can be non-equimolar. These mixtures form a wide variety of aggregates depending on surfactant mixing ratio and concentration, from elongated micelles and stable vesicles to liquid crystals in relatively complex phase equilibria.[6-12] Due to the presence of one of the individual ionic surfactants in excess, along with the inorganic counterions, they are at least four-component systems, with the consequence that their thermodynamic description is not straightforward.[9] On the other hand, a catanionic surfactant is a single net uncharged surfactant, with two (or more) alkyl chains, and when dispersed in water the system is a two-component one, of much easier handling. The catanionics are generally insoluble, in the sense that they give rise to swelling bilayers, closely resembling zwitterionic double-chained surfactants such as lecithin.[2,13-15] However, soluble micellar-forming catanionic surfactants have also been prepared, such as hexylammonium hexanoate[16] and short asymmetric chain compounds of the type n-alkylbromide dodecylsulfate.[17,18] The latter have been reported to self-organize into non-spherical micelles (cylinders and discs), depending on the degree of chain asymmetry, prior to liquid crystal formation.

We have recently shown that if the solubilities of the individual anionic and cationic parts of the surfactant are very different, the most soluble part will partition between aggregate and bulk, with the result that the aggregate surface becomes charged.[19] Thus, the solubility and swelling ability of the catanionic can be greatly enhanced. This is the case for hexadecyltrimethylammonium octylsulfonate, here designated by TASo.[19] The anionic octylsulfonate part is much more soluble than the longer hexadecyltrimethylammonium one. The fraction of charged TASo aggregates is concentration-dependent, and this has important consequences in the phase behavior, with some uncommon features arising as a function of both concentration and temperature. As will be discussed in further detail, on the concentrated side, two lamellar phases are observed with a coexistence region between them.[19] This is a rare phenomenon in binary systems, and has been observed for very few compounds, the most well known case being didodecyldimethylammonium bromide (DDAB).[20-23]

On the dilute side, at room temperature, vesicles are found dispersed in an isotropic (L_1) medium. When temperature is increased, the system forms a one-phase micellar region. This behavior is closely related to that of ionic surfactants that have their inorganic counterions replaced with hydrophobic ones (or hydrotropes), like salycilate[24] or hydroxynaphthalenecarboxylate, HNC^-.[25] Mixtures of cetyltrimethylammonium bromide (CTAB) with salycilate yield elongated micelles, while the CTA^+ HNC^- system yields vesicles (and possibly elongated micelles) solutions until concentrations as high as 300 mM.[25-28] In the latter case, a vesicle-to-micelle transition is also observed when temperature is increased. The HNC^- species can be regarded as weakly surface active, while octylsulfonate is a true anionic surfactant. On the other hand, salycilate on its own, does not self-assemble in water.[29]

The higher the difference in solubility and surface activity between the two individual parts of the catanionic pair, the higher the ionic character of the catanionic surfactant will be (i.e. the higher the extent of dissociation). Thus, CTA^+HNC^- has a higher ionic character than TASo. A good evidence for this is the ability of this system to form solution phases up to concentrations where TASo already shows a lamellar phase. On the other hand, catanionics with smaller differences between the solubilities of the individual ionic pairs, have a smaller ionic character than TASo. Hence, catanionic mixtures where the 1:1 charge ratio is reached readily precipitate from solution, yielding crystalline solids.[6-11] In some cases, salt-free catanionics can also form flat nanodiscs and icosahedra hollow capsules.[5,30] They also display rich thermal phase behavior, forming a large number of thermotropic liquid-crystalline phases prior to isotropization.[31-33]

6.2. Experimental and modeling details

Sodium octylsulfonate and hexadecyltrimethylammonium bromide were both purchased from Sigma and their purity checked from the CMCs obtained by surface tension. The catanionic compound TASo was prepared by mixing of two equimolar solutions of the oppositely charged surfactants and further purified (inorganic counterion removal) by 5 cycles of recrystalization, according to a procedure previously described.[32]

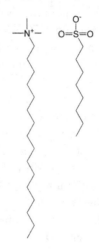

Fig. 6.1 The molecular structure of hexadecyltrimethylammonium octylsulfonate (TASo).

SAXS measurements were made at two different instruments. The dilute liquid-crystalline region was measured with synchrotron radiation SAXS at Max-Lab, beam-line I711. A wavelength λ of 1.08 $Å$ was used, at a sample-to-detector distance of 1456 mm. The concentrated liquid-crystalline region was studied using a Kratky camera, with wavelength of 1.54 $Å$, and sample-to-detector distance of 277 mm. SAXS structural data was obtained according to the following equation:

$$Q_1 = \frac{2\pi}{d} = 2\pi \frac{\phi_s}{\delta} = \pi \frac{\phi_s a_s}{V_s}, \tag{1}$$

where Q_1 is the scattering vector from the first order Bragg peak; d is the repeat distance; δ is the bilayer thickness; ϕ_s is the surfactant volume fraction; a_s is the average surface area per polar headgroup; and V_s is the molecular volume of TASo. This later value was obtained from density determination of the compound, which was found to be similar to that of water. Hence, the surfactant volume fraction ϕ_s and the weight percent wt% are similar.

Cryo-TEM measurements were also performed, in order to visualize the aggregates present at higher dilutions. The films were vitrified from 25°C by quick submersion in liquid ethane at −177°C. Subsequently, the samples were transferred in liquid nitrogen and kept at -180°C during transfer into the microscope, a Philips CM210 BioTWIN cryotransmission electron microscope, operating at 120 kV. Images were digitally recorded with a Gatan MSC791 CCD camera under low dose conditions. Turbidity measurements were performed in a Cary 300 UV/Vis spectrophotometer from Varian, operating at 400 nm, with water as reference. DSC measurements were carried out with a Setaram microDSCIII high-sensitivity calorimeter at a heating rate of 1.0 Kmin.$^{-1}$

The cell model calculations were made by use of the PBCell program (developed by Bengt Jönsson at Lund University).[34] A lamellar phase geometry is used with the obtained structural parameters from SAXS. The Poisson − Boltzmann equation is then solved for a given surface charge density, counterion concentration and d distance, at 25°C. Along with these calculations, the program also calculates the van der Waal's attractive force, for the given geometry and a given Hamaker constant H. Additionally, the hydration force was also calculated separately and added to the osmotic pressure obtained from the PBCell program. The hydration force is given by :

$$\Pi_{hyd} = a \exp\left(\frac{-(d-\delta)}{\lambda}\right). \quad (2)$$

The a parameter stands for the force amplitude, λ for the decay length, d and δ for the lamellar repeat distance and thickness, respectively. This expression is empirical as the origin of the hydration force is still not completely understood.[1] This force is highly repulsive at short ranges, and vanishes quickly at higher lamellar spacings. The parameters a and λ are generally situated in the range 10^6–10^8 Nm^{-2}, and 0.15−0.3 nm respectively.

6.3. Results and discussion

6.3.1. *Concentrated side: lamellar-lamellar coexistence and modeling*

The one-component catanionic surfactant system (neat anhydrous TASo) displays a rich thermotropic behavior, with soft-crystal and liquid-crystalline phases of the smectic type formed as the temperature is increased, until decomposition occurs at 218.6°C.[32] When it comes to the binary system TASo-water, in the concentrated side, from 2.6 to 80 wt%, two lamellar liquid crystalline phases are observed, with a miscibility gap between 15 and 54 wt%. This can be seen in a qualitative way from the birefringent lamellar textures detectable in a polarized light microscope (Fig. 6.2). SAXS measurements provide unambiguous proof and quantitative data (Fig. 6.3).

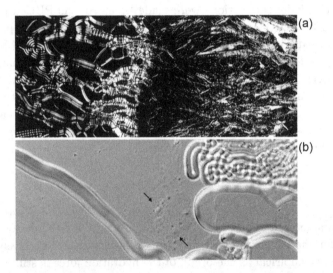

Fig. 6.2 Phase penetration scans in the polarized light microscope, with the surfactant concentration increasing from left to right: (a) two lamellar textures with an intermediate band or gap can be observed; (b) in the dilute side, together with the myelinic figures from the dilute lamellar phase, one sees an isotropic region, with some spontaneously detached vesicles (arrows).

In the phase penetration scan depicted in Fig. 6.2, a gradient of concentration exists from left to right and one clearly sees two different lamellar textures separated by a black gap. From SAXS measurements, one is able to see in a 20 wt% sample, superimposed Bragg peaks corresponding to the two lamellar spacings. On the other hand, when the Q values of the Bragg peaks are plotted against the surfactant volume fraction, linear swelling is observed, except for concentrations higher than 80 wt%. The miscibility gap exists in the middle region, where the structural parameters from both lamellar phases are maintained, and only their relative amounts are changed. These observations altogether can be summed up in the phase diagram displayed in Fig. 6.4.

As previously stated, this unusual behavior can be understood, in the case of TASo, if one considers that the short chain octylsulfonate ion (So^-) has a much higher solubility than the long alkyl chain hexadecyltrimethylammonium one (CTA^+). Let us consider then a section (or cell) from a lamellar phase, with repeat distance d, thickness δ and area A (Fig. 6.5).

The solubility of CTA^+ is set to zero, which should be a good approximation, compared with the solubility of So^-. The latter is denoted by C_{So^-} and considered to be constant with surfactant concentration. The volume of water between the two bilayers is obtained by the following simple expression:

$$V = A(d - \delta) \tag{3}$$

In this water volume, there are n ions of So^-, which stem from the two adjacent bilayers, leaving it with a charge density σ. The n molecules are obtained in a straightforward way from the solubility C_{So^-} and the water volume, according to:

$$n = C_{So^-} A(d - \delta) \tag{4}$$

Finally, the charge density σ of the bilayers can be obtained by dividing the total number of So^- molecules by a factor of 2 (since the n ions left from two bilayers), and by the area A, yielding:

$$\sigma = \frac{1}{2} C_{So^-} \delta (d - \delta) \tag{5}$$

which can be further rearranged to the more useful form

$$\sigma = \frac{1}{2} C_{So^-} z e \delta \left(\frac{1}{\phi_s} - 1 \right) \tag{6}$$

since $\phi_s = \delta/d$, and z is the So^- valence, and e the elementary charge.

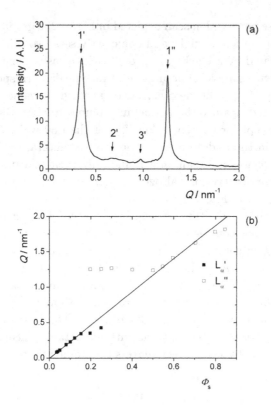

Fig. 6.3 SAXS results: (a) Diffractogram for a 20 wt% sample of TASo in water. Three reflection peaks for the L'_α phase can be seen and one peak which cannot be attributed to a fourth order reflection of L'_α, but instead belongs to L''_α. (b) Scattering vector values for both lamellar phases. At $\phi=0.20$ and 0.25 two Q values are obtained for the coexisting lamellar phases. Above 80 wt%, the swelling is no longer linear. From the slope of the fit, the lamellar thickness δ is 2.7 nm, and the area per molecule as is 0.59 nm^2. L'_α swells between $d=18\text{-}80$ nm, while L''_α between $d=3.2\text{-}4.9$ nm.

Eq. (6) bears a very important feature, regarding TASo phase behavior. It states that if the solubility C_{So^-} is assumed to be constant, regardless of concentration, then the charge density σ will increase when the volume fraction of surfactant ϕ_s decreases. This will have a pronounced effect on the phase behavior, since this charge gives rise to an electrostatic repulsive force. Taking into account this force, along with the attractive van der

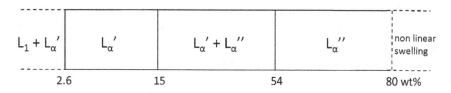

Fig. 6.4 Phase behavior for the concentrated region of the TASo-water binary system at 25°C.

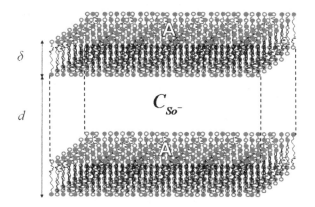

Fig. 6.5 Schematic representation of the lamellar phase cell used in the calculations.

Waals and repulsive hydration force, one can calculate the total pressure curve for TASo, by means of a cell model (see experimental section). In these calculations, one uses the structural parameters given by SAXS, and the adjustable parameter is the solubility C_{So^-}.

From Fig. 6.6, one can observe that there is a loop in the pressure, at small interbilayer separations. This loop implies phase separation (in this case, a miscibility gap in the lamellar phase) and originates from the fact that the electrostatic force arises from the partial solubilization of So^-. As a consequence, the force predominates only at higher d values, where the charge density σ is higher, and the other two forces are already very weak. Hence, this force is able to stabilize the swollen lamellar phase only until 15 wt% (the upper limit). At short distances, the hydration force dominates, and stabilizes the collapsed lamellar phase. At intermediate distances, there is a region where the van der Waals force dominates, creating a loop, which is the cause for the miscibility gap.

In this case, calculations were performed assuming a C_{So^-} value of 10 mM, which yielded an upper calculated limit of 9.9 wt%. By changing the solubility C_{So^-} (which is unknown) to 20 mM, one is able to fit very well the experimentally observed miscibility gap, with limits between 14.5 and 54 wt%. One should only state as a final note in the calculations, that the assumption that the C_{So^-} is independent of concentration is only approximate. In fact, this value is expected to decrease slightly when the concentration is diminished, since the charging of the lamellar surfaces will add an extra electrostatic penalty. However, this will not change the qualitative features or the physicality of the model.

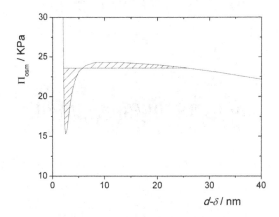

Fig. 6.6 Cell model calculations, using the charge density and counterion concentration given by eq. 2, and the lamellar structural parameters obtained by SAXS measurements. The Hamaker constant H is $6 \times 10^{-21} J; a = 3 \times 10^7 N.m^{-2}$ and $\lambda = 0.3$ nm. The enclosed region is a Maxwell construction, where the two areas are equal, signaling the phase boundaries of the swollen lamellar (d=24.8 nm, ϕ_s=0.099), and the collapsed lamellar phase (d=4.99 nm, ϕ_s=0.543).

Comparing this system with a fully ionic lamellar phase such as that of didodecyldimethylammonium chloride (DDAC), where a lamellar phase swells extensively in the phase diagram,[35] one easily understands that the miscibility gap here is originated by the partial solubilization of So^-. This is not sufficiently high to originate an electrostatic force strong enough to overcome the van der Waals force over a wide d range. On the other hand, lecithin, which is essentially a double-chained zwitterionic surfactant, displays a very limited swelling, promoted only by the hydration force. Considering these comparisons, the definition of ionic character to characterize these catanionic compounds seems adequate, since they indeed present an intermediate behavior between truly nonionic (or zwitterionic) and ionic surfactants. Finally, one should compare TASo with DDAB that displays a miscibility gap as well. In this case, however, the phase separation seems to originate from the high adsorption of Br^- ions on the charged DDA^+ surface, which creates an extra attractive force.[36]

6.3.2. Dilute side

Below 2.6 wt%, the dilute lamellar phase no longer swells and a new two-phase region is entered. This is what one generally observes for a lamellar phase, since at $d > 100$ nm it starts to be difficult to stabilize it further. In this regime, the lamellar phase is dispersed in L_1 in the form of vesicular aggregates (Fig. 6.7), which have a chain melting transition temperature of 15.9°C, as determined by DSC (Fig. 6.8).

The vesicles have been detected, at room temperature, even at very low concentrations (such as 0.05 wt%, from preliminary dynamic light scattering data), where the samples appear as completely transparent, colorless and slightly viscous. Due to this fact, and if we consider these bilayer aggregates to be a dispersion of L'_α, the experimental determination of the phase boundary $L_1/L_1 + L'_\alpha$ is not easily achieved. What is a fact is that TASo at room temperature shows a clear CMC at 0.035 wt%,[19] with the samples almost immediately starting to display some viscosity, indicating that large aggregates are present.

When the temperature is increased in the range 0.05-2.6 wt%, a gradual transition from the two-phase region to the viscous and clear L_1 phase is observed, with the transition temperature strongly depending on concentration. This can be observed in a clear way by turbidity measurements (Fig. 6.8). A phase map for this dilute region can thus be drawn (Fig. 6.9).

Again, this unusual behavior can be at least partially understood if one considers that the fraction of dissociated surfactant increases upon dilution (Eq. 6) and temperature. We note here that in such dilute solutions, potentiometric (EMF) measurements could be particularly useful to determine So^- activities and thus the extent of partitioning into the bulk, giving

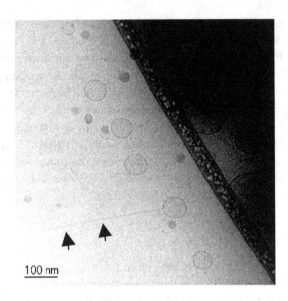

Fig. 6.7 Cryo-TEM image of a 1.2 wt% sample of TASo at 25°C. Unilamellar vesicles are seen together with an elongated micelle (arrows), probably a metastable aggregate. These micelles exist as equilibrium aggregates in the single phase L_1 at higher temperatures.

further evidence for our model. In any case, it is reasonable to assume that at sufficiently low concentration, the charge density σ (Eq. 6) increases. When the temperature is increased, within the two-phase region, the solubility C_{So^-} is expected to also increase (as with normal solutes) and, thus, σ will increase as well. This increase in σ (both with dilution and temperature) should reach a value high enough to favor higher curvature aggregates such as large non-spherical micelles. This mutual dependence of σ on the concentration and temperature in the dilute side, justifies the observed phase diagram in Fig. 6.9.

The gradual transition from a lamellar phase to the L_1 micellar region is then suggested to arise mainly from an increase in surfactant film curvature, driven by the increase in ionic character of the surfactant film (i.e. dissociation of the short chain), occurring both upon dilution and temperature rise.

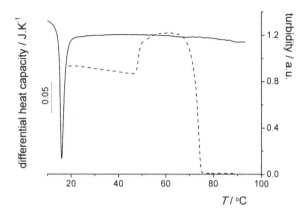

Fig. 6.8 DSC thermogram and turbidity temperature scan for a 1.2 wt% TASo sample. The chain melting peak for the vesicle bilayer occurs at 15.9°C. The transition of the vesicle region to a L_1 phase is hardly noticeable on DSC, but clearly observed by turbidity changes and naked eye around 70-75°C.

6.4. Conclusions

The phase behavior and aggregation structures of the salt-free catanionic surfactant TASo, which bears a large asymmetry in the chain lengths of the paired ions, has been discussed. The most striking features are the miscibility gap of the lamellar phase region, and the formation of vesicles and large micelles in the dilute one. Upon temperature increase, the vesicles undergo a transition into micelles. It is suggested that this highly unusual behavior is a result of the concentration dependence of the charge density of TASo aggregates, driven by a solubility mismatch between both individual ionic surfactant parts. Under this framework, it is possible to model and qualitatively rationalize the observed phase equilibria. It is suggested that this type of phase behavior may also be found in other types of catanionics bearing a large asymmetry in solubility of the anionic and cationic counterparts.

Acknowledgments: Financial support from the Portuguese Science Foundation (FCT) and FEDER Funds, through the research project

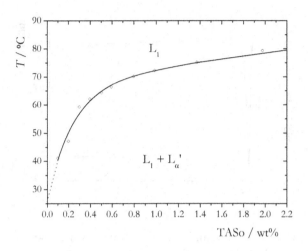

Fig. 6.9 Phase map for the TASo-water system on the dilute region. The dashed line at very low concentrations implies an uncertainty in the phase boundary.

POCTI/QUI/44296/2002, and from the Swedish Research Council (VR) is gratefully acknowledged. Centro de Investigao em Qumica(UP)- L5 is also acknowledged for financial support. BBS is grateful to F.C.T. for the PhD grant SFRH/BD/24966/2005.

References

1. D. F. Evans and H. Wennerström, *The Colloidal Domain : Where Physics, Chemistry, Biology and Technology Meet*, 2nd ed., (Wiley-VCH, New York, 1999).
2. P. Jokela, B. Jönsson and A. Khan, *J. Phys. Chem.* 91, 3291 (1987).
3. A. Khan and E. Marques, in *Specialists Surfactants*, I.D. Robb, Ed. (Blackie Academic and Professional, an imprint of Chapman & Hall, London, 1997), p. 37.
4. E. F. Marques, O. Regev, A. Khan and B. Lindman, *Adv. Colloid Interface Sci.*, 100, 83 (2003).
5. T. Zemb and M. Dubois, *Aust. J. Chem.*, 56, 971 (2003).
6. E. W. Kaler, A. K. Murthy, B. E. Rodriguez and J. A. N. Zasadzinski, *Science*, 245, 1371 (1989).
7. K. L. Herrington, E. W. Kaler, D. D. Miller, J. A. Zasadzinski and S. Chiruvolu, *Phys. Chem.*, 97, 13792 (1993).
8. L. L. Brasher, K. L. Herrington and E. W. Kaler, *Langmuir*, 11, 4267 (1995).

9. E. Marques, A. Khan, M. D. Miguel and B. Lindman, *J. Phys. Chem.*, 97, 4729 (1993).
10. E. F. Marques, O. Regev, A. Khan, M. D. Miguel and B. Lindman, *J. Phys. Chem. B*, 102, 6746 (1998).
11. E. F. Marques, O. Regev, A. Khan, M. D. Miguel and B. Lindman, *J. Phys. Chem. B*, 103, 8353 (1999).
12. E. F. Marques, *Langmuir*, 16, 4798 (2000).
13. P. Jokela, B. Jönsson, B. Eichmüller and K. Fontell, *Langmuir*, 4, 187 (1988).
14. P. Jokela, B. Jönsson, A. Khan, B. Lindman and A. Sadaghiani, *Langmuir*, 7, 889 (1991).
15. P. Jokela, B. Jönsson and H. Wennerström, *Progr. Colloid Polym. Sci.*, 70, 17 (1985).
16. A. Khan and C. Mendonca, *J. Colloid Interface Sci.*, 169, 60 (1995).
17. J. Eastoe, J. Dalton, P. Rogueda, D. Sharpe, J. F. Dong and J. R. P. Webster, *Langmuir*, 12, 2706 (1996).
18. J. Eastoe, P. Rogueda, D. Shariatmadari and R. Heenan, *Colloids Surf. A*, 117, 215 (1996).
19. B. F. B. Silva, E. F. Marques and U. Olsson, *J. Phys. Chem. B*, 111, 13520 (2007).
20. K. Fontell, A. Ceglie, B. Lindman and B. Ninham, *Acta Chem. Scand.*, 40, 247 (1986).
21. T. Zemb, D. Gazeau, M. Dubois and T. Gulikkrzywicki, *Europhys. Lett.*, 21, 759 (1993).
22. M. Dubois, T. Zemb, N. Fuller, R. P. Rand and V. A. Parsegian, *Biophys. J.*, 72, Th329 (1997).
23. M. G. Noro and W. M. Gelbart, *J. Chem. Phys.*, 111, 3733 (1999).
24. U. Olsson, O. Soderman and P. Guering, *J. Phys. Chem.*, 90, 5223 (1986).
25. P. A. Hassan, J. Narayanan, S. V. G. Menon, R. A. Salkar, S. D. Samant and C. Manohar, *Colloids Surf. A*, 117, 89 (1996).
26. P. A. Hassan, B. S. Valaulikar, C. Manohar, F. Kern, L. Bourdieu and S. J. Candau, *Langmuir*, 12, 4350 (1996).
27. E. Mendes, R. Oda, C. Manohar and J. Narayanan, *J. Phys. Chem. B*, 102, 338 (1998).
28. K. Horbaschek, H. Hoffmann and C. Thunig, *J. Colloid Interface Sci.*, 206, 439 (1998).
29. T. K. Hodgdon and E. W. Kaler, *Curr. Opin. Colloid Interface Sci.*, 12, 121 (2007).
30. M. Dubois, B. Deme, T. Gulik–Krzywicki, J. C. Dedieu, C. Vautrin, S. Desert, E. Perez and T. Zemb, *Nature*, 411, 672 (2001).
31. N. Filipovic–Vincekovic, I. Pucic, S. Popovic, V. Tomasic and . Tesak, *J. Colloid Interface Sci.*, 188, 396 (1997).
32. B. F. B. Silva and E. F. Marques, *J. Colloid Interface Sci.*, 290, 267 (2005).
33. E. F. Marques, R. O. Brito, Y. J. Wang and B. F. B. Silva, *J. Colloid Interface Sci.*, 294, 240 (2006).
34. V. Rajagopalan, H. Bagger-Jörgensen, K. Fukuda, U. Olsson and B. Jönsson, *Langmuir*, 12, 2939 (1996).

35. C. J. Kang and A. Khan, *J. Colloid Interface Sci.*, 156, 218 (1993).
36. J. Forsman, *Langmuir*, 22, 2975 (2006).

Chapter 7

Mixed Proteins/Surfactants Interfacial Layers as Studied by Drop Shape Analysis and Capillary Pressure Tensiometry

V.S. Alahverdjieva, D.O. Grigoriev, A. Javadi, Cs. Kotsmar, J. Krägel, R. Miller, V. Pradines

*Max Planck Institute of Colloids and Interfaces,
14424 Potsdam, Germany*

and

A.V. Makievski

*SINTERFACE Technologies,
12489 Berlin, Germany*

Models for the thermodynamics, adsorption kinetics and dilational rheology are presented for the quantitative analysis of adsorption layer formed from mixed solutions containing proteins and surfactants. The models are applied to experimental data obtained from experiments from drop profile analysis and capillary pressure tensiometry. A particular modification of drop profile tensiometry is presented, namely the real-time bulk exchange of the drop volume by using a co-axial double capillary, which allows for the adsorption studies of different components in a sequential way. The results indicate that proteins form complexes with surfactants in the bulk and at the interface. The properties of these complexes strongly depend on the composition of the solution in the bulk and the location of their formation. In particular, the dilational rheology is very sensitive for changes in mixed adsorption layers.

7.1. Introduction

Mixed protein/surfactant adsorption layers at liquid interfaces are wide spread in many modern technologies, for example in food processing or coating technologies. Until very recently, such mixed interfacial layers have been described essentially qualitatively. Now thermodynamic models have been derived for describing quantitatively the behaviour of interfacial layers formed by proteins and non-ionic or ionic surfactants. These models are needed as the basis also for a quantitative understanding of the respective adsorption dynamics and dilational rheology of liquid interfaces.

Tensiometry methods have been developed significantly during the recent years and became very versatile and efficient for the characterisation of liquid interfaces and their modification by surface active molecules. In particular, drop profile tensiometry and capillary pressure methods belong now-a-days to the most promising methodologies in surface science.

On the basis of a quantitative interpretation by the new theoretical models of experimental data obtained from tensiometry studies, a reliable picture can be drawn now about liquid interfacial layers.

7.2. Theoretical Models

Only recently, great progress was achieved in the development of theoretical models for describing adsorption layers of proteins at liquid interfaces.[1] These new thermodynamic models were based on the two-dimensional solution theory. This approach also allowed an expansion to mixed interfacial layers comprised by proteins and surfactants, such as briefly discussed in.[2] This paragraph deals with a brief summary of the thermodynamics of adsorbed layers of proteins alone and mixed with low molecular weight surfactants.

7.2.1. *Thermodynamic Model*

An equation of state for the surface layer was derived by Miller et al.[1] based on the assumption that protein molecules can adsorb in a number of states with different molar area, varying from a maximum area given by ω_{max} to a minimum area denoted by ω_{min}:

$$-\frac{\Pi\omega_0}{RT} = ln(1 - \Theta_P) + \Theta_P(1 - \omega_0/\omega_P) + \alpha_P \theta_P^2. \qquad (1)$$

Here R is the gas law constant, α_P is the intermolecular interaction parameter, ω_0 the area occupied by one segment of the protein molecule or the molar area of the solvent.

The total surface coverage Θ is given by the adsorption of proteins in all possible n states

$$\Theta_P = \omega_P \Gamma_P = \sum_{i=1}^{n} \omega_i \Gamma_{Pi}, \qquad (2)$$

where we use ω_P as the average molar area of the adsorbed protein, $\omega_i = \omega_1 + (i-1)\omega_0$ for $1 \leq i \leq n$ as the molar area in the adsorbed state i with $\omega_1 = \omega_{\min}$ and $\omega_{\max} = \omega_1 + (n-1)\omega_0$. The equations for the adsorption isotherm for each state (j) of the adsorbed protein molecules are:

$$b_{P_j} c_P = \frac{\omega_P \Gamma_{P_j}}{(1-\Theta_P)^{\omega_j/\omega_P}} \exp\left[-2\alpha_P (\omega_j/\omega_P)\Theta_P\right]. \tag{3}$$

with c_P being the protein bulk concentration and b_{P_j} the equilibrium adsorption constant for the protein in the jth state.

The thermodynamic model given by Eqs. (1)–(3) allows us to describe the change of the state of adsorbed protein molecules with increasing bulk concentration c_P, which agrees in many details with known experimental results.

For the description of simple surfactants' adsorption many models are known, among which the Langmuir and the Frumkin model are the most frequently used.[3,4] Especially from experiments on the limiting elasticity $E_0 = d\gamma/dln\Gamma$, (the derivative of the surface tension over the relative surface concentration) we know, however, that these models propose a continuous increase in the E_0 values, while experimental data rather level off or even pass through a maximum with increasing surfactant bulk concentration. A new interpretation was proposed recently, which explains this effect on the basis of a finite compressibility of the adsorbed surfactant molecules. The rigorous theoretical model given in[5] can be simplified by neglecting the contribution from the non-ideality of entropy, so that the equations of state and adsorption isotherm turn into equations for the ordinary Frumkin model:

$$b_S c_S = \frac{\Theta_S}{(1-\Theta_S)} \exp(-2\alpha_S \Theta_S), \tag{4}$$

$$\Pi = -\frac{RT}{\omega_{S_0}} \left[ln(1-\Theta_S) + \alpha_S \Theta_S^2\right], \tag{5}$$

with $\Theta_S = \omega_S . \Gamma_S$ being the surface coverage by surfactant molecules, Γ_S is the adsorption of the surfactant, b_S is the adsorption equilibrium constant, α_S is the interaction constant. The parameter $\Pi = \gamma_0 - \gamma$ is the surface pressure, which is an experimentally accessible quantity, given by the difference between the surface tension of the solvent γ_0 and that of the solution $\gamma(c_S)$. In contrast to the classical Frumkin model, however, the molar area of a surfactant molecule ω_S depends linearly on surface pressure

Π^5:

$$\omega_S = \omega_{S0}(1 - \epsilon\Pi\Theta_S). \tag{6}$$

In this relationship ω_{0S} is the molar area at $\Pi = 0$ and ϵ is the two-dimensional relative surface layer compressibility coefficient. This parameter characterises the intrinsic compressibility of the molecules in the surface layer. As a possible physical picture, we can assume that the intrinsic compressibility reflects the change of the tilt angle of the adsorbed molecules, entailing also an increase in the thickness of the surface layer.

The two sets of equations, for adsorbed protein and surfactant molecules were the starting point for the models describing the adsorption from mixed solutions. With the approximation, $\omega_0 \simeq \omega_S$, an equation of state for a protein/non-ionic surfactant mixtures was derived in :[6]

$$-\frac{\Pi\omega_0}{RT} = ln(1-\Theta_P-\Theta_S) + \Theta_P(1-\omega_0/\omega_P) + \alpha_P\Theta_P^2 + \alpha_S\Theta_S^2 + 2\alpha_{PS}\Theta_P\Theta_S, \tag{7}$$

where α_{PS} describes the additional interaction between the protein and surfactant molecules, while α_S and α_P are responsible for the interaction between the surfactant and protein molecules, respectively. The adsorption isotherms for the protein in state $j = 1$ and the surfactant, respectively, are given by :

$$B_{P1}c_P = \frac{\omega_P\Gamma_{P1}}{(1-\Theta_P-\Theta_S)^{\omega_1/\omega_P}} \exp\left[-2\alpha_P(\omega_1/\omega_P)\Theta_P - 2\alpha_{PS}\Theta_S\right], \tag{8}$$

$$b_Sc_S = \frac{\Theta_S}{(1-\Theta_P-\Theta_S)} \exp\left[-2\alpha_S\Theta_S - 2\alpha_{PS}\Theta_P\right]. \tag{9}$$

7.2.2. Adsorption Dynamics Modelling

The theoretical model for the adsorption kinetics is based on the equation of state for the surface layer and adsorption isotherm. Ward and Tordai derived the most general relationship between dynamic adsorption $\Gamma(t)$ and sub-surface concentration $c(0,t)$.[7] For a spherical drop or bubble surface (the sign "−" or "+" before the second term on the right hand side corresponds to diffusion inside a drop and outside a drop or bubble, having the radius r) this equation has the following form :[8]

$$\Gamma_{PS}(t) = 2\sqrt{\frac{D_{PS}}{\pi}} \left[c_{0P}\sqrt{t} - \int_0^{\sqrt{t}} c_P(0, t-t')d(\sqrt{t'})\right] \pm \frac{c_{0P}D_{PS}}{r}t, \tag{10}$$

and

$$\Gamma_S(t) = 2\sqrt{\frac{D_S}{\pi}} \left[c_{0S}\sqrt{t} - \int_0^{\sqrt{t}} c_S(0, t-t')d(\sqrt{t'}) \right] \pm \frac{C_{0S}D_S}{r}t, \quad (11)$$

for the adsorptions of protein and surfactant molecules, where D_{PS} and D_S are the diffusion coefficients for the protein/surfactant complex and the surfactant, respectively. Note, as we deal with a competitive adsorption process, both Eqs. (10) and (11) apply simultaneously, coupled with each other via the adsorption model, given by Eqs. (7) to (9). In Eqs. (10) and (11) we use t as the time, and t' as a dummy integration variable, and c_{0P} and c are the corresponding bulk concentrations.

7.2.3. Dilational Rheological Model

In the literature, two groups derived a relationship for the determination of the dilational elasticity of a mixed adsorption layer. Jiang et al. arrived at :[9]

$$E = \frac{1}{B}\left(\frac{\partial \Pi}{\partial n\Gamma_1}\right)_{\Gamma_2}\left[\sqrt{\frac{i\omega}{D_1}}a_{11} + \sqrt{\frac{i\omega}{D_2}}a_{12}\frac{\Gamma_2}{\Gamma_1} + \frac{i\omega}{\sqrt{D_1 D_2}}(a_{11}a_{22} - a_{12}a_{21})\right]$$
$$+ \frac{1}{B}\left(\frac{\partial \Pi}{\partial n\Gamma_1}\right)_{\Gamma_1}\left[\sqrt{\frac{i\omega}{D_1}}a_{21}\frac{\Gamma_1}{\Gamma_2}\right.$$
$$\left. + \sqrt{\frac{i\omega}{D_2}}a_{22} + \frac{i\omega}{\sqrt{D_1 D_2}}(a_{11}a_{22} - a_{12}a_{21})\right], \quad (12)$$

where c_1 and c_2 are the bulk concentrations of the two components, the coefficients $a_{ij} = (\partial \Gamma_i/\partial c_j)_{c_{i\neq j}}$ are the partial derivatives to be obtained from the adsorption isotherm equation, and

$$B = 1 + \sqrt{i\omega/D_1}a_{11} + \sqrt{i\omega/D_2}a_{22} + (i\omega/\sqrt{D_1 D_2})(a_{11}a_{22} - a_{12}a_{21}).$$

In his book, Joos[10] arrived at a similar equation, which transfers into Eq. (12) via some simple manipulations.[11]

The visco-elastic modulus can be presented as a complex number $E = E_r + iE_i$, where the real part is called storage modulus equal to the dilational elasticity, and the imaginary part, which is called loss modulus and is proportional to the dilational viscosity (i is the imaginary unit). From Eq. (12) we can determine the visco-elasticity modulus $|E|$ and phase angle ϕ between stress ($d\gamma$) and strain (dA) via the relationships

$$|E| = \sqrt{(R^2 + S^2)/(P^2 + Q^2)} \text{ and } \phi = \arctan(E_i/E_r), \quad (13)$$

with the following abbreviations[11]

$$P = 1 + \left(\sqrt{\omega/D_1 a_{11}} + \sqrt{\omega/D_2 a_{22}}\right)/\sqrt{2}$$
$$Q = \left(\sqrt{\omega/D_1 a_{11}} + \sqrt{\omega/D_2 a_{22}}\right)/\sqrt{2} + \left(\sqrt{\omega/D_1 D_2}\right).(a_{11}a_{22} - a_{12}a_{21}),$$
$$P_i = \left[\sqrt{\omega/D_i a_{ii}} + \sqrt{\omega/D_j a_{ij}}(\Gamma_j/\Gamma_i)\right]/\sqrt{2}$$
$$Q_i = P_i + \left(\omega/\sqrt{D_1 D_2}\right).(a_{11}a_{22} - a_{12}a_{21}),$$
$$R = P_1.(\partial\Pi/\partial ln\Gamma_1)/\Gamma_2 + P_2.(\partial\Pi/\partial ln\Gamma_2)/\Gamma_1,$$
$$S = Q_1.(\partial\Pi/\partial ln\Gamma_1)/\Gamma_2 + Q_2.(\partial\Pi/\partial ln\Gamma_2)/\Gamma_1.$$

This complex set of equations and parameters reduces to the theory of Lucassen and van den Tempel,[12,13] when applied to a solution of a single surfactant.

7.3. Experimental Technique

In the present study of the dynamic properties of mixed protein/surfactant layers we used methods based on single droplets, including the drop profile analysis and capillary pressure tensiometry, and also a new versatile tool allowing for drop volume exchange. In this paragraph we describe these methods very briefly.

7.3.1. *Drop Profile Analysis Tensiometry*

The drop and bubble profile method is the most frequently applied technique in tensiometry. The shape of a drop results from the balance of two oppositely acting forces - the surface tension, which tends to make drops spherical and the gravity which elongates them. The advantages of pendant and sessile drops or buoyant and captive bubbles are numerous and make the drop/bubble profile method a superior one over all other tensiometry methods. For example, it facilitates studies at both liquid/air and liquid-liquid interfaces. Also, these methods can be applied to materials ranging from organic liquids to molten metals and from pure solvents to concentrated solutions.

The mechanical equilibrium for two homogeneous fluids separated by an interface is given by the Gauss-Laplace equation,

$$\gamma\left(\frac{1}{R_1} + \frac{1}{R_2}\right) = \Delta P \tag{14}$$

which relates the pressure difference ΔP across a curved interface to the surface tension γ and the curvature of the interface, where γ is the surface tension, R_1 and R_2 are the two principal radii of curvature. In this way the shape of a drop or bubble is determined by the interfacial tension.

The instrument used in the present study is the Drop Profile Analysis Tensiometer PAT1 (SINTERFACE Technologies) as shown in Fig. 7.1. The set up consists of a dosing system with a capillary to form the drop, a video camera with an objective and a frame grabber to transfer the image into a PC. Pictures of a drop or bubble are acquired by the video system and the computer extracts the shape coordinates and fits the Gauss-Laplace equation to these coordinates.

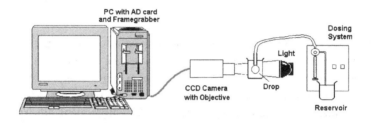

Fig. 7.1 Schematic of a drop/bubble profile analysis tensiometer (from ref. 14).

The first computer programme for this fitting routine was the so called ADSA methodology developed by Neumann et al.[15] Now, many commercial instruments are available using this methodology on the basis of an automated digital image acquisition and image analysis.

7.3.2. Bulk Exchange in Single Drops

Using the general set-up given above with a double dosing system, as proposed for the first time in,[16] a unique set of experiments becomes available. With two dosing systems, both connected to the double capillary (see Fig. 7.2), it becomes possible to exchange the bulk phase inside a drop without disturbing its surface layer.[17]

The operation of the drop exchange is based on a simple master-slave principle: one syringe, connected to the inner capillary, pumps small quantities of liquid into the drop, while the second syringe, connected to the outer capillary, controls a constant drop size, i.e. compensates any excess

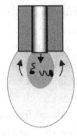 

Fig. 7.2 Schematic picture of the double capillary during the bulk exchange process; right - photo of the double capillary (SINTERFACE Technologies, Berlin, Germany)

liquid out of the drop. Thus, new liquid comes into the drop trough the inner glass capillary (1 mm diameter) and leaves the drop trough the outer plastic capillary (2 mm diameter), creating some convection/mixing within the drop. In this way the instrument allows for example to perform adsorption experiments of different types, such as the subsequent adsorption of different surface active compounds at the same surface/interface. Experimental data obtained from this methodology are presented below and show impressively its large potential.

7.3.3. *Capillary Pressure Tensiometry*

A special cell, equipped with a pressure sensor and a piezo drive represents the technical basis for capillary pressure experiments. The general principle of this method was developed by Passerone et al.[18] The main part is the ODBA-1 module developed as extra equipment for the instrument PAT-1 (SINTERFACE Technologies, Germany) the schematic of which is shown in Fig. 7.3.

The signal of the pressure sensor is passed to a high speed AD-board and monitored with a data acquisition rate of 10 MHz. The video technique available as standard technique of the PAT-1 is useful for controlling the wetting behaviour of the capillary tip. Moreover, the change of drop volume and drop area with time are directly available.[19–21]

The module is suitable also for drop and bubble oscillations experiments measuring the dilational visco-elasticity at frequencies up to few hundreds Hz.

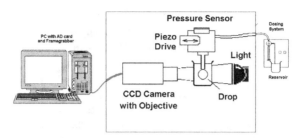

Fig. 7.3 Schematic of a drop and bubble profile analysis tensiometer equipped with a special capillary pressure cell ODBA-1 as additional module for the drop profile analysis tensiometer PAT-1 (SINTERFACE Technologies, Germany) (from ref. 21)

7.4. Adsorption Kinetics and Equilibrium Isotherms

This paragraph is dedicated to experimental results obtained by the introduced methodology for interfaces formed from mixed solutions containing proteins and surfactants. We studied many proteins, such as human and bovine serum albumin (HSA and BSA), β-lactoglobulin (β-LG), β-casein (β-CS), lysozyme, and mixed with various surfactants like sodium dodecyl sulphate (SDS), dodecyl or cetyl trimethylammonium bromide (DTAB or CTAB), decyl or dodecyl dimethyl phosphine oxides (C_{10}DMPO or C_{12}DMPO), or even technical surfactants like Tween 20. We present here only few selected results out of a huge pool of systematic studies.[22,23]

7.4.1. Dynamic Surface Tension

In Fig. 7.4 the dynamic surface tension for lysozyme are presented, as determined from the profile of a buoyant bubble. The tensions were recorded over 24 hours. As one can see from the graphs, no equilibrium has been reached and the dynamic surface tension continues to decrease over a prolonged period of time. It has been shown, that the use of drop and bubble shape analysis can lead to considerable differences of the dynamic surface tension, in particular at very low protein concentrations.[24] The use of buoyant bubbles makes sure that the real equilibrium surface tensions are obtained.[25] When pendant drops are used to measure very low protein concentrations, equilibrium data are reached which are significantly higher because of the depletion of molecules from the solution through the adsorption process.

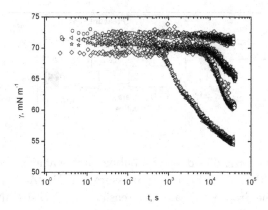

Fig. 7.4 Dynamic surface tension of lysozyme, buoyant bubble, at different bulk concentrations: ○ - 10^{-10}; ★ - 2.10^{-8}; ◊ - 5.10^{-8}; ◁ - 10^{-7} (from ref. 25)

From the comparison between both experiments one can conclude about the adsorbed amount at very low bulk concentrations,[26] which will, however, not be further discussed here.

The dynamic surface tensions shown in Fig. 7.5 demonstrate that the addition of surfactant to a 7×10^{-7} mol/l lysozyme solution leads to a significant change in the measured surface tensions. The dynamic surface tension of a pure 7×10^{-7} mol/l lysozyme solution is shown for comparison as well. For surfactant concentration above 10^{-5} mol/l significant effects are observed, although the surfactant alone would not yet significantly adsorb. At surfactant concentrations above 10^{-4} mol/l the adsorption kinetics looks essentially like the pure surfactant. At the CMC (9×10^{-3} mol/l) the equilibrium surface tension is around 29 mN/m and no further dynamics appears in the measured time interval.[27]

When SDS is added to the same 7×10^{-7} mol/l lysozyme solution, the picture is quite different, as shown in Fig. 7.6. As compared to the pure lysozyme a significant change in surface tension is observed. In comparison to C_{10}DMPO these dramatic changes appear at much lower surfactant bulk concentrations. At intermediate concentrations the dynamic surface tensions are characterised by both components, however the influence of the present surfactant becomes more evident with increasing concentration. At the highest studied concentration of 1×10^{-2} mol/l (just above the CMC of SDS) yields dynamic surface tension completely controlled by the surfactant.

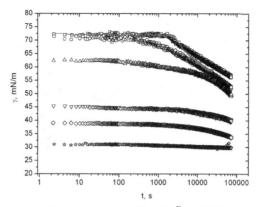

Fig. 7.5 Dynamic surface tension of 7×10^{-7} mol/l lysozyme-C_{10}DMPO solution at the air/water interface measured by pendant drop tensiometry at different C_{10}DMPO concentrations :
□ - 10^{-9} mol/l; ○ - 10^{-5} mol/l; △ - 10^{-4} mol/l; ▽ - 10^{-3} mol/l; ◊ - 2.10^{-3} mol/l; ⋆ - 9.10^{-3} mol/l (from ref. 27)

7.4.2. Equilibrium Surface Tension Isotherms

Based on the dynamic surface tension data, equilibrium isotherms can be derived. Such dependencies allow good access to thermodynamic quantities of the adsorption layer via application of models, such as those given in section 3.

An example is illustrated by Fig. 7.7, where the dependencies for the individual solutions of the non-ionic surfactant Tween 20 (polyoxyethylene 20 sorbitan monolaurate), and mixtures of 10^{-6} mol/l β-LG with different amounts of Tween 20 are presented.[28] As the value of ω_0 for β-LG was unknown, calculations were performed with various ω_0 values typical for proteins, and a satisfactory agreement between the experimental and theoretical dependencies of surface tension on surfactant concentration in the protein/surfactant mixture was achieved. Data obtained[29] for mixtures of β-CS and Tween 20 agreed well with the thermodynamic model given by the Eqs. (7) to (9).

In Fig. 7.8 the experimental surface tension isotherm for SDS and β-LG/SDS mixtures at a different SDS concentrations and a fixed β-LG concentration of 10^{-6} mol/l are illustrated. Assuming $\omega_S = 3.5 \times 10^5$ m^2/mol for SDS, taken equal to ω_0 for β-LG, the values for the other parameters in Eqs. (4) and (5) are $\alpha_S = 0.4$ and $b_S = 2.39 \times 10^4$ l/mol. The respective

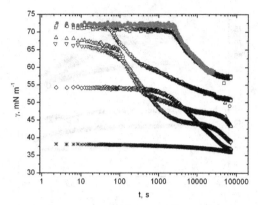

Fig. 7.6 Dynamic surface tension of 7×10^{-7} mol/l lysozyme/SDS at the air/water interface measured with the pendant drop method for different SDS concentrations

□ - 10^{-9} mol/l; ○ - 10^{-7} mol/l; △ - 10^{-6} mol/l; ▽ - 10^{-4} mol/l;
◊ - 5.10^{-4} mol/l; ⋆ - 10^{-3} mol/l; ∗ = 10^{-2} (from ref. 27).

calculated curve shown in Fig. 7.8 demonstrates a very good agreement with the theory given by Eqs. (7) to (9) using the parameters $\alpha_{SPS} = 0$, $\alpha_{PS} = \alpha_P$, and $b_{PS} = b_P$.[30] The adsorption behaviour of β-LG/SDS mixtures is similar to that found also in[30] for the HSA/CTAB mixtures. However, the experimental surface tensions are essentially lower than those for the systems containing SDS.

The best agreement between experimental and calculated values is obtained for three free positive charges (given by the parameter m) in the β-LG molecule, which is not in contradiction with data obtained by other methods.[31] The calculations made in[30] show that at a certain SDS concentration a maximum in the adsorption of protein/surfactant complex appears, i.e. it becomes significantly higher than the free surfactant. The existence of such a maximum was also found experimentally for mixtures of β-LG with non-ionic surfactants.[32]

7.4.3. Reversibility of Adsorption

A particular feature of proteins is their large adsorption energy when attached to liquid interfaces.[33] This energy, of the order of 20 RT, is so large that the protein adsorption appears as if it is irreversible, which means that once adsorbed, proteins would not desorb into the solvent when the

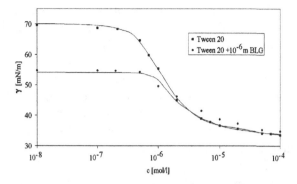

Fig. 7.7 Surface tension of mixtures of β-lactoglobulin (10^{-6} mol/l) with Tween 20 ($\circ$); the theoretical curve for the mixture was calculated from Eqs. (7) – (9) with $\omega_0 = 5 \times 10^5$ m^2/mol; individual Tween 20 ($\square$); (from ref. 28)

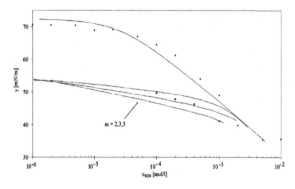

Fig. 7.8 Surface tension of mixtures of β-LG (fixed at 10^{-6} mol/l) with SDS in buffer ($\square$); theoretical curve for mixtures (from ref. 30); the individual SDS solution in the buffer ($\circ$).

solution in the bulk would be replaced by the pure buffer. This is really observed experimentally. The question, however, arises if surfactants can help to remove proteins from an interfacial layer.[34] The presented drop profile tensiometry equipped with a double capillary represents the optimum tool to study such phenomena. For this we pre-adsorbed a protein first, then exchanged the drop volume against a certain surfactant solution, let the surfactant diffuse to the interface, penetrate into the protein layer and adsorb, thereby eventually replace the protein molecules, and then wash all surfactants off the drop. As the adsorption energy of single surfactant molecules is rather low, they can be completely removed from the interface when the drop volume does not contain any surfactant molecules.

Fig. 7.9 shows the dynamic surface tensions after a protein was pre-adsorbed at a fixed concentration and subsequently the drop volume replaced by a solution with a respective surfactant concentration. The higher the surfactant concentrations used in the volume exchange, the lower are the final surface tensions.[35] We can assume that the surfactant strongly adsorbs at the drop surface and displaces an increasing number of protein molecules from the surface layer due to stronger competition.

As mentioned above, the experimental tool allows now to replace the surfactant from the drop volume by the pure buffer. For pure surfactant solutions, this leads to a complete wash-off of any surface active molecules and hence to a surface tension value close to that of the pure buffer. In Fig. 7.10 the curves are shown which correspond to the replacement of the previously injected C_{12}DMPO solution by the buffer. The larger the concentration of the temporarily injected surfactant solution was, the lower was the intermediate surface tension, but the higher is the final surface tension after washing with the buffer. We can conclude that a sufficiently high surfactant concentration is able to remove protein molecules from the interface. At low surfactant concentrations, the effect is negligible, while at very high concentrations, the protein is almost completely removed.

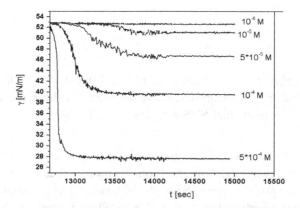

Fig. 7.9 Dynamic surface tensions during drop-bulk exchange measured after pre-adsorption of β-CS at a fixed concentration of 10^{-6} mol/l and a subsequent adsorption of C_{12}DMPO at different concentrations between 10^{-6} M and 5×10^{-4} M (from ref. 35)

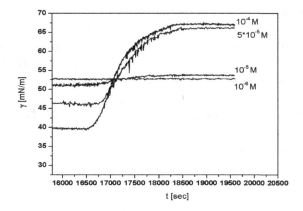

Fig. 7.10 Dynamic surface tensions during drop-bulk exchange processes after an initial pre-adsorption of β-CS at a fixed concentration of 10^{-6} mol/l, a subsequent adsorption of C_{12}DMPO at different concentrations and subsequently exchange with a buffer solution (washing out) (from ref. 35)

A quantification of how much protein is still remaining at the interface, is difficult to do. One option is a direct measurement of the adsorbed amount by surface ellipsometry. This, however, becomes less accurate with decreasing surface concentration. A second, however, not fully quantitative is to probe the amount of yet adsorbed protein molecules by relaxation experiments. As it will be shown in section 5, the frequency dependence of the dilational visco-elasticity is very sensitive to the type of molecules occupying the interface. Such experiments are presently under way and cannot be shown here yet.

7.5. Dilational Elasticity and Viscosity

The described drop and bubble profile tensiometry is suitable also to study the response of interfacial layers to dilation and expansion deformations.[36,37] Such experiments provide the elasticity modulus and the phase angle, or the dilational elasticity and viscosity, respectively.

The dilational elasticity modulus $|E|$ in dependence of the oscillation frequency for β-LG/C_{10}DMPO mixtures at different C_{10}DMPO concentrations are shown in Fig. 7.11, using the experimental conditions presented elsewhere.[11] For the relaxation experiments the solution drop was subjected to harmonic oscillations with a magnitude of about 7.5% and frequencies in the range between 0.01 and 0.2 Hz. With increasing C_{10}DMPO concentration, the elasticity modulus of the β-LG/C_{10}DMPO mixture decreases significantly, as one can see in Fig. 7.11. At $f = 0.1$ Hz, for instance, the modulus for β-LG mixed with 7×10^{-4} mol/l C_{10}DMPO is about 20 times lower than that for pure β-LG. The theoretical dependencies, shown in Fig. 7.11 as solid lines, were calculated from Eq. (12) using known parameters of the individual β-LG and C_{10}DMPO solutions, and a surfactant diffusion coefficient D_S between $10^{-10} m^2/s$ and 3×10^{-10} m^2/s. The agreement between the theory and experiment is quite satisfactory, especially in the low C_{10}DMPO concentrations range. At C_{10}DMPO concentrations above 10^{-4} mol/l the agreement is somewhat worse.

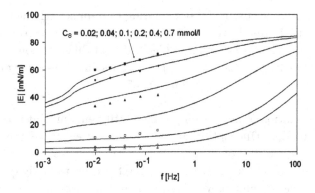

Fig. 7.11 Dilational elasticity modulus $|E|$ as a function of oscillation frequency at various C_{10}DMPO concentrations in mixed β-LG/C_{10}DMPO solutions; theoretical curves were calculated from Eq. (12) (from ref. 38)

In order to obtain better agreement with the experimental data, one could use a higher diffusion coefficients for the protein, e.g., $D_P = (10^{-10} - 10^{-11}) m^2/s$ instead of $10^{-12} m^2/s$. Probably in the presence of a surfactant the processes of protein reconformation and aggregation in the surface layer are accelerated, which increases the corresponding effective diffusion coefficient.

The surface dilational modulus for a fixed lysozyme concentration of 7×10^{-7} mol/l as a function of the surfactant concentration is shown for lysozyme/C_{10}DMPO mixtures in Fig. 7.12, and for lysozyme/SDS mixtures in Fig. 7.13, respectively. Lysozyme belongs to the rarely studied proteins due to its rather difficult structure.[39] These data were measured at a fixed oscillation frequency of 0.08 Hz. Similar data were presented in[25] for other frequencies (0.01, 0.04, 0.16 and 0.4 Hz). The elasticity modulus increases slightly with the oscillation frequency for both systems, while the phase angle decreases but remains below 12°. For lysozyme/C_{10}DMPO mixtures the $|E|$ decreases with increasing surfactant concentration as shown in Figs. 12 and 13, calculated with the theoretical model of Eq. (12). For the theoretical calculations a set of model parameters was used in[25], including the diffusion coefficients $D = 10^{-12} m^2/s$ for lysozyme, and $D = 410^{-10} m^2/s$ for the surfactant molecules. The theoretical curve given in Fig. 12 describes the experimental results very well. As it is seen from the surface tension isotherm,[25] at C_{10}DMPO concentrations above 10^{-4} mol/l, a partial hydrophilisation of the protein/surfactant complexes can take place, which decreases the protein adsorption and causes a considerable decrease of the dilational elasticity (the bold dotted line in Fig. 7.12).

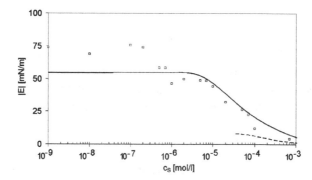

Fig. 7.12 Concentration dependence of the surface dilational modulus $|E|$ of 7×10^{-7} mol/l lysozyme/C_{10}DMPO at a fixed oscillation frequency of 0.08 Hz (□), lines are theoretical values (see text) (from ref. 25)

For lysozyme/SDS mixtures, however, there is a surfactant concentration interval, in which the elasticity of the mixed layer is increased. In agreement with the situation shown in Fig. 7.13, the arrows indicate the transitions from non-associated lysozyme (A-B) to the hydrophobic complex (B-C-D) and then to the hydrophilized complex (D-E). Note that the transition points B and D, shown in Fig. 7.13, correspond to approximately the same SDS concentrations found in the respective isotherms.

It is essential that the calculations with the theoretical model allow us to explain the maximum in the observed elasticity dependence —E—(c_S) for lysozyme/SDS mixed adsorption layers. This maximum can be attributed essentially to the adsorption of hydrophobized lysozyme/SDS complexes. The step by step transition to significantly less hydrophobic complexes in point E, are due to the hydrophobic interaction. This in turn, is the reason for a decreased activity of the complex, as comparable to the pure protein, by a factor of about 10 or more, accompanied by a decreased visco-elasticity modulus $|E|$.

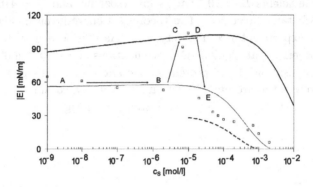

Fig. 7.13 Surface dilational modulus of 7×10^{-7} mol/l lysozyme/SDS versus SDS concentration at an oscillation frequency of 0.08 Hz (□), lines are theoretical values (see text) (from ref. 30)

7.6. Conclusions

The presented material demonstrates that the topic of mixed interfacial layers, comprised of proteins and surfactants, is a great challenge and requires good theories as well as excellent experimental tools. Still one can conclude that only some very first ideas exist on the overall interfacial behaviour and many effects are only qualitatively understood.

For example, the adsorption dynamic data, as shown in Figs. 4–6 are not well described by the Ward and Tordai Eqs. (10) and (11). Thus, for the presented systems obviously another mechanism in addition to simple diffusion plays a role for the establishment of the surface layer. Recently orogenic displacement is discussed as a mechanism of replacing proteins by surfactants.[40–42] The given thermodynamic models also partially cover such phenomena. Additional information can be expected from simulations,[43,44] however, these require a good basis of experimental data, which is established only slowly now. One of the most important set of information can be provided by the interfacial rheology as emphasised recently.[45]

References

1. V.B. Fainerman, E.H. Lucassen-Reynders and R. Miller, *Adv. Colloid Interface Sci.*, 106, 237 (2003)
2. R. Miller, V.B. Fainerman, M.E. Leser and M. Michel, *Current Opinion in Colloid Interface Sci.*, 9,350 (2004)
3. I. Langmuir, *J. Amer. Chem. Soc.*, 39, 1848 (1917)
4. A.N. Frumkin, *Z. Phys. Chem. Leipzig*, 116, 66 (1924)
5. V.B. Fainerman, V.I. Kovalchuk, E.V. Aksenenko, M. Michel, M.E. Leser and R. Miller, *J. Phys. Chem. B.*, 108, 13700 (2004)
6. V.B. Fainerman, S.A. Zholob, M. Leser, M. Michel and R. Miller, *J. Colloid Interface Sci.*, 274, 496 (2004)
7. A.F.H. Ward and L. Tordai, *J. Phys. Chem.*, 14, 453 (1946).
8. R. Miller, V.B. Fainerman, E.V. Aksenenko, M.E. Leser and M. Michel, *Langmuir*, 20, 771 (2004).
9. Q. Jiang, J.E. Valentini and Y.C. Chiew, *J. Colloid Interface Sci.*, 174, 268 (1995).
10. P. Joos, *Dynamic Surface Phenomena* (VSP, Dordrecht, The Netherlands, 1999).
11. E.V. Aksenenko, V.I. Kovalchuk, V.B. Fainerman and R. Miller, *J. Phys. Chem. C*, 111, 14713 (2007).
12. J. Lucassen and M. van den Tempel, *Chem. Eng. Sci.*, 27, 1283 (1972)
13. J. Lucassen and M. van den Tempel, *J. Colloid Interface Sci.*, 41, 491 (1972)
14. G. Loglio, P. Pandolfini, R. Miller, A.V. Makievski, P. Ravera, M. Ferrari and L. Liggieri, In: *Novel Methods to Study Interfacial Layers*, D. Möbius and R. Miller (Editors), (Elsevier Science, Amsterdam, 2001).
15. Y. Rotenberg, L. Boruvka and A.W. Neumann, *J. Colloid Interface Sci.*, 93, 169 (1983)
16. H.A. Wege, J.A. Holgado-Terriza, A.W. Neumann and M.A. Cabrerizo-Vilchez, *Colloids Surfaces A*, 156, 509 (1999)
17. R. Miller, D.O. Grigoriev, J. Krägel, A.V. Makievski, J. Maldonado-Valderrama, M.E. Leser, M. Michel and V.B. Fainerman, *Food Hydrocolloids*, 19, 479 (2005).
18. A. Passerone, L. Liggieri, N. Rando, F. Ravera and E. Ricci, *J. Colloid Interface Sci.*, 146, 152 (1991)
19. L. Liggieri, F. Ravera and A. Passerone, *J. Colloid Interface Sci.*, 169, 226 (1995)
20. L. Liggieri and F. Ravera, *Capillary Pressure Tensiometry and Applications in Microgravity*, in "Drops and Bubbles in Interfacial research", Studies in Interface Science, D. Möbius and R. Miller (Eds.), Vol. 6, (Elsevier, Amsterdam, 1998), pp. 239-278
21. A.V. Makievski, J. Krägel, P. Pandolfini, G. Loglio, L. Liggieri, F. Ravera, E. Santini, M.E. Leser, M. Michel, R. Miller, *Microgravity - Science and Technology Journal*, 18, 108 (2006)
22. V.S. Alahverdjieva, PhD Thesis, Potsdam, 2007
23. Cs. Kotsmar, Ph.D. Thesis, Potsdam, 2008.

24. A.V. Makievski, G. Loglio, J. Krägel, R. Miller, V.B. Fainerman and A.W. Neumann. *J. Phys. Chem.*, 103, 9557 (1999)
25. V.S. Alahverdjieva, V.B. Fainerman, E.V. Aksenenko, M.E. Leser and R. Miller, *Colloids Surfaces A*, 317, 610 (2008)
26. R. Miller, V.B. Fainerman, A.V. Makievski, M. Leser, M. Michel and E.V. Aksenenko, *Colloids Surfaces B*, 36, 123 (2004)
27. V.S. Alahverdjieva, D.O. Grigoriev, V.B. Fainerman, E.V. Aksenenko, R. Miller and H. Möhwald, *J. Phys. Chem.*, DOI:10.1021/jp074753k (2008).
28. R. Miller, V.B. Fainerman, M.E. Leser and M. Michel, *Colloids Surfaces A*, 233, 39 (2004)
29. J. Krägel, R. Wüstneck, F. Husband, P.J. Wilde, A.V. Makievski, D.O. Grigoriev and J.B. Li, *Colloids Surfaces B*, 12, 399 (1999).
30. V.B. Fainerman, S.A. Zholob, M.E. Leser, M. Michel and R. Miller, *J. Phys. Chem.*, 108, 16780 (2004).
31. S. Magdassi, Y. Vinetsky and P. Relkin, *Colloids Surfaces B*, 6, 353 (1996)
32. E. Dickinson, D.S. Horne and R.M. Richardson, *Food Hydrocolloids*, 7, 497 (1993)
33. V.B. Fainerman, R. Miller, J.K. Ferri, H. Watzke, M.E. Leser and M. Michel, *Adv. Colloid Interface Sci.*, 123-126, 163 (2006)
34. T.F. Svitova, M.J. Wetherbee and C.J. Radke, *J Colloid Interface Sci.*, 261, 170 (2003)
35. Cs. Kotsmar, D.O. Grigoriev, A.V. Makievski, J.K. Ferri, J. Krägel, R. Miller and H. Möhwald, *Colloid Polymer Sci.*, 286, (2008).
36. R. Miller, R. Sedev, K.-H. Schano, Ch. Ng and A.W. Neumann, *Colloids Surfaces A*, 69, 209 (1993).
37. J. Benjamins, A. Cagna and E.H. Lucassen-Reynders, *Colloids Surfaces A*, 114, 245 (1996).
38. E.V. Aksenenko, V.I. Kovalchuk, V.B. Fainerman and R. Miller, *Adv. Colloid Interface Sci.*, 122, 57 (2006).
39. S. Sundaram, J.K. Ferri, D. Vollhardt and K.J. Stebe, *Langmuir*, 14, 1208 (1998).
40. A.R. Mackie, A.P. Gunning, P.J. Wilde and V.J. Morris, *J. Colloid Interface Sci.*, 210, 157 (1999).
41. A.R. Mackie, A.P. Gunning, M.J. Ridout, P.J. Wilde and V.J. Morris, *Langmuir*, 17, 6593 (2001).
42. N.C. Woodward, P.J. Wilde, A.R. Mackie, A.P. Gunning, P.A. Gunning and V.J. Morris, *J. Agric. Food Chem.*, 52, 1287 (2004).
43. L.A. Pugnaloni, R. Ettelaie and E. Dickinson, *Langmuir*, **19**, 1923 (2003).
44. E. Dickinson, *Soft Matter*, 2, 642 (2006).
45. Interfacial Rheology, Vol. 1, Progress in Colloid and Interface Science, R. Miller and L. Liggieri (Eds.) Brill Publ., Leiden, 2009.

Chapter 8

Factors Affecting Mixed Aggregation

Pablo C. Schulz

Departamento de Química,
Universidad Nacional del Sur,
Bahia Blanca, Argentina

Some characteristics of surfactants were commonly not considered when mixed micellization is studied. We have studied some of them. The definition of the interaction parameter in the regular solution theory of mixed micelles was taken as $\beta = \beta_{ph} + \beta_{core}$, in which β_{ph} is the excess energy of interaction between polar head groups and β_{core} is the core excess of energy, caused by the interaction of the hydrophobic tails in the mixed micelle core, when compared with the situation in pure component micelles. In the case of ionic surfactants, β_{ph} is considered as an electrostatic contribution β_{elec}, associated with electrostatic interactions between the charged hydrophilic groups of components. The effect of the difference in chain length between components, steric constraints originated in the structure of aggregates or/and components, and the presence of double bonds in one of the surfactants in mixed micelles or vesicles on both components of β is discussed.

8.1. The Regular Solution Theory of Mixed Micelles

Regular solution theory combined with the pseudophase separation model, which is a thermodynamic one, has been very widely used to model the thermodynamic nonidealities of mixed micelles; it has been shown to accurately model critical micelle concentration (CMC) values[1] and monomer-micelle equilibrium compositions[2] in surfactant systems exhibiting negative deviations from ideality. However, it must be pointed out that the theoretical validity of using regular solution theory to describing nonideal mixing in mixed surfactant micelles has been questioned.[3] Although this theory assumes that the excess entropy of mixing is zero, it has been demonstrated that in some surfactant mixtures this assumption is not true.[4,5] However, this model remains as a very widely used and convenient method for analyzing experimental data. An application of the theory on the aqueous system sodium dodecyl sulfate-n-octanol, in which the water solubility of octanol was taken as equivalent to its CMC, gave results which were in agreement with those obtained using other approaches, giving some support to the regular solution theory for mixed micelles.[6]

To study the non-ideality in mixed micelles it is commonly used the interaction parameter (in $k_B T$ units, k_B being the Boltzmann constant and T the absolute temperature) $\beta = \beta_{ph} + \beta_{core}$, in which β_{ph} is the excess energy of interaction between polar head groups in the mixed system, compared with that of the polar heads in pure components micelles (*i.e.*, micelles having only one of the components of the mixture). In case of ionic surfactants β_{ph} considered as an electrostatic contribution β_{elec}, associated with electrostatic interactions between the charged hydrophilic groups of surfactants 1 and 2. The core excess of energy, β_{core}, is caused by the interaction of the hydrophobic tails in the mixed micelle core, when compared with the situation in pure component micelles.[7] It is commonly accepted that β_{core} is typically equal to zero for mixtures of two hydrocarbon based (or fluorocarbon based) surfactants,[8,9] but is larger than zero for a binary mixture of hydrocarbon and fluorocarbon surfactants due to the repulsive interactions in the micelle core.[8-13]

The parameter β is related to the molecular interactions in the mixed micelle by[14]:

$$\beta = N_A(W_{11} + W_{22} - 2W_{12}), \qquad (1)$$

where W_{11} and W_{22} are the energies of interaction between molecules in the pure micelle and W12 is the interaction between the two species in the mixed micelle. N_A is the Avogadro's number.

8.2. Discussion

8.2.1. *Factors Affecting β_{core}*

Accordingly to the above approach, it is commonly supposed that the mixed micellization of homologous surfactants is ideal, because the polar head groups are the same and then β_{ph} is not affected, and the mixture of homologous hydrocarbons is also ideal, so β_{core} must not change. Actually, mixtures of linear hydrocarbons with different chain length are not strictly ideal, showing a small repulsive interaction.[17] We have found that mixed micelles of homologs surfactants are not ideal (except if the difference in chain length Δn_C between both components is only of about one carbon atom), irrespective of the nature of the polar head groups. This nonideality is attractive, and increases linearly with Δn_C. For $\Delta n_C \geq 5.31 \pm 0.48$ there is an increase in slope, i.e, the dependence of β on Δn_C becomes stronger than below this limiting Δn_C[16], as it can be seen in Figure 8.1.

The relationships found are:

At low Δn_C values the equation of the straight line is:

$$\beta = -(0.526 \pm 0.049)\Delta n_C + 0.403 \pm 0.021 \qquad (2)$$

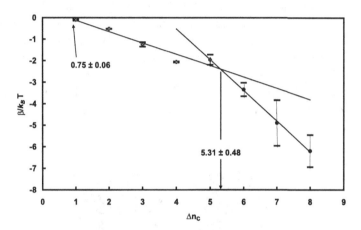

Fig. 8.1 Dependence of the average interaction parameter β_M on Δn_C.[16]

giving $\beta = 0$ at $\Delta n_C = 0.75 \pm 0.06$, and the equation of the line at high Δn_C values is:

$$\beta = -(1.426 \pm 0.044)\Delta n_C + 5.18 \pm 0.28 \qquad (3)$$

The intersection of both lines occurs at $\Delta n_C = 5.31 \pm 0.48$.

There is a change in the nature of the nonideal interaction between the two components below and above $\Delta n_C = 5.31 \pm 0.48$. Below this value, $\partial \beta / \partial \Delta n_C = -0.526 k_B T$ i.e., $-(1.30 \pm 0.12)$ kJ.mol at 25° C, and above the intersection, $\partial \beta / \partial \Delta n_C = -1.43\ k_B T [-(3.55 \pm 0.11)$ kJ.mol at 25°C].

The study of the micellized surfactants activity coefficients shows that in all cases, both components affect each other. Strictly speaking, no one of the components can be considered as the solvent or the solute.

The origin of this non-ideality in mixed micelles of homologs surfactants was attributed to the difference between the structure of liquid hydrocarbon mixtures and liquid hydrocarbon tails mixture in micelles. The main difference is that the hydrocarbon chains in micelles are anchored by one of their ends to the Stern layer in a spherical structure, whilst the hydrocarbon molecules are free to move without restrictions. Spherical micelles are formed by a mixture of molecules having stretched chains and other having their chains folded to fill the spaces near the surface created by the radial disposition of the stretched molecules. This packing reduces the hydrocarbon/water contact at the micelle surface. Then, the inclusion of a shorter surfactant into a micelle of a longer homolog may produce stabilization because the shorter surfactant may fill these spaces with a reduced folding (Figure 8.2). The advantage must be small if Δn_C is small, but must increase with increasing Δn_C. The increase in stabilization above $\Delta n_C =$ 5.31 probably reflects that the need of folding of the shorter surfactant is reduced when the difference in chain length exceeds this value. Since the micelle core is of liquid nature,[17] the folding of a chain must reduce its freedom of motion in comparison with the stretched ones.

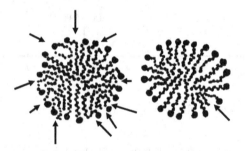

Fig. 8.2 left: proposed scheme of a pure surfactant micelle of a surfactant having 13 carbon atoms in the hydrocarbon chain showing by arrows the hydrocarbon surface exposed to water; right: mixed micelle composed by two homologous surfactants, one of them having 13 and the other 6 carbon atoms in the alkyl chain, i.e. with $\Delta n_C = 7$. The amount of hydrocarbon surface exposed to water was substantially reduced.[16]

In the light of this interpretation, both components are affected by their inclusion in the mixed micelles. In a pure surfactant micelle some of the chains are extended while other must be folded to fulfill the spaces between the extended surfactant molecules in order to reduce the hydrocarbon/water interface (Figure 8.2, left). Since it is impossible to cover the entire micelle surface with the polar headgroups, some part of the chain, about 3-4 methylene groups by each micellized surfactant molecule, remains in contact with water.[18-25] A shorter molecule may be intercalated between the extreme of the folded chain of the longest component and the micelle surface, replacing part of the hydrocarbon/water interface by a polar headgroup. This reduces the surface free energy of micelles (Figure 8.2, right). The energetic advantage must increase with (nC, because when (nC is large, the longest component can fold its distal end more easily in the space within the centre of the micelle and the end of the shorter component chain. This situation must be more favorable when that space is of five methylene groups or more (i.e., about 0.66 nm or more). Both the longer and the shorter chains in the mixed micelle expose less hydrocarbon surface to water than in their respective pure surfactant micelles. This is reflected in their activity coefficients in micelles[16].

When the mixed micelles composition was computed it was seen that larger the Δn_C, lower the proportion of the shorter surfactant in the mixed micelle. This means that the inclusion of longer surfactant molecules in the mixed micelles is favored by the presence of a shorter homolog and that effect increases with the difference in chain length.

Two factors govern the value of α, the micelle ionization degree, if the polar head group and the counterion in mixed micelles are the same: the ionic strength (I) in the intermicellar solution and the electrostatic surface potential of the micelle polar heads layer (without counterions) (Ψ_0). An increase of I or Ψ_0 must reduce the value of α. Since the values of α were computed at the CMC, a reduction of that concentration is equivalent to a reduction in I and then α must increase. However, the inclusion of the shorter molecule replaces micelle hydrocarbon surface by polar head groups, and then the surface charge density of micelles must increase, producing an increase of Ψ_0, which in turn, captures more counterions in the micelle Stern layer. As a consequence, the inclusion of the shorter component must produce an increase of α by reduction of the CMC, but an increase in the efficiency in the accommodation of the components must work in the opposite direction reducing the value of the micelle ionization degree. A more efficient packing of the components produces a more high Ψ_0 and a reduction of α. In the light of the preceding results, the values of α in mixed micelles must increase with Δn_C between 1 and 5, and then α must decrease for $\Delta n_C > 5$, as it was experimentally observed.[16]

Herrington et al.[26] found that surfactant geometry strongly affects the microstructures present in anionic/cationic surfactant mixtures. When the surfactants are branched and /or contain a bulky group (e.g. a benzene group) in the tail, the solid phase stability is reduced relative to the micellar and vesicular phases. As a result, wide regions of vesicle phase stability were observed. The system is strongly non-ideal.[27-30] Steric constraints were also found in mixtures of single chain/twin tailed surfactants such as dodecyltrimethylammonium bromide (DTAB) and didodecyldimethylammonium bromide (DDAB).[31] Despite the chains are both hydrocarbon with the same n_C, and the polar head groups (including counterions) are closely similar, the system was strongly nonideal. The model predicts that micelles must form for DTAB:DDAB proportions higher than 1:1, and vesicles and/or liposomes at lower proportions. The β parameter was positive indicating a repulsive interaction. The steric hindrance interaction parameter of Huang and Somasundaran theory[32] P^*, which describes the non-random mixing of components in mixed aggregates, indicated that the interaction of DTAB into DDAB layers was favored at very low DTAB content, but became highly unfavorable when the proportion of DTAB increased. The aggregates were systematically much richer in DDAB than the inter-aggregates solution. The Israelachvili−Mitchell−Ninham[33] packing parameter values (IMN = v/al, where v and l are the volume and length of the non-polar group and a the area occupied by the polar head group, computed with Evans et al. procedure[34]) were systematically compatible with flat surfaces, i.e., lamellar liquid crystals and vesicles. Then, vesicles form at DTAB:DDAB proportions below 9:1. Micelles were detected for a 9.5:0.5 proportion (Figure 8.3). The same result was found by Viseu et al.[35] in DDAB:DTAC mixtures. Steric hindrances must be taken into account in dealing with complex structures such as liquid crystals and mixed micelles having very dissimilar components. The concept of sterically stabilized vesicles and liposomes, caused by asymmetric bilayers with non-zero spontaneous curvature, which may originate from differences in internal and external monolayers, is known.[36] The results may be explained by the different composition of the inner and outer monolayers in the bilayer structure of vesicles.

Viseu et al.[35] proposed that the main factor which ruled the appearance of vesicles in the system DDAB-DTAC-water was the difference in the spontaneous curvature (or IMN packing parameter) of the two surfactants. They said that an asymmetric bilayer with adequate inner (c_{in}) and outer (c_{out}) curvatures may be reached spontaneously when DDAB is the main (even exclusive) component of the inner layer, but it is mixed with DTAC in the outer layer. Pure DDAB also forms vesicles, so the addition

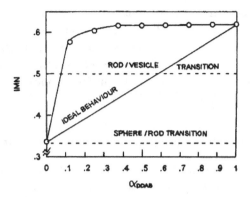

Fig. 8.3 IMN parameter for the aggregated DDAB-DTAB mixture versus the composition of the mixture ((DDAB). The straight line corresponds to ideal behavior[31].

of DTAC (or DTAB) should favour vesicle formation. Vesicles form at a concentration much lower than micelles and capture the adequate proportion of DTAB molecules to form the outer monolayer with an appropriate curvature. The appropriate composition to obtain spontaneous vesicles seems to be $X_{DDAB} = 0.994 \pm 0.003$.[31] By assuming that half of the aggregated DDAB molecules are in the inner monolayer of a vesicle, this means that the mole fraction of DTAB molecules in the outer monolayer is about $X_{DTAB,out} = 0.012 \pm 0.003$, and the packing parameter of the mixture in the outer monolayer is IMN $= 0.617 \pm 0.006$.

Since DTAB is much more water soluble than DDAB, the excess of unaggregated DTAB molecules remains in monomer form until its own concentration reaches the DTAB CMC. This situation may explain the presence of globular micelles together with vesicles in DDAB-DTAC systems.[35] A posterior study of DDAB:DTAB mixtures at higher concentration also showed micelles in equilibrium with vesicles.[37]

The above results indicate that the process of obtaining single tailed -double tailed mixed aggregates by simply mixing the surfactants is not so easy as it was assumed, because of the non-ideal interaction arising from steric packing constraints. However, it must be pointed out that Weers and Scheuing[38] studied DDAB-DTAB mixtures and they found micelles and ideal mixing behavior. They suggested that as DDAB is added to DTAB micellar solutions, spherical micelles undergo a transition from spheres to disks to rods as a function of composition.

Small β values were found for some particular anionic - cationic surfactant aqueous mixtures, as CTAB-sodium deoxycholate[39] ($\beta = -2.7$), CTAB - sodium cholate39 ($\beta = -4.0$), and dodecyltrimethylammonium bromide - disodium dodecanephosphonate[40] ($\beta = -1.66$). All these systems have some structural characteristics that are different to the most commonly studied cationic-anionic mixtures.

8.2.2. *The interaction between π-electrons and water*

Another factor to be taken into account in interpreting the behavior of mixtures of surfactants is the presence of double bonds in the chain. Since the π-electrons may form hydrogen bonds with water,[41–46] thus strongly reducing the free energy of interaction of the hydrocarbon surface of micelles with water, there is a preference to situate the double bonds at the micelle core/water interface giving rise to a preferential micelle composition and the formation of a natural curvature that difficult precipitation. This effect was found in hexadecyltrimethylammonium bromide (CTAB)/ sodium oleate (SOL) mixtures.[47] It is very interesting that in this system (has its minimum at $X_{CTAB} = 0.75$ ($\beta = -6\ k_B T$). Except for this composition, the β value is closer to that of ionic-nonionic surfactant mixtures than for typical anionic - cationic ones (about -20).[48] The oleate chain has a double bond in the middle.

The double bond has the tendency to remain at the micelle surface, explaining the difference between the structure of SOL micelles and those of CTAB. CTAB chains have a tendency to avoid the contact with water. It is impossible to make a spherical or cylindrical micelle of CTAB or SOL with a surface completely covered with hydrophilic groups: a certain fraction of the surface consists of hydrocarbon exposed to water. Since the hydrocarbon chains are in liquid-like state,[17] some of them are folded and about four carbon atoms for each micellized hydrocarbon chain are exposed to water.[18–25] For a chain having 16 carbon atoms, this means that the 25% of the hydrocarbon core is exposed to water. Then, the substitution of 25% of the fully saturated CTA^+ chains in micelles by OL^- ions having a double bond with water affinity must diminish the interfacial Gibbs' free energy of micelles. This diminution is of the order of 13-16 mJ/m^2 with respect to a saturated hydrocarbon-water interface.[49,50] All the CTA^+ chains could remain in the hydrophobic core interior, whereas all the OL^- double bonds could remain in contact with water. The energetic situation of CTA^+ ions in mixed micelles having about 25% of OL^- will be more favorable than in other proportions, which may explain the observed minimum of the micellized CTAB activity coefficient (f_{CTAB}). However, the energetic situation of the OL^- ions is not considerably affected by the mixed micellization, provided that all (or almost all) of them have their double bond at the micelle surface.

The inclusion of the $-CH = CH-$ groups in the Stern layer can also favor the formation of spherical micelles at low concentration. This inclusion gives an appropriate curvature and some hydrophilicity to the micelle-water interface which favors the spherical shape against cylindrical or plane structures. The folding of the oleate chain to expose the double bond to water causes a hydrocarbon crowding at the proximity of the surface. There is a loop of the portion of the chain between the carboxylate and the $-CH = CH-$ groups, both anchored at the surface, and the short terminal fraction of the chain which submerges into the hydrocarbon core. Then, this system may be considered similar to a mixture of a long-chained surfactant (CTAB) with a surfactant having a bulky, short hydrophobic group (SOL). In this case, Edlund et al.[51] explained this behavior on the basis of geometrical effects caused by the different values of nC. According to Edlund et al.,[51] the inclusion of the shortest surfactant molecule in the longest surfactant micelle gives rise to a reduction of the effective IMN of the surfactant mixture, by augmentation of a, whereas l remains constant and v increases slightly. A $IMN < 1/3$ value is compatible with the formation of spherical micelles. Increasing surfactant mixture concentration produces a reduction of the value of a, caused by the reduction of the electrostatic repulsion between the charged headgroups in excess in non-equimolecular micelles, and the possible change in the micelle composition (the values of X_I given by the mixed micelles theory are valid at the CMC). This leads to an increase of IMN. If $1/3 < IMN < 1/2$, the surface curvature of aggregates is compatible with cylindrical micelles. This situation may be extrapolated to the CTAB-SOL mixture.

The effect of double bonds was also studied extensively in mixtures of DTAB and sodium 10-undecenoate (SUD). The aqueous SUD-DTAB catanionic system was studied at low concentration.[52] The system did not precipitate, even at 1:1 SUD:DTAB proportion, but showed the formation of a coacervate in a range of surfactant mixture compositions. Micelles have a preferential composition of 0.37-mole fraction of SUD (X_{SUD}). This behavior is attributed to the presence of the double bound at the distal extreme of the SUD molecule, which can form hydrogen bonds with water. As a consequence, the vinyl groups ($-CH = CH_2$) are situated at the interface between the hydrocarbon micelle core and water (Figures 8.4 and 8.5), reducing the interfacial free energy. Structural computations demonstrated that the mentioned SUD proportion produces complete coverage of the micelle surface by the double bonds.[52] The interaction between the surfactant components in the mixed micelle was non-ideal.

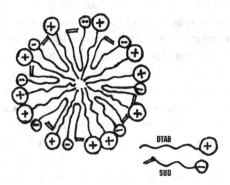

Fig. 8.4 Representation of a DTAB:SUD mixed micelle.

The mixed surfactant system SUD-DTAB was also studied by computational simulation to determine the composition and structure of the mixed microstructures.[53] Results were contrasted with experimental data obtained from literature and our own laboratory. The modellisation predicts spherical or cylindrical micelles with a preferential composition of SUD:DTAB about 1:2, while the system predicts a lamellar structure with a proportion 1:1 when SUD is replaced by the saturated soap sodium undecanoate. The model also predicts the inclusion of bromide ions deeply in the mixed micelle Stern layer. All predictions were in agreement with the previous experimental results.

It was also found that the partial molar volume [Figure 8.6] and micelle hydration [Figure 8.7] of both micellized and unmicellized DTAB:SUD mixtures are non-ideal, depend on the mixture composition and are related to structural changes in micelles. These phenomena are also caused by the presence of the double bond at the distal extreme of the SUD molecule and its affinity with water. In particular, as far as we know, this is the first reported non-ideal behavior of the partial molar volume in mixed micelles.[54]

Figure 8.7 shows that the hydration behavior is different in the one-phase region at the right-hand side of the two-phase region (coacervate) than in the monophasic region at the left - hand side of the biphasic region.

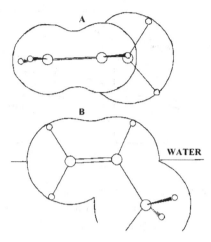

Fig. 8.5 Localization of the vinyl group at the micelle hydrocarbon/water interface (A) upper view, (B) lateral view.

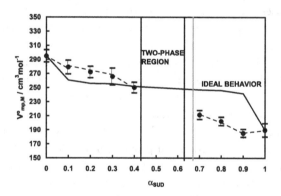

Fig. 8.6 The micellized surfactant partial molar volume at the CMC, $V_{mp,M}$, and the ideal value, $V^o_{mp,M,ideal}$, as a function of the micelle composition XSUD. Lines are eye guides (taken from ref. 55).

Micelles having lower SUD content show dehydration when compared with the ideal one. There is a minimum of hydration at $X_{SUD} = 0.37$, when the interaction parameter β shows a minimum and the micelle ionization degree (α) is maximum.[52]

These facts suggest that when $X_{SUD} = 0.37$, micelles compactness is maximal, with the opposite charged groups close each other. This situation

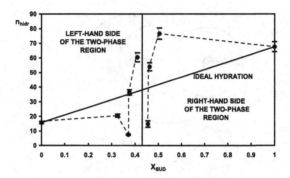

Fig. 8.7 The hydration of micelles (n_{hydr} = water molecules per surfactant molecule) represented as a function of X_{SUD}. Lines are eye guides (taken from ref. 55).

must produce dehydration in both the polar head groups and the double bonds of SUD included in micelles. When X_{SUD} increases, β increases and α decreases. The accumulation of the double bonds at the micelle hydrocarbon-water interface may explain the increase in hydration. The increase in β indicates that these micelles are less stable than that having $X_{SUD} = 0.37$. The capture of counterions indicates that the opposite polar head groups are far each other, retaining more water molecules among them. The same effect was seen in the SOL-CTAB mixed system.

Micelles in the right - hand monophasic side have almost the same α value but a large increase in β. This situation may be caused by an excess of double bonds over the ideal proportion ($X_{SUD} \approx 0.39$).[52] Probably the compactness of micelles is reduced when X_{SUD} increases, causing an increase in hydration by inclusion of water among the ionic head groups that are separated by the hydrophilic $-CH = CH_2$ groups covering the hydrocarbon -water interface. Moreover, if there is water intrusion into dehydrocarbon core, as suggested by some authors,[19,20,55-58] that intrusion will be facilitated by the inclusion of the double bonds in the hydrocarbon core of both the SUD micelles and in the mixed micelles. This inclusion of water in the micelle core may also contribute to the reduction in partial molar volume.

The formation of a coacervate in mixtures of DTAB-SUD aqueous mixtures was studied by light scattering, zeta potential measurements, and electronic microscopy. The coacervate appears when the zeta potential goes to zero, and the energy barrier against agglomeration disappears, promoting the agglomeration of micelles. Rod-like micelles agglomerate in bundles in the DTAB-rich side of the phase diagram [Figure 8.8], while spherical or globular micelles agglomerate in clusters in the SUD-rich side [Figure 8.9].

This difference explains the differences in the transition micelles - coacervate in the opposite sides of the two-phase region.[59] In general, ζ absolute value increased when the surfactant concentration increased. Since this is the opposite behavior that one can expect by the increase in the ionic strength while increasing C (as it can be seen in the pure SUD micelles), this means that mixed micelles change in composition by incorporating molecules of the component in excess when the total concentration increases.

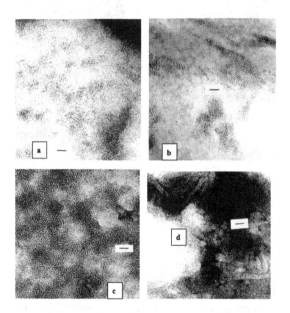

Fig. 8.8 TEM of the uranyl stained aggregates in the DTAB-rich side of the composition diagram, as a function of α_{SUD}: a: $\alpha_{SUD} = 0$, $C = CMC$, b: $\alpha_{SUD} = 0$, $C = 5\ CMC$, c : $\alpha_{SUD} = 0.3$, $C = CMC$, d : $\alpha_{SUD} = 0.4$, $C = CMC$. In the photograph for $\alpha_{SUD} = 0.4$ it can be seen the large bundles of rod-like micelles. Bars represent 21 nm (taken from ref. 60).

Pure SUD micelles ($\alpha_{SUD} = 1$) are small and nearly spherical. When DTAB was added (($\alpha - SUD = 0.9$ to 0.7) aggregates formed. These aggregates seem multilamellar vesicles. However, there are not true vesicles, since no birefringence was detected with the polarizing microscope. True vesicles are structures associated with lamellar liquid crystals. As it can be easily seen in the TEM photographs of the (SUD = 0.7 and 0.8 [Figure 9], near the formation of two phases, micelles aggregate in unordered clusters, which in turn undergo self-accommodation giving strings of globular micelles concentrically ordered.

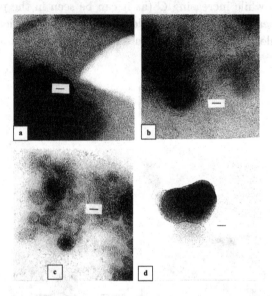

Fig. 8.9 TEM of the uranyl stained aggregates in the SUD-rich side of the composition diagram, as a function of α_{SUD} : a : $\alpha_{(SUD} = 1$, $C = CMC$, b : $\alpha_{SUD} = 0.8$, $C = CMC$, c : $\alpha_{SUD} = 0.8$, $C = 2CMC$ d : $\alpha_{SUD} = 0.7$, $C = CMC$. It can be seen unordered clusters, clusters partially ordered and clusters completely ordered. Bars represent 21 nm (taken from ref. 60).

Figure 8.8 shows the variation of the structure of surfactant aggregates when SUD is added to DTAB micelles. Pure DTAB micelles ($\alpha_{SUD} = 0$) are globular or spherical and become short rods as the concentration increased. These rods may self-organize in clusters looking as fingerprints. When increasing SUD content (from $\alpha_{SUD} = 0.1$ to 0.4), the rods increased their length and aggregated in clusters. In the $\alpha_{SUD} = 0.4$ system, near the formation of two phases, the rod-like micelles aggregated in bundles. However, there is no formation of hexagonal mesophase since no birefringence was detected. Thus, the sudden formation of two phases when SUD is added is preceded by the formation of bundles of rod-like micelles, whereas the gradual formation of the coacervate when DTAB is added to SUD, is preceded by the formation of agglomerations of spherical or globular micelles. This difference in the mechanism of formation may explain the different transition from monophasic to biphasic at the two sides of the coacervate region. In both, upper and lower layers in the coacervate region, the structure is an agglomeration of micelles, as it was supposed on the basis of the polarized light microscope observations.

Between $\alpha_{SUD} = 0.44$ and 0.674 the mixtures of SUD and DTAB form a coacervate. When approaching to the two-phase region from both sides of the composition diagram, micelles increase in size, reduce their ζ potential and aggregate in highly polydisperse clusters with increasing size. The coacervate appears when ζ approaches to zero, because the repulsion energy barrier between micelles disappears. The structure of micelles and the mechanism of aggregation of them are different at the two sides of the composition diagram, which explains the different behavior of the two biphasic / monophasic borders.

To corroborate that the double bonds are responsible of the nonideality, we have also studied the mixed system sodium dodecanoate (SDD) -SUD. The system was strongly non–ideal.[60]

Figure 8.10 depicts the micelle composition (X_{SUD}) as a function of the total mixture composition (α_{SUD}). It is clearly visible that there is a preferential composition, which is $X_{SUD} \approx 0.35 - 0.48$, similar to that obtained in previous work dealing on a SUD - dodecyltrimethylammonium bromide mixture, with X_{SUD} lying between 0.33 and 0.4752. The sample with $\alpha_{SUD} = 0.4$ has $X_{SUD} = 0.4$, which indicates that this proportion is the most favorable in micelles.

Micelles in samples having $\alpha_{SUD} < 0.4$ are richer in SUD than the intermicellar solution, which is not the expectable behavior because SUD is much more water soluble than SDD, and micelles above this proportion are richer in SDD. This indicates the existence of a mechanism which leads to a preferential micelle composition.

Figure 8.11 represents the intra-micellar interaction parameter β as a function of the micelle composition (X_{SUD}). All the mixtures show negative β values higher than that related to the difference in chain length $((-0.129 \pm 0.088)k_BT)$.[16] The values lie between ionic-nonionic mixtures such as the system SDS- poly (oxyethylene) (4) dodecylether

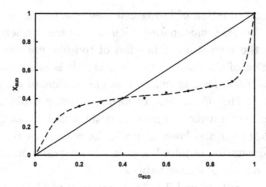

Fig. 8.10 The composition of SDD-SUD mixed micelles as a function of the total composition of the surfactants mixture. The full line represents the ideal composition. Dotted line is an eye guide (taken from ref. 61).

$(-3.9k_BT)^{61}$ and that of a cationic-anionic mixture: sodium decylsulfate-decyltrimethylammonium bromide $(-13.2k_BT)$.[14] This indicates an unexpected strongly nonideal system. There is a maximum in $\beta(=-4.22k_BT)$ at $X_{SUD} = 0.397$. This maximum seems strange because it is at about the preferential composition. However, the explanation comes from Figure 8.11: this micelle composition coincides with the overall total surfactant composition, i.e., when $(SUD = X_{SUD} = 0.4$, the system behaves as ideal and β must decrease. The interaction parameter cannot be zero because the conformation of both surfactants in the mixed micelles is not equal to that in the pure surfactants micelles. In pure SDD micelles, some portion of the saturated chains will remain in contact with water, and in pure SUD micelles, some of the unsaturated groups must remain in a hydrocarbon medium. These situations are represented in Eq. (1) by W_{11} and W_{22}. In the mixed micelle at the preferential composition, SDD chains are all in a hydrocarbon medium, and vinyl groups are all in contact with water. This means that W_{12} is different from W_{11} or W_{22}. This gives the value of β at the maximum, which is similar to that expected in mixtures of ionic-nonionic surfactants.

Figure 8.12 shows the ionization degree of micelles (α) as a function of the composition of micelles. It may be seen that all mixed micelles have α values higher than that of the pure surfactant micelles. This effect may be caused by the inclusion of the hydrated vinyl groups among the carboxylate ones producing a reduction in the micelle surface charge density, which in turn produces a reduction in the attraction of counterions to the Stern layer

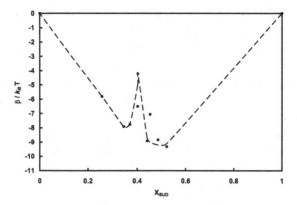

Fig. 8.11 The micelle interaction parameter as a function of the mixed micelle composition in the SDD- SUD aqueous system. Line is an eye guide (taken from ref. 61).

and an increase in α. This effect must be considered as a β_{elec} contribution, which must be added to the effect discussed in the preceding paragraph and explains why β is not zero (or $\approx -0.13 k_B T$) at the preferential composition.

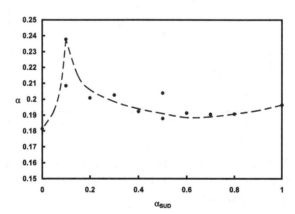

Fig. 8.12 The mixed micelle ionization degree as a function of the SDD-SUD mixed micelle composition. Line is an eye guide (taken from ref. 61).

The effect of double bonds can also be detected in mixed monolayers at the air/solution interface. The aqueous mixed system sodium dehydrocholate (NaDHC)-SOL mixed micellization and mixed monolayer adsorbed at the air-water interface was studied.[62] The molecular area at the cmc in pure surfactant solutions suggest that the adsorbed oleate chain was folded to allow the double bond in the middle of the molecule to keep in contact with water, and that the NaDHC molecule was situated with its plane lying parallel to the water surface, allowing the three carbonyl groups in the hydrocarbon backbone to form hydrogen bonds with water. The interaction was repulsive at the surface, and in the mixed monolayer some molecules must move away the less hydrophilic groups from water (double bond of SOL, carbonyl groups of $NaDHC$). The interaction in the mixed micelles was strongly attractive showing a preferential composition roughly equimolar (Figure 8.13) with $\beta = -16k_BT$, which is typical of catanionic mixtures. Since the polar head groups are the same, this situation must reflect the different hydrophobicity of both non-polar groups.

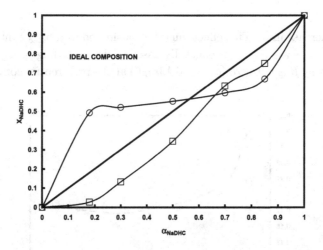

Fig. 8.13 Mole fraction of NaDHC in mixed micelles (circles) and in mixed monolayer adsorbed at the air-water interface (squares) versus composition of the SOL-NaDHC mixture (taken from ref. 63).

The behavior was attributed to the oleate location in the interstices of the open structure of NaDHC micelles, thus reducing the hydrocarbon-water contact via both, the carboxylate and the double bond. The micellar ionization degree α fell from the pure NaDHC micelles (0.54) to an almost constant value of 0.2, similar to that of pure SOL micelles, which indicates that the polar head groups of both components locate in the same zone increasing the charge density of the carboxylate layer, thus capturing counterions.

8.3. Factors affecting β_{ph}

8.3.1. *The Size of Polar Head Groups*

Another commonly accepted supposition is that the interaction in the Stern layer of micelles in anionic-cationic surfactants mixtures is predominantly electrostatic (i.e., that $\beta_{ph} = \beta_{elec}$). We have found that when the two dissimilar polar heads are bulky enough, the steric hindrance produces interesting effects. Catanionic mixtures having the proportion 1:1 do not precipitate. The DTAB/disodium n-dodecanephosphonate (DSDP) aqueous mixed system does not precipitate even at the 2:1 proportion which ought to generate uncharged aggregates. At all DTAB:DSDP proportions, the system is highly soluble. There is a wide domain of existence of spherical micelles, followed by a much larger domain of existence of rod-like micelles. At high concentration, only hexagonal liquid crystal exists. Neither vesicle nor lamellar mesophase was found. The explanation of this unusual behavior was proposed on the basis of steric effects due to the size of both headgroups ($-PO_3^=$ and $-N(CH_3)_3^+$; whose radius in micelles is 0.295 nm and 0.298 nm, respectively, without considering hydration.[63] This size causes a) the reduction of the electrostatic interaction intensity, and b) the production of a curvature at the aggregates' surface which is incompatible with planar structures at low concentration, giving rise to the formation of micelles, instead of solids or lamellar mesophases. Nevertheless, the interaction was attractive as expected in a catanionic system, though the β value ($\approx -2.5 k_B T$ as a maximum) was in absolute value much less that that of common catanionic systems, which is of the order of about $-20 k_B T$. The same behavior was found in aqueous dodecyltrimethylammonium hydroxide (DTAOH): dodecanephosphonic acid (DPA), having ($\approx -1.4 k_B T$.[64]

The air/solution monolayer structure of the DTAOH/DPA was also studied, and the interaction parameter at the surface was extremely low ($\beta_{surface} \approx -8x10^{-10} k_B T$) compared with that of common catanionic mixtures, which are between -30 and $-40 k_B T$, and is more close to that of ionic / nonionic surfactant mixtures.[65]

8.3.2. The Change in the Structure of the Polar Layer

Another supposition is that the interaction parameter must be constant for all proportions of both components. This lies on the supposition that the nature of the interaction between the micellized components does not change with micelle composition. However, this is not a general situation. The dependence of β on the micelle composition was previously observed by several authors.[66,67] In particular, there are situations in which it is evident that the nature of the interaction between components changes with micelle composition. In the case of mixtures of DPA-DTAOH the local effective dielectric constant at the mixed micelle polar surface was ($\epsilon_{eff} \approx 30$ from $X_{DPA} = 0$ (i.e., pure DTAOH) up to $X_{DPA} = 0.33$, (i.e., the neutral proportion DTA_2DPA,[68] which is a common value for the Stern layer of ionic micelles. However, when DPA is in excess ($X_{DPA} > 0.33$), (ϵ_{eff} increased up to about 65) [Figure 8.14].

This indicated that the micelle surface is structured by hydrogen bonds [Figure 8.15], as it was also detected in the lamellar liquid crystalline dodecane- and decanephosphonic acids polar layer (both acids form lamellar liquid crystals when contacted by water).[69] This situation is also reflected in the dependence of the CMC [Figure 16] and β [Figure 8.17] on mixed micelle composition.[70] The CMC behavior for DTAOH -rich micelles is similar to that of common ionic micelles, whilst in DPA - rich micelles the CMC is similar to that of pure DPA solutions. The value of β was between -0.4 and $-0.7k_BT$ when the hydrogen-bonded structure does not exist, fell up to $-1.4k_BT$ at $X_{DPA} \approx 0.4$ and then rose up to about zero in micelles having an excess of DPA. This behavior reflects a change in the interaction between components inside the mixed micelle. The hydrogen-bonded polar structure was destroyed between $X_{DPA} = 0.5$ and 0.33.

The behavior of the related mixed system DTAB-DSDP was more conventional, showing a statistically composition-independent β value.[65] When the system has an excess of DPA, lamellar mesophase appears at high concentration. When the DTAOH:DPA ratio is 1:1 an hexagonal mesophase appears, and increasing the DTAOH proportion the lamellar liquid crystal domain is reduced while that of the hexagonal one increases, and finally the lamellar mesophase disappears.[63]

Another interesting effect of these interactions is that when the DTAOH-DPA mixed system is dried to obtain crystals, their composition changes during their growing. Then, it was impossible to obtain mixed crystals with a well-determined composition.[64]

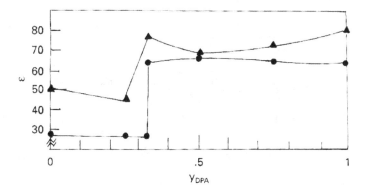

Fig. 8.14 Effective dielectric constant of the styrene vinyl (•) and aryl (△) microenvironments in micellar solutions in the DTAOH-DPA mixed micelles vs. the mole fraction of DPA in the surfactant mixture (y_{DPA}) (taken from ref. 69).

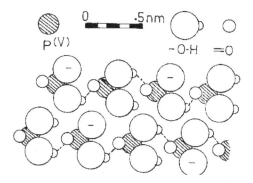

Fig. 8.15 Probable hydrogen-bonded structure of the surface of DPA micelles (taken from ref. 70).

8.4. Conclusions and Outlook

When the design of mixed surfactants systems is planned, the following factors must be taken into account to obtain some specific properties: the difference in the chain length of components, the presence of double bonds in the chain of one of the components, the steric hindrances between components having different hydrophobic structures, the size of polar headgroups, and the formation of hydrogen bonds between polar heads. Using these

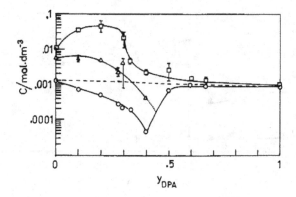

Fig. 8.16 Critical aggregation concentrations in aqueous DTAOH-DPA mixtures as a function of the DPA mole fraction in micelles (y_{DPA}). ○: aggregation of surfactant ions, △: change in structure of aggregates, □: counterion aggregation, - - - - : ideal aggregation. Vertical lines indicate 90% probability errors (taken from ref. 70).

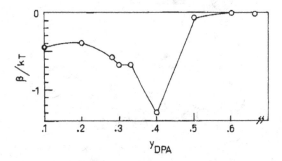

Fig. 8.17 β values of the DTAOH-DPA mixed micelles as a function of micellised DPA mole fraction y_{DPA}.[70]

factors the composition and stability of micelles, the kind of aggregates (micelles, vesicles, and liquid crystals), the solubility of catanionic systems, and the properties of the micelle surface can be tuned.

The possible applications are very broad: catanionic emulsifiers for the petroleum industry, the possibility of use water soluble catanionic mixtures for surgical washing profiting the anti-microbiological properties of cationic surfactants and the fungicidal properties of some unsaturated soaps, the production of vesicles and micelles with polar interfaces tailored for specific catalytic applications, among others. Research on these subjects seems to be promising.

References

1. J.C. Eriksson, S. Ljunggren, U. Henriksson, *J. Chem. Soc. Faraday Trans. 2*, 81, 833 (1985).
2. H. Hoffmann, G. Possnecker, *Langmuir* 10, 381(1994).
3. T. Forster, W. von Rybinski, M.J. Schwuger, *Tenside Surf Deterg*, 27, 254 (1990).
4. I.W. Osborne-Lee, R.S. Schechtere, in J.F. Scamehorn, Ed., *Phenomena in Mixed Surfactant Systems*, ACS Symp Series 311, ACS, Washington DC, p. 30 (1986).
5. J.F. Scamehorn, R.S. Schechter, W.H. Wade, *J. Disp. Sci. Technol.*, 3, 261(1982)
6. S.E. Moya, P.C. Schulz, *Colloid Polym. Sci.*, 277, 735 (1999)
7. J.F. Scamehorn, in J.F. Scamehorn, Ed. *Phenomena in Mixed Surfactant Systems*, ACS Symp. Series 311, American Chemical Society, Washington (1986).
8. M. Ghogomu, M. Bourouka, J. Dellacherie, D. Balesdent, M. Dirand, *Thermochim. Acta 306*, 69, (1997).
9. P. Mukerjee, T. Handa, *J. Phys. Chem.*, 85, 2298 (1981).
10. T. Handa, P. Mukerjee, *J. Phys. Chem.*, 85, 3916 (1981).
11. R.M. Clapperton, R.H. Ottewill, B.T. Ingram, *Langmuir*, 10, 51 (1994).
12. S.J. Burkitt, B.T. Ingram, R..H Ottewill, *Progr. Colloid Polym. Sci.*, 76, 247 (1988).
13. Lange H, Beck KH *Kolloid Z u Z Polym* 251, 424 (1973).
14. P.M. Holland, D.N. Rubingh, *J. Phys. Chem.*, 87, 1984 (1983).
15. J.M. Prausnitz, R.N. Lichtenthaler, E. Gomes de Azevedo, *Molecular Thermodynamics of Fluid-Phase Equilibria*, Prentice Hall, New Jersey, 3rd Ed., p.313-317 (1999).
16. P.C. Schulz, J.L. Rodrguez, R.M. Minardi, M.B. Sierra. M.A. Morini, *J. Colloid Interface Sci.*, 303, 264 (2006).
17. D. Stigter, *J.Colloid Interface Sci.*, 23, 379 (1967).
18. D. Stigter, *J. Phys. Chem.*, 78, 2480 (1974).
19. J. Clifford, B.A. Pethica, *Trans. Faraday Soc.*, 61, 182 (1965).
20. J. Clifford, *Trans. Faraday Soc.*, 61, 1276 (1965).
21. M. Muller, J.M. Pellerin, W.W. Chen, *J. Phys. Chem.*, 76, 3012 (1972).
22. T. Drakenberg, B. Lindman, *J. Colloid Interface Sci.*, 44, 184 (1973).
23. T. Walker, *J. Colloid Interface Sci.*, 45, 372 (1973).
24. G.H. Griffith, P.H. Dehlinger, S.P. Van, *J. Membrane Biol.*, 15, 159 (1974).
25. P. Ekwall, L. Mandell, P. Solyom, *J. Colloid Interface Sci.*, 35, 519 (1971).
26. K.L. Harrington, E.W. Kaler, D.D. Miller, J.A.N. Zasadzinski, S. Chirovolu, *J. Phys. Chem.*, 97, 13792 (1993).

27. E.W. Kaler, K.L. Harrington, D.D. Miller, A.N. Zasadzinski, in S.H. Chen, J.S. Huang, P. Tartaglia (Eds.), *Structure and Dynamics of Strongly Interacting Colloids and Supramolecular Aggregates in Solution*, Kluwer Academic, Dordrecht, The Netherlands, pp. 571-577 (1992).
28. E.W. Kaler, A.K. Murthy, B. Rodriguez, J.A.N. Zasadzinski, *Science*, 245, 1371 (1989).
29. E.W. Kaler, L.K. Herrington, A.K. Murthy, J.A.N. Zasadzinski, *J. Phys. Chem.*, 96, 6689 (1991).
30. G.X. Zhao, W.L. Yu, *J. Colloid Interface Sci.*, 173, 159 (1995).
31. Z.E. Proverbio, P.C. Schulz, J.E. Puig, *Colloid Polym Sci.*, 280, 1045 (2002).
32. Huang L. Somasundaran P., *Langmuir*, 13, 6683 (1997).
33. D.J. Mitchell, B.W. Ninham, *J. Chem. Soc Faraday Trans.*, 2, 77, 776 (1980).
34. H.C. Evans, *J. Chem. Soc.*, Pt 1, 579 (1956).
35. M.I. Viseu, K. Edwards, C.S. Campos, S.M.B. Costa, *Langmuir*, 16, 2105 (2000).
36. D.D. Lasic, *Angew. Chem. Int. Ed. Engl.*, 33, 1685 (1994) and references therein.
37. Z.E. Proverbio, P.V. Messina, J.M. Ruso, G. Prieto, P.C. Schulz, F. Sarmiento, *J. Argentine Chem. Soc.*, 94(4/6), 19-30 (2006).
38. J.G. Weers, D.R. Scheuing, *J. Colloid Interface Sci.*, 145, 563 (1991).
39. M. Swanson-Vethamuthu, M. Almgren, P. Ansson, J. Zheo *Langmuir*, 12, 2186 (1996).
40. P.C. Schulz, R.M. Minardi, B. Vuano *Colloid Polym. Sci.*, 277, 837 (1999).
41. S. Furutaka, S-I. Ikawa, *J. Phys. Chem.*, 108, 1347 (1998).
42. S. Furutaka, S-I. Ikawa, *J. Chem. Phys.*, 108, 5159 (1998).
43. M. Goldman, R.O. Crisler, *J. Org. Chem.*, 23, 751 (1958).
44. M. Oki, H. Iwamura, *Bull. Chem. Soc. Japn.*, 33, 717 (1960).
45. P.R. Rablen, J.W. Lockman, W.L. Jorgensen, *J. Phys. Chem. A*, 102, 3782 (1998).
46. J.L. Atwood, *Nature*, 349, 683 (1991).
47. N. El Kadi, F. Martins, D. Clausse, P.C. Schulz, *Colloid. Polym. Sci.*, 281, 353 (2003).
48. I. Liu, M.J. Rosen, J. Colloid Interface Sci., 179, 454 (1996).
49. F.M. Fowkes, *J. Adhes. Sci. Technol.*, 1, 7 (1987).
50. F.M. Fowkes, *J. Phys. Chem.*, 67, 2538 (1963).
51. H. Edlund, A. Sadaghiani, A. Khan, *Langmuir*, 13, 4953 (1997).
52. M.B. Sierra, M.A. Morini, P.C. Schulz, *Colloid Polym. Sci.*, 282, 633 (2004).
53. M.L. Ferreira, M.B. Sierra, M.A. Morini, P.C. Schulz, *J. Phys. Chem. B*, 110, 17600 (2006).
54. M.B. Sierra, M.A. Morini, P.C. Schulz, M.L. Ferreira, *Colloid Polym. Sci.*, 283, 1016 (2005).
55. N. Muller, R.H. Birkhahn, *J. Phys. Chem.*, 71, 957 (1967).
56. N. Muller, H. Simsohn, *J. Phys. Chem.*, 75, 942 (1971).
57. N. Muller, R.H. Birkhahn, *J. Phys. Chem.*, 72, 583 (1968).
58. R. Svens, B. Rosenholm, *J. Colloid Interface Sci.*, 21, 634 (1966).
59. M.B. Sierra, P.V. Messina, M.A. Morini, J.M. Ruso, G. Prieto, P.C. Schulz,

F. Sarmiento, *Colloids and Surf. A*, 277, 75 (2006).
60. M.B. Sierra, M.A. Morini, P.C. Schulz, E. Junquera, E. Aicart, *J. Phys. Chem. B*, 111, 11692-(2007).
61. D.N. Rubingh, in *Solution Chemistry of Surfactants*, Ed. K.L. Mittal, Plenum Press, New York, 1, p. 337 (1979).
62. P. Messina, M.A. Morini, P.C. Schulz, *Colloid Polym. Sci.*, 281, 1082 (2003).
63. R.M. Minardi, P.C. Schulz, B. Vuano, *Colloids and Surfs. A*, 197, 167 (2002).
64. R.M. Minardi, P.C. Schulz, B. Vuano, *Colloid Polym. Sci.*, 276, 589 (1998).
65. P.C. Schulz, R.M. Minardi, B. Vuano, *Colloid Polym. Sci.*, 277, 837 (1999).
66. C. Treiner, A. Amar Khodja, M. Fromon, *Langmuir*, 3, 729 (1987).
67. A. Shiloach, D. Blankschtein, *Langmuir*, 14, 1618 (1998) and references therein.
68. P.C. Schulz, R.M. Minardi, B. Vuano, *Colloid Polym. Sci.*, 276, 278 (1998).
69. P.C. Schulz, M. Abrameto, J.E. Puig, F.A. Soltero-Martnez, A. Gonzlez-lvarez, *Langmuir*, 12, 3082 (1996).
70. R.M. Minardi, P.C. Schulz, B. Vuano, *Colloid Polym. Sci.*, 274, 669 (1996).

Chapter 9

Micellization Characteristics of Sodium Dioctylsulfosuccinate: An Overview

S. Chanda, O.G. Singh and K. Ismail

*Department of Chemistry,
North-Eastern Hill University
Shillong - 793022, India*

The counter ion binding constant of sodium dioctyl sulfosuccinate (AOT) abruptly increases by two fold at about 0.02 mol kg^{-1} NaCl. However, such a sudden change in the counter ion binding constant of AOT does not take place in the presence of salicylate ion. The deviation from the Corrin–Harkins (CH) relation that takes place in AOT is not due to salting-out effect. It has been shown that at this particular concentration of NaCl the aggregation number of AOT and the polarity of the micellar interface change suddenly. Shift in the polarity of the interface occurring at about 0.02 mol kg^{-1} NaCl is found to be dependent on the probe used. An increase in the dielectric constant of the micellar interface is observed on the basis of the intensity ratio, I_3/I_1 of pyrene and $E_T(30)$ value of Reichardt's dye, whereas $\lambda_{\max}$ value of pyrenecarboxaldehyde indicate a decrease in the interfacial dielectric constant. Pyrene and Reichardt's dye are, therefore, considered to reside initially at the micellar interior and they move to the interface when the NaCl concentration reaches about 0.02 mol kg.$^{-1}$ At this particular concentration of NaCl the shape of the AOT micelle is expected to change.

9.1. Introduction

Sodium dioctyl sulfosuccinate (AOT) is known to be a special surfactant owing to its ability to form microemulsions without the presence of any cosurfactant.[1-4] Recently we reported two more special properties of AOT, which are: (1) a different type of counter ion binding behaviour of AOT micelle[5,6] and (2) a different type of electrical conductance behavior of AOT.[7] In this paper we highlight the special counter ion binding behaviour of AOT.

Counter ion binding ability is one of the important characteristics of ionic micelles. Counter ions control not only the critical micelle concentration[8] (*cmc*) and aggregation number[9] (N) of ionic surfactants, but also the reactions[10] that take place in the presence of ionic surfactants. Generally, the counter ion binding behaviour of ionic micelles is quantified in terms of counter ion binding constant, β. β is equal to $1-\alpha$, where α is the degree of dissociation of AOT micelles. The definition of α is the same as that reported by Fletcher and coworkers[11] and is given by the relation,

$$\alpha = (c_{Na,aq} - cmc - c_{NaCl})/(c_{AOT} - cmc) \qquad (1)$$

In Eq. (1) $c_{Na,aq}$, c_{AOT} and c_{NaCl} refer to the concentrations of total free counter ion, total AOT and of added $NaCl$ solution, respectively. To estimate β Corrin–Harkins (CH) relation[8] has been widely used, which is represented by the equation

$$ln(cmc) = A - \beta ln C \qquad (2)$$

Eq. (2) can be derived by applying the mass-action model to the micellization equilibrium and the constant A provides the value of standard free energy of micellization per mole of monomer. In Eq. (2) C indicates the total counter ion concentration at the *cmc*. C and cmc are expressed in either mole fraction or molar units. Relations that are similar to the CH equation have also been reported by others[12–15] signifying the importance of counter ion binding in the micellization process.

Normally, for ionic surfactants the CH plot has been found to be linear. On the other hand, in the case of AOT deviation from CH plot occurs at about 0.02 mol kg^{-1} of added 1:1 electrolyte containing sodium counter ion. The value of β below and above $\sim$ 0.02 mol kg^{-1} concentration of added 1:1 electrolyte was found to be about 0.4 and 0.8, respectively.[5,6] An abrupt two-fold increase in the counter ion binding constant of AOT micelle at about 0.02 mol kg^{-1} of added sodium ion is an interesting characteristic of AOT and warrants further investigation to establish the driving force responsible for such a sudden change in β. In this work we report other experimental evidences, which also indicate that at about 0.02 mol kg^{-1} of added sodium ion a sudden change occurs in the AOT micelle - solution interfacial property.

9.2. Experimental

AOT (Sigma, 99%), sodium dodecyl sulfate (Aldrich, 99%), sodium chloride (Merck, GR grade), pyrene (Fluka), pyrenecarboxaldehyde (Aldrich), cetylpyridinium chloride (Fluka) and Reichardt's dye (Sigma) were used without further purification. Surface tension measurements were made by the Wilhelmy plate method using K11 Krüss tensiometer. Fluorescence emission spectra were recorded using Hitachi F4500 FL spectrophotometer. The excitation wavelengths were 335 nm for pyrene and 380 nm for pyrenecarboxaldehyde. For determining aggregation number pyrene and cetylpyridinium chloride were used as probe and quencher, respectively. Solution of Reichardt's dye in water was made by dissolving the dye first in a small amount of methanol. Absorption spectra of this dye in surfactant solution were recorded using Beckmann DU 650 spectrophotometer and maintaining the pH at 11. All measurements were made at 25°C.

9.3. Results and Discussion

9.3.1. *Counter ion binding constant*

Using the cmc values obtained from the surface tension method CH plots were drawn for AOT and SDS as shown in Fig. 9.1. From Fig. 9.1 it is obvious that β has a single value for SDS equal to 0.72, whereas for AOT it has two values equal to 0.39 (in the region where $c_{NaCl} < 0.016$ mol kg^{-1}) and 0.81 (in the region where $c_{NaCl} > 0.016$ mol kg^{-1}). We have confirmed the two values of β for AOT in various aqueous electrolytic media.[5,6] CH plots were also drawn for AOT + SDS mixed system and are shown in Fig. 9.2. In the mixed surfactant system deviation from the CH relation starts occurring above 0.163 mole fraction of AOT (α_{AOT}). Deviation from the CH plot has been reported[16-21] in a few other surfactants also. However, all these reported deviations occurred at fairly high concentration of the added electrolyte (above 0.1 mol dm^{-3}) and it is attributed to the salting-out effect.[21] Surprisingly, the deviation from the CH plot occurs in the case of AOT at a much lower concentration of added electrolyte, i.e., about 0.02 mol kg^{-1}. Since the salting-out effect would be very negligible at this low electrolyte concentration, the cause for the deviation from the CH plot in AOT must be other than the salting-out effect.

9.3.2. *Polarity of the micellar interface using pyrene probe*

From the fluorescence emission spectra of pyrene in AOT solution, the values of the intensity ratio (I_3/I_1) of peak 3 (384 nm) to peak 1 (373 nm) were calculated as a function of NaCl concentration at fixed ($c_{AOT} - cmc$) values. The term ($c_{AOT} - cmc$) denotes the number of moles of AOT that

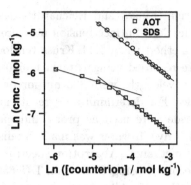

Fig. 9.1 CH plots for AOT and SDS

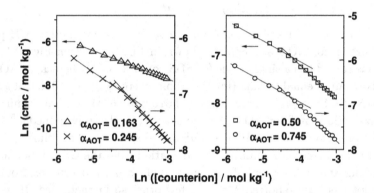

Fig. 9.2 CH plots for AOT + SDS mixed system

had undergone micellization. The intensity ratio I_3/I_1 is known as the polarity index and its values are shown in Fig. 9.3 as a function of total sodium ion concentration, which is equal to $(c_{NaCl} + c_{AOT})$. From Fig. 9.3 it is evident that the polarity index versus $(c_{NaCl} + c_{AOT})$ plot passes through a maximum when the total sodium ion concentration becomes nearly equal to or more than 0.02 mol kg^{-1}. Such a trend in the variation of polarity index shows that the dielectric constant of the environment around the pyrene probe decreases as the concentration of sodium ion increases and at around 0.02 mol kg^{-1} of sodium ion concentration the dielectric constant suddenly increases, since the value of I_3/I_1 varies inversely with the dielectric constant.[22] At a fixed NaCl concentration, dielectric constant around the pyrene probe has been found to decrease by increasing the concentration of AOT and when the c_{NaCl} is nearly equal to or more than 0.02 mol kg^{-1} the dielectric constant increases suddenly at a particular $(c_{AOT} - cmc)$ value.

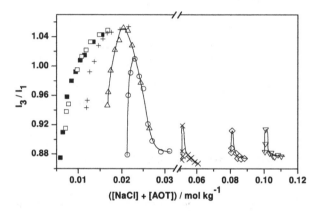

Fig. 9.3 Variation of polarity index of AOT with total Na^+ ion concentration in the solution. The concentrations of NaCl in mol kg^{-1} are 0.003 (●), 0.005 (□), 0.01 (+), 0.015 (△), 0.02 (○), 0.05 (×), 0.08 (◇) and 0.10 (▽)

9.3.3. Aggregation Number

The aggregation number of ionic surfactants is a function of counter ion concentration and increases as functions of both surfactant concentration and added electrolyte concentration.[9] Therefore, corresponding to 0.02 mol kg^{-1} NaCl concentration at which I_3/I_1 is maximum it is expected that a sudden change in the aggregation number occurs. Such a sudden change in the aggregation number around 0.02 mol kg^{-1} NaCl concentration is evident from the measured aggregation numbers of AOT using steady-state fluorescence quenching method (Fig. 9.4). This sudden change in aggregation number may lead to a change in the shape or surface morphology of AOT micelle,[5] which in turn may be responsible for the sharp increase in the value of the counter ion binding constant of AOT micelle.

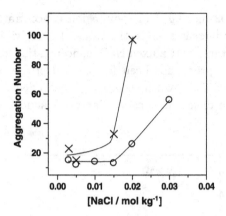

Fig. 9.4 Aggregation number of AOT in the presence of NaCl. The values of $c_{AOT} - cmc$ are 4.5×10^{-4}(o) and 0.002(×) mol kg^{-1}

9.3.4. *Polarity of the micellar interface using pyrenecarboxaldehyde probe*

The λ_{max} values of the fluorescence emission band of pyrenecarboxaldehyde in the AOT micellar solution as a function of NaCl concentration are shown in Fig. 9.5. λ_{max} of pyrenecarboxaldehyde is sensitive to the dielectric constant of the medium and increases with increase in the dielectric constant.[23] From Fig. 9.5 it can be seen that λ_{max} suddenly decreases at about 0.02 mol kg^{-1} NaCl concentration. From the λ_{max} versus dielectric constant correlation reported by Kalyanasundaram and Thomas,[22] we estimated the dielectric constant values corresponding to the λ_{max} values of pyrenecarboxaldehyde, which are shown in Fig. 9.6. Surprisingly, at about 0.02 mol kg^{-1} NaCl concentration, pyrenecarboxaldehyde probe reveals a sudden decrease in the dielectric constant of the AOT micellar medium, while pyrene probe reveals on the contrary a sudden increase in the dielectric constant. The sites of residence of pyrene and pyrenecarboxaldehyde in the micelle are therefore different. Pyrenecarboxaldehyde appears to reside at the micellar surface with its CHO group projecting towards water and an increase in the counter ion binding to the AOT micelle decreases the dielectric constant at the micelle - solution interface. Using pyrenecarboxaldehyde probe a decrease in the dielectric constant at the micellar interface by the addition of NaCl has been reported for SDS micelles.[23] Pyrene, on the other hand, seems to reside at the micellar interior initially and moves

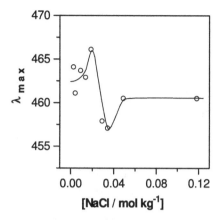

Fig. 9.5 Variation of λ_{max} (in nm) of the fluorescence emission band of pyrenecarboxaldehyde in aqueous AOT micellar medium as a function of added $NaCl$ concentration $((c_{AOT} - cmc) = 4.5 \times 10^{-4} mol kg^{-1})$

towards the surface when the NaCl concentration becomes nearly equal to or more than 0.02 mol kg^{-1} thereby sensing an increase in the dielectric constant.

9.3.5. *Polarity of the micellar interface using Reichardt's dye probe*

The measured λ_{max} values of the absorption band of Reichardt's dye in AOT micellar solution as a function of NaCl concentration are shown in Fig. 9.7. A sudden decrease in the λ_{max} value of this dye occurs at about 0.02 mol kg^{-1} NaCl. The values of $E_T(30)$ calculated from the λ_{max} values are shown in Fig. 9.8 and an abrupt increase in $E_T(30)$ takes place at about 0.02 mol kg^{-1} NaCl, which is opposite to the trend reported[24] for SDS as a function of NaCl concentration. From the $E_T(30)$ values (Fig. 9.8) it can be seen that a sudden increase in the dielectric constant of the micellar interface takes place at about 0.02 mol kg^{-1} NaCl since $E_T(30)$ value is directly proportional to the dielectric constant.[24] This observation is similar to that made by using the pyrene probe. It therefore seems that Reichardt's dye resides initially at the micellar interior and moves towards the surface when the NaCl concentration becomes nearly equal to or more than 0.02 mol kg^{-1} as the case with pyrene.

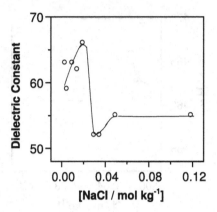

Fig. 9.6 Estimated dielectric constant at the AOT micelle - solution interface as a function of added NaCl concentration $((c_{AOT} - cmc) = 4.5 \times 10^{-4}$ mol $kg^{-1})$

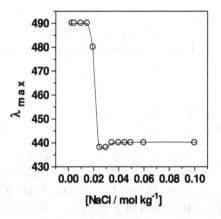

Fig. 9.7 Variation of $\lambda_{\max}$ (in nm) of Reichardt's dye in AOT $(c_{AOT} = 3.2 \times 10^{-3}$ mol $kg^{-1})$ micellar solution with concentration of added NaCl $(pH = 11.0)$

From the above investigations it has become clear that at about 0.02 mol kg^{-1} NaCl concentration (i) the counter ion binding constant of AOT increases abruptly by two fold, (ii) aggregation number of AOT increases sharply and (iii) a sudden change in the polarity of the AOT micellar interface occurs.

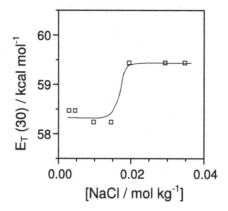

Fig. 9.8 $E_T(30)$ values of AOT ($c_{AOT} = 3.2 \times 10^{-3}$ mol kg^{-1}) micellar solution at different NaCl concentrations

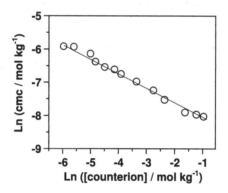

Fig. 9.9 CH plot for AOT in aqueous sodium salicylate solution

9.3.6. *Phase behaviour of AOT - water system and shape of AOT micelle*

From the small angle neutron scattering study Sheu *et al.*[25] reported that AOT forms spherical micelles near the cmc and as the AOT concentration increases the micellar shape changes from spherical to oblate spheroid with a continuous increase in the axial-to-polar-radius ratio. At [AOT] ≈ 0.02 mol dm^{-3} the axial-to-polar-radius ratio of the oblate spheroid micelle was reported to be 1.64. The reported[26–28] phase diagram of AOT - water system exhibits an isotropic solution phase ([AOT] < 0.03 mol kg^{-1}), a lamellar phase (0.23 < [AOT] < 1.6 mol kg^{-1}), a cubic phase (1.64 < [AOT] < 1.8 mol kg^{-1}) and a hexagonal phase ([AOT] > 1.84 mol kg^{-1}). A miscibility gap exists between isotropic solution phase and lamellar phase (0.03 < [AOT] < 0.23 mol kg^{-1}) and in this region appearance of polydispersed vesicles has also been reported.[27] On addition of NaCl the lamellar phase extends to lower AOT concentration and at about 0.04 mol kg^{-1} of NaCl the lamellar phase starts appearing at nearly 0.1 mol kg^{-1} of AOT. A new disordered cubic phase appears when [NaCl] ≈ 0.26 mol kg^{-1} and [AOT] ≈ 0.1 mol kg^{-1}. From the above reported phase behaviour of AOT - water system in the presence and absence of NaCl it is evident that the concentration range of AOT covered in our study falls within the region of isotropic solution phase. Therefore, a shape change of the AOT micelle, rather than a phase change of the AOT - water - NaCl system, may be responsible for the sudden shift in β.

9.3.7. Change of cmc of AOT in the presence of sodium salicylate

The CH plot for AOT in the presence of sodium salicylate is shown in Fig. 9.9 based on our earlier reported work[29] and it is interesting to note that the plot has only one slope. Therefore, AOT has a single value for β equal to 0.47 in the presence of sodium salicylate, which is a hydrotrope. We reported[29] that in the Stern layer salicylate ions might be forming ion-pairs with the bound sodium ions resulting in only the lower value of β for AOT.

9.4. Conclusions

AOT micelle, unlike other ionic micelles, exhibits a two fold increase in its counter ion binding capacity when the concentration of the added NaCl becomes about 0.02 mol kg^{-1}. This unusual behaviour of AOT is not attributable to the salting-out effect of the electrolyte. It has been demonstrated that at this particular concentration of NaCl sudden change also occurs in the aggregation number of AOT, I_3/I_1 value of pyrene, $E_T(30)$ value of Reichardt's dye and λ_{max} value of pyrenecarboxaldehyde. The observed abrupt change at around 0.02 mol kg^{-1} of NaCl in the property of AOT micelle and micellar interface is not due to a phase change of AOT. At this particular NaCl concentration the shape of AOT micelle is, however, expected to change. Since counter ions play important role in the performance of ionic surfactants, further investigations are required to have a better understanding of the special counter ion binding behaviour of AOT.

Acknowledgments

We acknowledge the financial support received from the Department of Science and Technology, New Delhi for carrying out this work. Equipment grant received from the DST under the FIST program is also acknowledged. SC acknowledges the UGC, New Delhi for providing a research fellowship.

References

1. S. Nave and J. Eastoe, *Langmuir*, 16, 8733 (2000).
2. S. Nave and J. Eastoe, *Langmuir*, 16, 8741 (2000).
3. S. Nave, J. Eastoe, R. K. Heenan, D. Steytler, and I. Grillo, *Langmuir*, 18, 1505 (2002).
4. S. Nave, A. Paul, J. Eastoe, A. R. Pitt and R. K. Heenan, *Langmuir*, 21, 10021 (2005).
5. I. M. Umlong and K. Ismail, *J. Colloid Interface Sci.*, 291, 529 (2005).
6. I. M. Umlong, J. Dey, S. Chanda and K. Ismail, *Bull. Chem. Soc. Jpn.*, 80, 1522 (2007).
7. O. G. Singh and K. Ismail, *J. Surf. Deterg.*, 11, 89 (2008).
8. M. L. Corrin and W. D. Harkins, *J. Am. Chem. Soc.*, 69, 683 (1947).
9. N. V. Lebedeva, A. Shahine and B. L. Bales, *J. Phys. Chem.*, B109, 19806 (2005) and references therein.
10. L. S. Romsted, *Langmuir*, 23, 414 (2007) and references therein.
11. R. Aveyard, B. P. Binks, P. D. I. Fletcher, C. E. Rutherford, P. J. Dowding and B. Vincent, *Phys. Chem. Chem. Phys.*, 1, 1971 (1999).
12. T. Sasaki, M. Hattori, J. Sasaki and K. Nukina, *Bull. Chem. Soc. Jpn.*, 48, 1397 (1975).
13. L. Gaillon, J. Lelievre and R. Gaboriaud, *J. Colloid Interface Sci.*, 213, 287 (1999).
14. R. Palepu, D. G. Hall and E. Wyn-Jones, *J. Chem. Soc. Faraday Trans.*, 86, 1535 (1990).
15. H. Maeda, *J. Colloid Interface Sci.*, 241, 18 (2001).
16. M. Shigeyoshi, H. Kurimoto, Y. Ishihara and T. Asakawa, *Langmuir*, 11, 2951 (1995).
17. M. Shigeyoshi, H. Suzuki and T. Asakawa, *Langmuir*, 12, 2900 (1996).
18. M. Shigeyoshi, W. Akasohu, T. Hashimoto and T. Asakawa, *J. Colloid Interface Sci.*, 184, 527 (1996).
19. M. K. Franchini and J. T. Carstensen, *J. Pharma. Sci.*, 85, 220 (1996).
20. E. K. Mysels and K. J. Mysels, *J. Colloid Sci.*, 20, 3718 (1965).
21. P. Mukerjee, *Adv. Colloid Interface Sci.*, 1, 241 (1967).
22. K. Kalyanasundaram and J. K. Thomas, *J. Am. Chem. Soc.*, 99, 2039 (1977).

23. K. Kalyanasundaram and J. K. Thomas, *J. Phys. Chem.*, 81, 2176 (1977).
24. K. A. Zacharlasse, N. V. Phuc and B. Kozanklewicz, *J. Phys. Chem.*, 85, 2676 (1981).
25. E. Y. Sheu, S.-H. Chen and J. S. Huang, *J. Phys. Chem.*, 91, 3306 (1987).
26. P. Ekwall, L. Mandell, K. Fontell, *J. Colloid Interface Sci.*, 33, 215 (1970).
27. A. Caria and A. Khan, *Langmuir*, 12, 6282 (1996).
28. J. Rogers and P. A. Winsor, *Nature*, 216, 477 (1967).
29. I. M. Umlong and K. Ismail, *J. Surf. Sci. Technol.*, 22, 101 (2006).

Chapter 10

Phase Separation Study of Surface-Active Drug Promazine Hydrochloride in Absence and Presence of Organic Additives

Kabir-ud-Din, Mohammed D.A. Al Ahmadi, Andleeb Z. Naqvi, Mohd. Akram

*Department of Chemistry,
Aligarh Muslim University,
Aligarh-202 002, India*

The present work focuses on the clouding phenomenon and dye solubilization results in a surface-active drug promazine hydrochloride (PMZ, a tricyclic tranquillizer) solution, in presence of various organic additives (*viz.*, alcohols, sugars, amino acids). Higher chain alcohols, due to their hydrophobic nature, decrease the cloud point (CP) while short chain alcohols show a small increase. With diols (ethane-diol and propane-diol) the cloud point (CP) remains almost constant, but cycloalkanols decrease it. Sugars hinder micellization and cause a decrease in CP. The effect of amino acids on the CP of PMZ solution depends upon their acidic/basic as well as polar/non-polar characteristics. Some dye solubilization experiments are also performed on the drug + sugar systems as well as on the drug at different pH values. The overall behaviour is interpreted in terms of variations in drug-aggregate morphology and modifications in the background solvent.

10.1. Introduction

Surface active behaviour of a large number of drugs has been reported[1,2] and therefore, attempts have been made to correlate surface and biological activities.[3] Self-association of amphiphilic compounds is a possible way of eliminating the energetically unfavourable contact between the non-polar part and water while retaining the polar part in an aqueous environment. Manifestation of changes in some physical properties of the amphiphilic drugs in aqueous solutions indicates aggregation of drugs and the discontinuity is taken as the critical micelle concentration (cmc). Although the surface activity of a drug does not explain its mechanism of action, but the presence of a hydrophobic portion in the drug molecule can be responsible for phenomenon like formation of hydrophobic bonds with proteins or with other compounds present in pharmaceutical formulations, affecting the bioavailability and stability of a product.[3]

Promazine hydrochloride (PMZ), a tranquillizer, though possesses a rigid, planar hydrophobic ring system (Fig. 10.1), associates in water to form micelles.[4] Like surfactants, phase transitions (dependent upon concentration, temperature, and pH) occur in this drug also. The temperature of phase transition is called the cloud point (CP).[5-7] Above the CP, solutions spontaneously separate into two distinct phases; one phase is surfactant rich and the other is surfactant lean. The CP phenomenon is reversible and a single phase reappears if the temperature is lowered below the CP. Among several factors responsible for the CP phenomena, viz., structure of drug molecule, concentration, temperature and a third component (additive), and the last factor has shown greater sensitivity in a system, even at a very low concentration. The additives modify the drug-solvent interactions, change the cmc, size of micelles and phase behaviour in the solution. Since the phase transition temperature is determined by a very delicate balance between opposing effects, a slight change in the intermicellar interaction potential may affect the cloud temperature considerably. We, therefore, report herein a study on the effect of addition of various alcohols and diols, sugars, and amino acids on the CP behaviour of PMZ solutions prepared in aqueous buffer (10 mM sodium phosphate). The pK_a value of PMZ (Fig. 10.1) is 9.4.[8] At low pH values the tertiary nitrogen atom acquires a positive charge (i.e., PMZ exists in cationic form) while it becomes neutral at high pH values.

10.2. Materials

The details of manufacture and purity of the chemicals, which were used as received, are given below; PMZ hydrochloride ($\geq$ 98%, P 6656), L-(+)- arabinose ($\geq$ 98%), Sigma, USA; ethanol (C_2OH, $\geq$ 99.8%), L-leucine ($\geq$ 99%), lysine. HCl ($\geq$ 99%), Merck, Germany; 1-propanol (C_3OH, $\geq$ 99.9%), 1-hexanol (C_6OH, 99%), 1-heptanol (C_7OH, $\geq$ 99%), L-histidine. HCl ($\geq$ 99%), BDH, England; n-butanol (C_4OH, $\geq$ 99.9%), Sarabhai, India; 1-pentanol (C_5OH, $\geq$ 99%), 1-octanol (C_8OH, > 97%), L(−)sorbose (98%), cyclopentanol ($\geq$ 98%), L-arginine (99.5%), Fluka, Switzerland; ethanediol (> 99%), propane-1, 2-diol (95%), cyclohexanol (> 98%), BDH India; dextrose ($\geq$ 99%), D(−)fructose ($\geq$ 99%), D(+)mannose (99%), Merck, India;

allyl alcohol (> 95%), Duchem Lab, India; D(+)xylose (≥ 98%), alanine (≥ 99%), threonine (≥ 99%), S.D. Fine, India; Sudan III, L-arginine. HCl (≥ 99%), Loba Chemie, India; glutamic acid (≥ 99%), glycine (≥ 99.5%), phenylalanine(≥ 99%), DL-aspartic acid (≥ 99%), SISCO, India.

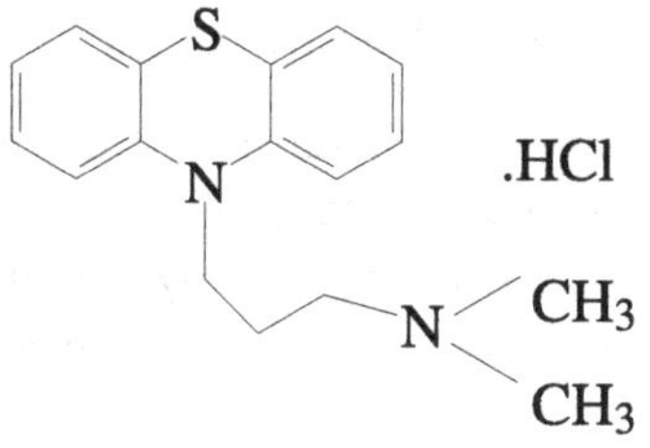

Fig. 10.1 Molecular structure of promazine hydrochloride (PMZ).

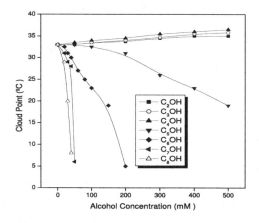

Fig. 10.2 Effect of addition of alcohols on the cloud point of 50 mM PMZ in a 10 mM sodium phosphate buffer solution (pH = 6.67).

Trisodium phosphate dodecahydrate and sodium dihydrogen phosphate monohydrate were obtained from Merck (India). 10 mM sodium phosphate (SP) buffer (prepared from SP monobasic monohydrate, 5 mM, and tribasic dodecahydrate, 5 mM was used as solvent. Deionized and doubly-distilled water (sp. cond. = 1-2 μS cm^{-1}) was used to prepare buffer solutions. The pH of the solutions was measured with an ELICO pH meter (model LI 120) using combined electrode (CL 51B).

10.3. Methods

10.3.1. *Cloud Point (CP) Measurements*

To determine the CP, the sample solutions were taken in securely stoppered Pyrex glass tubes which were then placed in a controlled stirring and heating device. The temperature was slowly raised. The onset of sudden clouding in the solution was taken as CP.[9-11] The heating was discontinued until the sample became clear again. The temperature was cycled (at least twice) in this way to obtain the CP temperature (reproducibility 0.5°C). The CMC of promazine hydrochloride is 36 mM,[12] hence the drug concentration was fixed at 50 mM. The pH was kept fixed at 6.67 (except where effect of pH variation was studied).

10.3.2. *Dye Solubilization Measurements*

Experiments of Sudan III dye solubilization in the PMZ solutions were performed at room temperature by vigorously shaking 20 cm^3 of PMZ with 20 mg Sudan III for 5 min., separating the insoluble dye by filtering the mixture and then measuring the UV-visible absorbance in the wavelength range of 430-650 mM on a Shimadzu 1240 UV-visible spectrophotometer.

10.4. Results and Discussion

The effect of alcohol concentration on the CP of 50 mM PMZ solution prepared in 10 mM SP buffer is shown in Fig. 10.2. The behaviour can be explained by taking cognizance of the nature of alcohols. As alcohols contain a hydrophilic -OH group and a hydrocarbon alkyl part, their distribution between aqueous and micellar phases depends upon these two factors. Longer chain alcohols are hydrophobic in nature (the hydrophobicity increases with increase in chain length).[13] Therefore, penetration of the alkyl chain of alcohols will increase the volume of hydrophobic part (V_c) of drug molecule whereas intercalation of -OH group between the charged head groups reduces the repulsion among them which leads to a decrease in area per head group (a_o). Consequently, the value of R_p increases[14] ($R_p = V_c/a_o l_c$, the so-called Mitchell-Ninham packing parameter, l_c is the

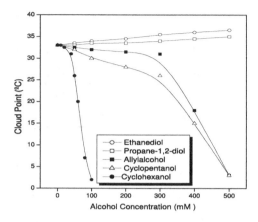

Fig. 10.3 Variation of cloud point of 50 mM PMZ in a 10 mM sodium phosphate buffer solution (pH = 6.67) with addition of alcohols.

maximum chain length), and thereby micellar size increases. The factors discussed above help in decreasing the CP of the drug system progressively in presence of C_5OH-C_8OH with the steepest decrease obviously being with the most hydrophobic alcohol, viz. C_8OH. The alkyl chain penetration toward the micellar core replaces water from the head group/ palisade layer and thus result in CP decrease with increase in the alcohol concentration. The short chain alcohols are hydrophilic molecules and are known to be adsorbed on the micelle- water interface.[15] They would thus hinder micellar aggregation with an effect of slow CP increase.

Figure 10.3 shows the effect of added diols on the CP of PMZ. With ethanediol and propane-1, 2-diol, CP remains almost constant. Diols being hydrophilic and highly miscible in water (as they contain two -OH groups on hydrophilic ethane or propane molecules), remain in aqueous phase at all concentrations and would not affect the micelle hydration. However, cycloalkanols show different behaviour; with allylalcohol and cyclopentanol, unlike cyclohexanol, the CP decrease is not pronounced in the beginning. The results are in conformity to the relative solubility of the cycloalkanols and prove cyclohexanol to be the highest on hydrophobicity scale in the present system.[16]

Results of adding different amino acids on the CP of PMZ solutions are shown in Figures 10.4 and 10.5. The nature and molecular structure of the amino acids seem to play as sharp increase in CP is observed a

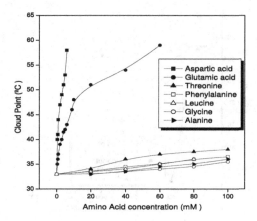

Fig. 10.4 Variation of cloud point of 50 mM PMZ in a 10 mM sodium phosphate buffer solution with addition of amino acids. The initial pH of PMZ solution was 6.67.

role as sharp increase CP is observed with the acidic amino acids (aspartic and glutamic), while non-polar and uncharged polar amino acids remain much less effective (Fig. 10.4). The negatively charged side chain of acidic amino acids would interact with tertiary amine of the drug. This will allow further hydration of the micelles, hence the CP of the system increases. Hydrophobic non-polar and uncharged polar amino acids, on the contrary, would partition in micellar interior or bulk water, respectively. In either case the hydration of micelles is not affected and the CP as well.

Figure 10.5 illustrates the effect of basic amino acids and their hydrochloride salts on the drug CP. The basic amino acids, being polar, partition in the head group region with the result that certain amount of water near the head group region is replaced; the observance of CP phenomenon at a lower temperature is justified this way. Hydrochloride salts bear a positive charge on them and their interaction with the PMZ micelles would result in increased micelle-micelle repulsion. The CP would then increase which indeed is the case (Fig. 10.5).

Figure 10.6 shows the variation of CP of PMZ solutions with the addition of sugars wherein a decrease in the CP is observed with each compound. Figure 10.7 illustrates the UV-visible spectra of Sudan III in the presence of sugars (200 mM dextrose and fructose).

One can see that the absorbance decreases indicating decreased solubility of the dye. The results imply that the size of the micelles is reduced.

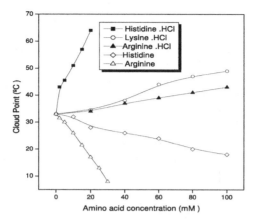

Fig. 10.5 Variation of cloud point of 50 mM PMZ in a 10 mM sodium phosphate buffer solution with addition of basic amino acids and their salts. The initial pH of PMZ solution was 6.67.

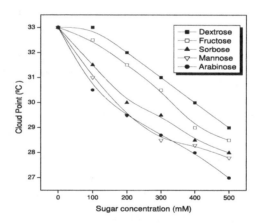

Fig. 10.6 Variation of cloud point of 50 mM PMZ in a 10 mM sodium phosphate buffer solution (pH = 6.67) with addition of sugars.

The reduction in micellar size would give rise to reduced repulsion between the drug micelles with the result of CP appearing at lower temperatures. Apparently, adsorption of sugars at the water- micelle interface hinders micellization giving rise to smaller micelles and lower CP.

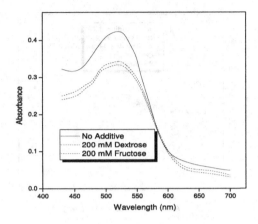

Fig. 10.7 UV-visible spectra of Sudan III in the presence of 75 mM PMZ in water containing 200 mM sugars.

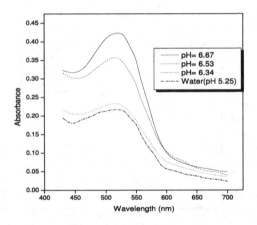

Fig. 10.8 UV-visible spectra of Sudan III in the presence of 75 mM PMZ in a 10 mM sodium phosphate solution at different pH value.

Figure 10.8 illustrates the influence of pH on the UV-visible spectra of Sudan III in 50 mM PMZ solutions. The absorbance was found to increase with increasing pH from 5.25–6.67. Increase in solution pH results in deprotonation of nitrogen atom of tert-amine portion of the drug molecules which reduces the micellar surface charge. Consequently, the micelle-micelle repulsion reduces and the micellar size increases at higher pH values.

The resulting larger micelles dissolve more dye and hence the UV-vis absorbance increases with increasing pH.

Variation of CP with arabinose concentration at different fixed pH's as well as fixed drug concentration is shown in Figs. 10.9 (a) and (b). Increase

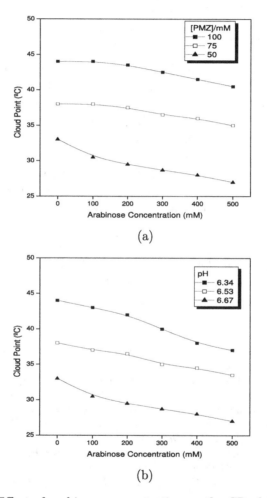

Fig. 10.9 (a) Effect of arabinose concentration on the CP of different fixed concentration (50-100 mM) of PMZ solution prepared in 10 mM sodium phosphate buffer (pH =6.67). (b) Effect of arabinose concentration on the CP of 50 mM PMZ solution prepared in 10 mM sodium phosphate buffer at different pH values.

in pH decreases the CP while increase in drug concentration works in an opposite way. As explained for Fig. 10.8, increase in pH decreases the intermicellar repulsion and the CP as well. Increase in drug concentration increases both the number and size of micelles increasing the repulsions among the micelles which increase the CP.

10.5. Conclusions

The effects of various organic additives on the cloud point and dye solubilization of amphiphilic drug promazine hydrochloride solutions were studied. The UV-visible absorbance of the dye increases with the increase in pH. Alcohols decrease the CP by their screening effect and palisade layer penetration. The behaviour of amino acids depends upon the nature of their side chain. Acidic amino acids and hydrochloride salts of basic amino acids increase the CP while basic amino acids decrease the CP and nonpolar, polar uncharged ones are less effective. Sugars hinder micellization due to adsorption and decrease the CP as well as the UV-Vis absorbance of the dye.

Acknowledgments: One of the authors, Mohd. D. A. Al-Ahmadi, thanks Aden University, Yemen, for grant of leave.

References

1. W.M. Gelbert, A.Ben Shaul and D. Roux, *Micelles, Membranes, Microemulsions, and Monolayers*, Springer, New York, 1994.
2. D. Attwood and A. T. Florence, *Surfactant Systems*, Chapman & Hall, London, 1987.
3. P.M. Seeman and H. S. Baily, *Biochem. Pharmacol.*, 12, 1181 (1963).
4. J.M. Ruso, D. Attwood, P. Taboada, M. J. Suarez, F. Sarmiento and V. Mosquera, *J. Chem. Eng. Data.*, 44, 941 (1999).
5. Z. Wang, Z. Fengsheng and D. Li, *Colloids Surf.* **A216**, 207 (2003).
6. W.N. Maclay, *J. Colloid Sci.*, 11, 272 (1956).
7. P. Yan, J. Huang, R. C. Lu, C. Jin, J.X. Xiao and Y.M. Chen, *J. Phys. Chem.*, B109, 5237 (2005).
8. A.L. Green, *J. Pharm. Pharmcol.*, 19, 10105 (1966).
9. S. Kumar, D. Sharma and Kabir-ud-Din, *Langmuir*, 16, 6821 (2000).
10. S. Kumar, D. Sharma, Z. A. Khan and Kabir-ud-Din, *Langmuir*, 17, 5813 (2001)
11. S. Kumar, D. Sharma, Z. A. Khan and Kabir-ud-Din, *Langmuir*, 19, 3539 (2003).

12. S. Schreier, S. V. P. Malheiros and E. de. Paula, *Biochim. Biophys. Acta.*, 1508, 210 (2000).
13. S. Kumar, D. Bansal and Kabir-ud-Din, *Langmuir*, 15, 4960 (1999).
14. J. N. Israelachvili, D. J. Mitchell and B. W. Ninham, *J. Chem. Soc., Faraday Trans.*, 272, 1525 (1976).
15. T. Gu and P. A. Galera-Gomez, *Colloids and Surf.*, A147, 365 (1999).
16. The Merck Index, 13th ed., Merck, New Jersey, USA, 2001.

Chapter 11

Effect of Urea on Surfactant Aggregates: A Comprehensive Review

Silvia M.B. Souza

Fundao Educacional do Municpio de Assis - FEMA, Brazil

and

E.B. Alvarez, Mario J. Politi

*Departamento de Bioqumica,
Instituto de Química,
Universidade de São Paulo, Brazil*

The effect of urea on surfactant aggregates is reviewed. Previous and new data are presented. The main focus is to demonstrate the salting in and salting out effects of urea on these supra molecular aggregates. The data are rationalized in terms of the effect of the water-urea mixture being a more polar medium leading to stronger electrostatic interactions as well as stronger hydrophobic repulsions. These properties lead to a simple model of the urea effect as a result of a simple increase in the dielectric permittivity with no minimum energy interactions of the additive with solutes or even with water molecules.

11.1. Introduction

The effect of urea at high concentrations (> 0.5 mol dm^{-3}) as a denaturating or as a chaotropic agent has been known for a very long period.[1] In general this effect has been attributed to either an indirect effect on the water structure or to a direct effect via hydrogen-bond with the solutes.[1,2] More recently an alternative explanation through a general effect of an increase of the solvent polarity (water/urea) that leads to a poorer ion pairing has been proposed.[3] This effect will lead to an increased hydrophobic effect via the increase in the electrostatic fields, in other words the salting out effect would originate from a better solvation of isolated hydrophilic species, such as simple ions, and also towards a more drastic hydrophobic repulsion. In parallel it has been also shown that urea in low concentrations (< 0.2 mol dm^{-3}) works as a salting in agent[4] both with proteins and with surfactant aggregates. It has been pointed out that the effect of urea is quite similar for association colloids and for proteins.[5] Given the easiness of working with surfactants where the hydrophobic and hydrophilic moieties are clearer defined, in this review we will focus on selected data that lends support to the simple interpretation of the "Urea Effect" on the association colloids both at the low and high concentration levels. The data will be reported in the sequence of the effect of urea in liquid water, aqueous solution, monolayers, micelles, reversed micelles, and vesicles.

11.2. Liquid Water

Theoretical analyses show that urea has little effect upon the 3-D structure of water.[6] Molecular dynamics simulations by Wallqvist et al.[6] show that urea stabilizes the methane-methane contact pair, a renaturation effect, and that urea is preferentially adsorbed by a fictitious partially charged methane (M^+, M^-) ion pair molecule, which destabilizes their contact. The authors concluded that these results are consistent with the concept that a large excess of urea must be added (6–8 mol dm^{-3}) before adsorption of urea on surface residues of proteins, rather than enhanced dissolution of the hydrocarbon components, leads to denaturation. The slightly enhanced solubilization of butane by urea found by Wetlaufer et al.[1] some years ago and attributed to the altering of water structure should be contrasted with recent molecular dynamics simulations of Mountain and Thirumalai[7a] Their results show that hydrocarbons from methane to octane undergo fluctuations in their conformations without significantly perturbing local water structure and that the hydrogen-bonding network is essentially preserved.[7]

In parallel, Grubmuller[8] have proposed that hydrogen bonds between urea and water are significantly weaker than those between water molecules, which drives urea self-aggregation due to the hydrophobic effect. From the reduction of the water exposed urea surface area, urea was found to exhibit an aggregation degree of ca. 20% at concentrations commonly used for protein denaturation.[8] Structurally, three distinct urea pair conformations were identified and their populations were analyzed. Urea was found to strengthen water structure in terms of hydrogen bond energies and population of solvation shells. The author's calculations also lead to the conclusion that as urea replaces water in the structure it has both a kosmotropic and a chaotropic effects. They also found that there is no further evidence of urea-water special interactions, except for the self-association of urea at higher concentrations starting from 0.1 mole fraction.

An almost ideal behavior of urea solutions from the relatively constant molar activity was also shown. Analysis of the simulations suggested that the reason for the ideal behavior of urea solutions is the similar average interaction energies between water molecules and urea and water molecules, with a smaller average urea self interaction energy.[9]

To conclude the modeling data in this section we mention the work by Rezus and Bakker[10] on measurements of the orientational dynamics of water molecules in mixtures of water and urea. They observed that even high concentrations of urea do not alter the reorientation time of the majority of the water molecules and concluded that urea does not change the strength of the hydrogen-bond interactions between water molecules.[10]

An important aspect is the effect of urea on the aqueous dielectric constant (ϵ). It has being shown that ϵ increases steadily when urea is added.[11]. ϵ values go from 78.54 at 0% to 96.58 at 42.97% weight percent urea (T= 25°C). These values also have being corroborated by more recent investigations.[12] It is thus clear that even relatively low amount of urea $\sim 0.1 mol\ dm^{-3}$ leads to an increase in ϵ, thereby enhancing ionic dissociations.

11.3. Aqueous Solutions

In this section the effect of urea on aqueous solutions is reviewed. NMR studies of the effect of urea on aqueous solutions of tert-butanol by Mayele and Holz resulted in the conclusion that low urea concentrations enhance the hydrophobic self-association in the water-rich region.[13] With low tert-butanol concentrations enhancement reaches a pronounced maximum. At higher urea concentrations a destabilization of the hydrophobic interaction is observed, thus pointing strong evidences towards the salting in and salting out effects.[13]

In studies of the effect of photoacids dissociation in water/urea mixtures Politi and Chaimovich[14] found no differences in the rate of the prototropic dissociation. They showed that the mixed solvent (3 $moldm^{-3}$ urea) has the same proton affinity as pure water. However, the effect on proton re-association has not been studied so far. It is tempting to anticipate a decrease in the rate constant due to the destabilization of ion pair formation in water urea mixtures. This effect should appear as well in the dissociation step but this is a very fast step and could be detected only with a very fast kinetics facility. Anyhow, it was concluded that urea-H_2O clusters have properties similar to pure water.[14] In parallel, determinations of the Gibbs energy of transfer of H^+ ion from water to urea - water mixtures (0–8 $mol\ dm^{-3}$) by measuring the dissociation of uronic acid (UH^+) show that the energy can be properly correlated with the variation of dielectric constant of the urea + water mixture. A good correlation exists between the variation of basicity of the solvent mixture and the dissociation constant of uronic acid[15].

Conductivity experiments on NaI in ternary solvents of water, alcohols (3–4 carbons), and urea are consistent with added urea competitively hydrating alcohol clusters and Na^+ and $I.^-$ [16] In related results, in 4–8 $mol\ dm^{-3}$ urea, the solubility of β-cyclodextrin is orders of magnitude greater than in water, without the inclusion of urea in/on the host cavity.[17a] Transfer heat capacities of the nucleic acid bases, nucleosides, and nucleotides from water to concentrated aqueous urea solutions suggest stronger hydrophilic/ionic group interactions with the solvent,[17b] in the same direction as urea stabilization of hydrophobic components in bulk solution.

The effects of added urea on the aggregation of an organic nickel complex and changes in the precipitation threshold of oppositely charged polyelectrolyte amphiphile mixture were studied. Dimerization of nickel tetrasulfophthalocyanine (NiPCS4) macrocycles is a result of partial charge interaction between adjacent macrocycles.[18,19]

Dimerization constants (K_D) and derived enthalpies and entropies of dimerization as a function of temperature and urea concentration are listed in Table 11.1.[3] Addition of urea decreases K_D by orders of magnitude, and the change is primarily enthalpic. These results are consistent with an earlier study of the effect of added urea on dye association, such as the dimerization of methylene blue in water or its association with perchlorate.[20]

Table 11.1. Dimerization Constant (K_D), Enthalpy (ΔH), and Entropy (ΔS) for NiPCS$_4$ in Water and Urea at Various Temperatures.[3]

[urea] ($mol\ dm^{-3}$)	$K_D(35)^a x$ 107	$K_D(45)^a x$ 107	$K_D(55)^a x$ 107	$K_D(65)^a x$ 107	ΔH (kcal/mol)	ΔS(kcal/°C mol)
0	5.47	2.14	0.86	0.79	0.219	0.018
3	0.43	0.41	0.22	0.18	0.12	0.017
5	0.0084	0.0070	0.0064	0.0067	0.028	0.017

[a] Numbers in parentheses are the temperature in Celsius

Enhanced solvation of ions also explains qualitatively why added urea reduces the rate of nickel complexation with *trans*-pyridine-2-azo-*p*-dimethylaniline in sodium decyl sulfate, micelles[21a] and the alkaline hydrolysis of cobalt complexes in SDS micelles.[21b] Urea also increases the solubility of organic salts such as potassium picrate, potassium tetraphenyl borate, tetraphenyl arsonium picrate[22] and inorganic salts such as NaCl, NaI, KCl, KBr, and CsCl.[23]

11.4. Monolayers

The effect of urea on insoluble surfactant monolayers has been studied with cationic (dioctadecyldimethylammonium), anionic (fatty acids) and zwiterionic (DPPC, and a sulphobetaine) surfactants. In the case of cationic systems, in the absence of urea, the minimum head group area increases in the order $Br^- > Cl^- \sim NO_3^- > OH^-, F^-, CH_3CO_2^-$. Urea causes film expansion, loss of defined phase transitions, and an increase in the calculated minimum head group area. The ion specificity observed without urea is lost upon increasing the concentration of the additive, suggesting the absence of a Stern layer at the interface. The effect of urea was interpreted in terms of urea association to the interface rather than as a general effect on the properties of the solvent.[24] These results support both theoretical calculations and experimental results, indicating direct interaction of urea with amphiphile aggregates in which urea displaces water molecules from the interface. It was shown, however, by ion-trapping studies that there is no differential affinity of urea over water on interfaces as that of alkylammonium surfactants. The general effect of increasing the medium polarity increase by addition of urea is thus again a simpler and more concise explanation.[25]

For stearic acid and polydimethyldiallylammonium (PDDA) the same effect of urea is observed. The minimum area per molecule of surfactant or polyelectrolyte (A_0) increases as function of urea concentration, indicating once again the effect of decreased ion-pairing which determines the extent of ionic screening at the monolayer level (Figs. 11.1 and 11.2).

In the case of PDDA, a simple determination of the cloud point induced by addition of SDS shows once again the retarding effect of urea. In other words, the increase in the medium polarity leads to a decrease in the ion pairing manifested by a shift towards larger concentrations of urea and as well by an overall spreading of the curve compared to absence of the additive (Fig. 11.3).

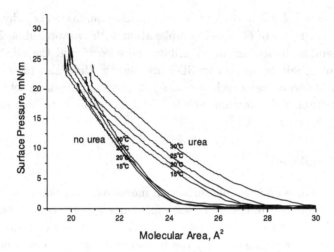

Fig. 11.1. Plot of surface area vs. molecular area for steric acid. color figs monolayer in the absence of urea; Red, monolayer in the presence of the urea.

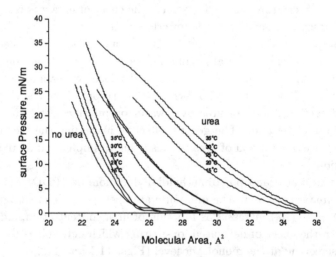

Fig. 11.2. Plot of surface area vs. molecular area for PDDA. Black, monolayer in the absence of urea; Red, monolayer in the presence of the urea.

Table 11.2. CMC and Dissociation Deree (α) of Various Amphiphiles as a Function of Urea Concentration ($T = 25°C$).

Amphiphile	[urea]	CMC (M)	α^a
CTACl[b]	0	1.60	0.33
	0.5	1.64	0.36
	1.0	1.68	0.37
	3.0	2.31	0.42
	5.0	3.03	0.44
CTABr	0	0.95	0.26
	3.0	1.57	0.29
	5.0	2.43	0.33
TTABr	0	3.82	0.23
	3.0	6.60	0.38
	5.0	9.48	0.31
SDS	0	8.33	0.36
	3.0	10.0	0.55
$C_{12}E_6^c$	0	0.0012	
	2.0	0.0021	
	4.0	0.0035	
TX-100[d]	0	0.33	
	1.0	0.41	
	2.0	0.50	
	3.0	0.72	
	4.0	1.00	

[a] α values were obtained from the ratio from the slopes of the lines before and after the cmc according to Sepulveda, L.; Cortez, J. *J Phys. Chem.* **89**, 5322 (1985). [b] Conductometric data. [c] Data from ref. 8a in ref. 3. [d] Data from ref. 7h in ref. 2.

11.5. Micelles

The effect of added urea on the CMC's on selected surfactant systems and as well on the dissociation degree (α) of ionic systems is presented in Table 11.2.[3]

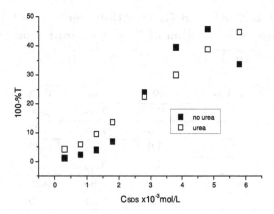

Fig. 11.3. Percentage transmission of PDDA in the absence and presence of $3 \, mol \, dm^{-3}$ urea as function of SDS concentration. $T = 25°C$, PDDA Mw $\sim$ 200 Kd, 5 gm litre^{-1}.

Clearly both an increase in the CMC for ionic and nonionic surfactants and in the dissociation degree (α) is observed. These data go in the direction of more ionized micelles having higher CMC values. Considering again the lack of differential partitioning or "binding" of urea against water,[25] the data are in agreement with the medium polarity increase. In a close study by Ruiz,[26] micelle formation and microenvironment properties of sodium dodecyl sulfate (SDS) in aqueous urea solutions were investigated. The effect of urea on the hydrophobic region of SDS micelles was clearly reflected in steady-state fluorescence anisotropy measurements, which indicated a more rigid microenvironment around the probe as urea concentration increased. The results obtained from conductivity (Fig. 11.4 and Table - 11.3) support a direct mechanism of urea action.[26] However, the more separation of hydrophobic and hydrophilic due to more polar water is a better and simpler explanation.

In another series of experiments dealing with SDS sphere to rod transitions by the Kabir-ud-Din et al. employing urea and alkylureas the effect of urea as a "re-naturant" ("salting effect") at low concentrations were very cleverly demonstrated.[27] They reported that at low [additive], the micellar association (or hydrophobic interactions) is facilitated by the presence of urea (or other alkyl ureas). Contrary to this, micellar destabilization predominates at higher [additive]. A similar understanding applies to amphiphilic biomolecules (e.g., proteins) and it is not unlikely that the above additives may renature the protein structure at lower concentrations. As stated, in this review the threshold between low and high additive concentration lies around $\sim 0.2\ mol\ dm^{-3}$.

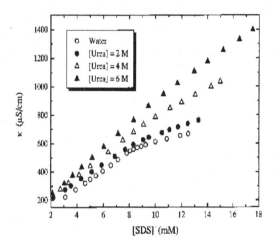

Fig. 11.4. Specific conductivities vs. concentration of SDS at different aqueous mixtures of urea. (from ref. 26.)

Table 11.3. CMC and degree of counterion dissociation (α) of SDS indifferent aqueous solutions of urea at 25°C, as compared with literature values[a].

Urea [mole]	CMC [mmol]	α
0	8.3	0.39
	8.0	0.30
	8.2	
2	8.7	0.49
	9.0	0.40
	9.2	
4	9.6	0.66
	9.6	0.40
6	11.2	0.77
	11.0	0.42
	12.0	

[a] Reference 26.

11.6. Reversed Micelles and Microemulsions

Initial studies with reversed micelles (RM) focusing the effect of urea were conducted by Amaral and co-workers.[28,29] It was shown that increasing the concentration of urea induced percolation in the system and, at sufficiently high urea concentration, even to a gel like system (Fig. 11.5). This effect was attributed to a larger polarization of the RM droplets and their contact stabilization. Several studies followed including RM from cationic systems containing ammonium based surfactants (Fig. 11.6).[3]

In a related study on urea in AOT / CCl_4 systems it was found that the additive stays confined in the drops. Even when drying the system, it stays nanoencapsulated.[30] This data indicates the confinement of urea in the aqueous pool of the droplets.

Another set of data shows the salting in effect of urea.[31] In this work evidence for a stabilizing effect of a low concentration (< 1 mol dm^{-3}) of urea incorporated in the central pool of AOT/n-heptane/water reverse micelles is presented. Static light-scattering experiments were performed to measure (ω_0), the molar ratio of water to AOT beyond which the micelles become unstable as a function of the concentration of urea in the central

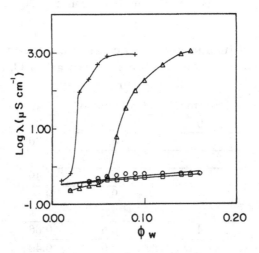

Fig. 11.5. Electrical conductivity (λ) as a function of the water volume fraction (ϕ_w) for urea/water / AOT/hexane mixtures : [urea] = 0mol dm^{-3} (o), 1.4mol dm^{-3} ($\square$), 3mol dm^{-3} ($\triangle$), and 5mol dm^{-3} (+), T = 25°C (from ref. 27).

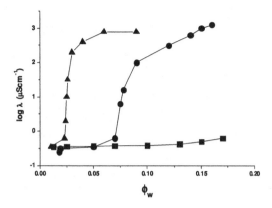

Fig. 11.6. Electrical conductance of 0.1mol dm^{-3} DDBABr in chlorobenzene as a function of the volume fraction, % v/v, of the hydrophilic phase: water (■), 1mol dm^{-3} urea (•), 4mol dm^{-3} urea (△), T = 50°C (from ref. 3).

water pool. The stabilizing effect of urea is reflected in an increase in the value of (ω_0), at low urea concentrations over that in the absence of urea. Dynamic light-scattering experiments show that the hydrodynamic radius of the micelles is smaller at low urea concentrations ($< 1 \ mol \ dm^{-3}$) than in the absence of urea. Size-distribution analysis shows that for $\omega_0 = 20$ the microemulsion containing 0.5 $mol \ dm^{-3}$ urea in its pool is significantly more monodisperse than that containing no urea.

In another study thin films on quartz substrates show the effect of urea on a larger scale. In the Figs. 11.7 and 11.8, the effect on a microscopic scale is depicted.[32]

It can be observed that at low volume fractions the films droplets merge on themselves in presence of urea whereas at higher volume fractions a supra structure is formed and appears as a tube like (percolative) one.

11.7. Vesicles

Initial studies of urea effect on DPPC vesicles showed the urea does not interact with the interface whereas ethylurea and butyl urea does, due to the methylene groups.[33]

Light scattering and fluorescence data on ammonium surfactants showed that urea at high concentrations ($\sim 3 mol \ dm^{-3}$) induced a destabilization

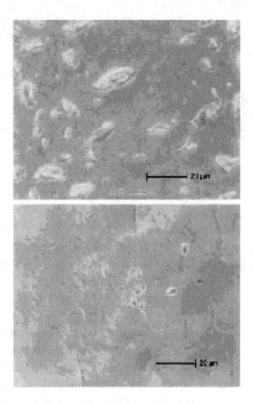

Fig. 11.7. Photographs of films deposited on hydrophilic surface for AOT/isooctane/water (top) and AOT / isooctane / water / urea 3M films (bottom). $\omega_0 = 15$ and $\phi = 0.06$. [AOT] $= 0.1$mol dm^{-3} (from ref. 31).

of the vesicles. A decrease in anion selectivity was also demonstrated, as found for monolayer (see above). These effects can be rationalized by the increase in the ion dissociation degree and thus towards larger fluctuations or ondulations on the bilayers. Once again an effect assigned to a more polar water.[34]

11.8. Conclusions

A set of evidences is presented to show the overall effect of urea as an additive that increases the medium polarity. This effect results in an increase in the electrostatic interactions that leads to a poorer ion pairing (better solvation of ions) and stronger hydrophobic repulsion. It should be added

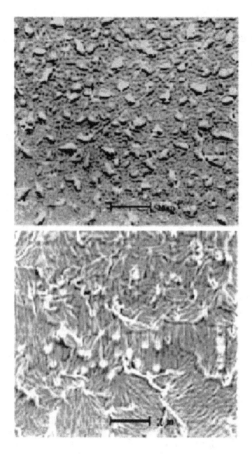

Fig. 11.8. "SEM" micrographs of the AOT/isoocatane / water (top) and AOT/isooctane/water / urea 3mol dm^{-3} films (bottom). [AOT] = 0.1mol dm^{-3}; $\phi = 0.06$ (from ref. 31).

that the hydrophobic forces are correlated to transfer free energy of methylene groups from hydrocarbons to liquid water. Such transfer is not very distinct from water to urea-water. Thus the main urea effect is within the bulk aqueous medium by favoring ionic dissociantions. In other words, the capital forces that drive aggregation of amphiphile species shows up latter. This effect, at low urea concentrations is similar to the salting in effects usually observed with proteins whereas at higher urea levels the effect is similar to the salting out.

Acknowledgments: We wish to express our gratitude to the Brazilian granting agencies FAPES, CNPQ and Capes. We also thank the ISSCS committee for the invitation and for their patience with the manuscript.

References

1. (a) D.R. Robinson and W.P. Jencks, *J. Am. Chem Soc.* **87** 2462 (1965), (b) D.B. Wetlaufer, S.K. Malik, L. Stoller and R.L. Coffin, *J. Am. Chem. Soc.* **86** 508, (1964), (c) H.S. Frank, and F. Franks, *J. Chem. Phys.* **48**, 4746 (1968), (d) M. Roseman and W.P. Jencks, *J. Am. Chem. Soc.* **97**, 631 (1975), (e) O. Enea and C. Jollcoeur, *J. Phys. Chem.* **86**, 3870 (1982).
2. (a) E. Liepinsh and G. Otting, *J. Am. Chem. Soc.* **116**, 9670 (1994), (b) P. Baglioni, E. Ferroni and L. Kevan, *J. Phys. Chem.* **94**, 4296 (1990), (c) C.C. Ruiz, *Colloids Surf. A.*, **147**, 349 (1999), (d) X.H. Shen, M. Belletete and G. Durocher, *J. Phys. Chem.* **101**, 8212 (1997), (e) S. Kim, D. Sarathchandra and D. Mainwaring, *J. Appl. Polym. Sci.* **59**, 1979 (1996).
3. L.G. Dias, F.H. Florenzano, W.F. Reed, M.S. Baptista, S.M.B. Souza, E.B. Alvarez, H. Chaimovich, I.M. Cuccovia, C.L.C. Amaral, C.R. Brasil, L.S. Romsted, and M.J. Politi, *Langmuir*, **18**, 319 (2002).
4. N. Gull, P. Sen, Kabir-ud-Din and R.H. Khan, *J. Biochem.* **141**, 261 (2007).
5. N. Gull, S. Kumar, B. Ahmad, R.H. Khan and Kabir-ud-Din, *Colloids and Surfaces B., Biointerfaces* **51**, 10 (2006).
6. A. Wallqvist, D.G. Covell and D. Thirumalai, *J. Am. Chem. Soc.* **120**, 427 (1998).
7. (a) D.R. Mountain and D. Thirumalai, *PNAS*, **95**, 8436 (1998). (b) W. Hinze, D.Y. Pharr, Z.S. Fu and W.G. Burkert, *Anal. Chem.* **61**, 422 (1989). (c) N. Kishore and J.C. Ahluwalia, *J. Chem. Soc., Faraday Trans.* **86**, 905 (1990).
8. M.C. Stumpe and H. Grubmuller, *J. Phys. Chem. B.* **111** 6220 (2007).
9. Weerasinghe, W. Samantha, and E. Paul Smith, *Journal of Physical Chemistry B.* **107**, 3891 (2003).
10. Y.L.A. Rezus, and H.J. Bakker, *PNAS* **103**, 18417 (2006).
11. J. Jr. Wyman *Journal of the American Chemical Society* **53**, 3292 (1931).
12. (a) F. Frank, 1963, *Water-A Comprehensive Treatise*, Plenum Press, N.Y., (b) B. John Hasted, 1973, *Aqueous Diletric*, Chapman and Hall.
13. M. Mayele, and M. Holz, *Phys. Chem. Phys.*, **2**, 2429 (2000).
14. M.J. Politi, and H. Chaimovich, *J. Solution Chemistry*, **18**, 1055 (1989).
15. B.P. Dey and S.C. Lahiri, Zeitschrift fuer Physikalische Chemie (Muenchen), **15**, 214 (2000).
16. E. Hawlicka and T.Z. Lis, *Naturforsch.* **49a**, 623 (1994).
17. (a) W. Hinze, D.Y. Pharr, Z.S. Fu and W.G. Burkert, *Anal. Chem.* **61**, 422 (1989). (b) N. Kishore and J. Ahluwalia, *C J. Chem. Soc., Faraday Trans.* **86** 905 (1990).
18. C.L.C. Amaral and M.J. Politi, *Langmuir* **13**, 4219 (1997).
19. E. Reddi and G. Jori, *Rev. Chem. Intermed.* **10**, 241 (1988).
20. P. Mukerjee and A.K. Ghosh, *J. Phys. Chem.* **67**, 193 (1963).

21. (a) K.A. Berberich and V.C. Reinsborough, *Langmuir* **15**, 966 (1999). (b) G. Calvaruso, F.P. Cavasino, C. Sbriziolo and M.L.T. Liveri, *J. Chem. Soc., Faraday Trans.*, **89**, 373 (1993).
22. H. Talukdar, S. Rudra and K.K. Kundu, *Can. J. Chem.* **67**, 321 (1989).
23. M. Woldan, Ber. Bunsen-Ges. *Phys. Chem.* **93**, 782 (1989).
24. S.M.B. Souza, H. Chaimovich, and M.J. Politi, *Langmuir* **11**, 1715 (1995).
25. L.S. Romsted, J. Zhang, I.M. Cuccovia, M.J. Politi and H. Chaimovich, *Langmuir* **19**, 9179 (2003).
26. C.C. Ruiz, *Colloids and Surfaces A : Physicochemical and Engineering Aspects*, **147**, 349 (1999).
27. (a) M. Abu-Hamdiyyah, L.J. Al-Mansour, *Phys. Chem.* **83**, 2236 (1979). (b) B. Kumar, N. Parveen and Kabir-ud-Din, *J. Phys. Chem. B.* **108**, 9588 (2004). (c) S. Kumar, D. Sharma, G Ghosh, and Kabir-ud-Din *Langmuir* **21**, 9446 (2005). (d) S. Kumar, D. Sharma, G. Ghosh, and Kabir-ud-Din *Colloids and Surfaces A : Physicochem. Eng. Aspects* **264**, 203 (2005). (e) S. Kumar, Z.A. Khan, N. Parveen, N. and Kabir-ud-Din, *Colloids and Surfaces A : Physicochem.Eng. Aspects*, **45**, 268 (2005).
28. C.L.C. Amaral, O. Brino, H. Chaimovich, and M.J. Politi, *Langmuir* **8**, 2417 (1992).
29. (a) R. Itri, C.L.C. Amaral and M.J. Politi, *J. Chem. Phys.* **111**, 7668 (1999). (b) C.L.C. Amaral, R. Itri and M.J. Politi, *Langmuir* **12**, 4638 (1996). (c) F.H. Florenzano and M.J. Politi, *Braz. J.Med. Biol. Res.* **30**, 179 (1997). (d) L. Garcia-Rio, J.R. Leis, J.C. Mejuto, M.E. Pena and E. Iglesias, *Langmuir* **10**, 1676 (1994). (e) M.S. Baptista, Tran, C. D. *J. Phys. Chem. B.* **101**, 420 (1997).
30. A. Ruggirello, and V.T. Liveri, *Chemical Physics* **288**, 187 (2003).
31. A. Chakraborty, M. Sarkar and S. Basak, *Journal of Colloid Interface Science*, **287**, 312 (2005).
32. C.L.C. Amaral and M.J. Politi, *Thin Solid Films* **468**, 250 (2004).
33. T. Inoue, K. Miyakama, and R. Shimozawa, *Bull.Chem. Soc. Jpn.*, **60**, 4148 (1987).
34. F.H. Florenzano, L.G.C. Santos, I.M. Cuccovia, M.V. Scarpa, H. Chaimovich, and M.J. Politi, *Langmuir* **12**, 1166 (1996).

Chapter 12

Specific Ion-Pair/Hydration Model for the Sphere-To-Rod Transitions of Aqueous Cationic Micelles. The Evidence from Chemical Trapping

Laurence S. Romsted

Department of Chemistry and Chemical Biology
Wright and Rieman Laboratories
Rutgers, The State University of New Jersey
New Brunswick, New Jersey 08903

"If there is Magic on the Planet, It is Contained in Water."
Loren Eisley, *The Immense Journey*, 1957

Models for the balance-of-forces controlling micelle formation, size and shape are briefly reviewed, including the hydrophobic effect which drives surfactant aggregation and headgroup repulsion/electrostatic screening that provides balance, the concept of ion specific interactions in the Stern layer, and the packing parameter for the relationship between surfactant structure and aggregate shape. The basic assumptions of the ion-pair/hydration model are introduced. This model is an alternative explanation for the balancing force and is a new approach for interpreting specific ion effects on the sphere-to-rod transitions of ionic micelles and potentially other structural transitions of association colloids. In this approach sphere-to-rod transitions occur because headgroups and counterions form ion-pairs in the micellar interfacial region and release water into the surrounding aqueous pseudophase. The concentration at which the transition occurs depends on headgroup structure, counterion type and the tendency to ion-pair correlates with the Hofmeister series. The model is used to interpret the sphere-to-rod transition concentrations of four different cationic micellar systems: gemini micelles with variable spacer lengths; cationic micelles with 3,5-dichlorobenzoate and 2,6-dichlorobenzoate counterions; and cetyltrimethylammnonium chloride, and cetyltrimethylammonium bromide and cetyltri-n-propylammonium bromide micelles. The strengths and limitations of the model are summarized and future conceptual and experimental applications are introduced.

12.1. Introduction

The aims of this review are several fold: (a) to summarize the current concepts and their limitations that are used to describe the balance-of-forces controlling the sizes and shapes of micelles and other association colloids such as vesicles, and microemulsions; (b) to show that the ion-pair/hydration model provides a natural explanation for ion specific effects on the formation of rod-like from spherical-like surfactant aggregates in aqueous micellar solutions with increasing surfactant concentration and counterion added as salt; and (c) to apply the ion-pair/hydration model to a series of chemical trapping experiments published over the last few years.

The ion-pair/hydration model is grounded in the pseudophase model of micelle formation and based on experimental estimates of the changes in interfacial concentrations of counterions and water obtained by using the chemical trapping method. This method is based on quantitative determination of competitive product yields from trapping of an arenediazonium ion by weakly basic nucleophiles in the interfacial regions of association colloids over large ranges in stoichiometric surfactant, salt and additive concentrations, such as urea and alcohols, in solution.[1-3] In the ion-pair/hydration model, specific counterion and headgroup effects are a consequence of changes in ion-pair and hydration interactions within the interfacial region of aqueous micelles caused by increasing surfactant and/or counterion concentration. The model provides an alternative to the "headgroup repulsion/electrostatic screening" interpretation of the balancing force to the hydrophobic effect that determines the equilibrium size and shape of surfactant micelles. The results demonstrate that micellar interfacial regions are "laboratories" for studying changes in ion and water interactions because modest changes in headgroup structure and counterion type have sometimes dramatic and easily observed effects on aggregate structure.

Hofmeister first reported specific ion effects on the solubilities of proteins in water 1888.[4] Since then a plethora of reports have appeared on specific effects on the solution properties of not only proteins, but also ionic association colloids such as micelles, microemulsions and vesicles, and on solution properties of polyelectrolytes and biomembranes. Nevertheless, a consensus view on interpreting specific ion effects at a molecular level remains stubbornly unrealized.[5] The wide variation in physical and chemical properties of ionic association colloids that are sensitive to both the types of cations and anions present in solution have been described in papers and reviews.[5-19] The focus of this review and examples given are on ion specific effects on sphere-to-rod transitions of ionic micelles, but many of the interpretations apply to structural transitions of association colloids and perhaps the specific ion sensitive properties of polyelectrolytes, proteins, and biomembranes as well.

The typical Hofmeister series order for halide ion effects on the properties of cationic micelles correlates qualitatively with the radius of the ions: $I^- > Br^- > Cl^- > F^-$.[19] In micellar solutions, the larger the halide ion, the lower the critical micelle concentration (cmc, the concentration above which micelles form); the higher the Krafft temperature (the temperature above which homogeneous micellar solutions form and below which the surfactant exists primarily as monomer in solution or hydrated solid); and the larger the micellar aggregation number.[20] In addition, increasing the anion radius decreases the sphere-to-rod transition concentration (see Table 12.1 below); and increases the inhibition of a chemical reaction between a micellar bound organic substrate and a reactive counterion such as OH^-.[21] The effect of cations such as alkali metal ions on the properties of anionic micelles typically follow the large to small cation radius order, but the effects are generally smaller.[22] Other anions or cations of the same valence, e.g., OH^-, NO_3^-, $CH_3CO_2^-$, ClO_4^- and $(CH_3)_4N^+$, are sometimes included in Hofmeister series for anions and for cations. Typically the larger, less strongly hydrated, more polarizable ions have larger effects on micellar solution properties.

Given that such qualitative orders of ion specific effects for micelles, association colloids, proteins and biomembranes are plentiful and have increased in number for more than a century, why is a quantitative interpretation still absent? A number of theoretical and experimental reasons make quantitative interpretation of the Hofmeister series one of the more difficult problems in chemistry.[5] The strengths of the interactions between the surfactant tails and water and surfactant headgroups with counterions and water are difficult to measure or calculate because the differences in strengths of the intermolecular and interionic forces controlling these interactions are small, on the order of the strengths of hydrogen bonds in solution, ca. 5 kcal/mole. Although the free energies of hydration of ions are large, the rate of exchange of water molecules on and off ions is fast, which means free energy of exchange of water of hydration is small. In addition, the free energies of ion-pair formation between simple ions (i.e., the typical, uncomplicated ions commonly used as headgroups and counterions of single chain anionic and cationic surfactants) in water are seldom strong.

If they were, we might not have many water soluble soaps.

12.2. Balance-of-forces Controlling Micelle Size and Shape

The distributions of ions around association colloids were classically modeled by using electrical double layer theory, in which Coulombic interactions between ions are treated as point charges and are screened by water modeled as a uniform dielectric.[23,24] However, treating ions as point charges cannot account for the ion specific on the properties of micelles and association colloids as represented by the Hofmeister series, which has been recognized for some time. However, ion specific effects are difficult to model from first principles,[25] as is including specific solvation in treating ionic interactions.[26] The properties of micellar and association colloid solutions are also modeled thermodynamically by including extra thermodynamic assumptions to interpret ion specific effects and that approach is used here.

12.2.1. Balance-of-forces/Free Energy of Micelle Formation

A delicate balance-of-forces determines the distributions of sizes and aggregation numbers of ionic micelles in dilute, homogeneous, aqueous solutions. The forces must be in balance when the micellar sizes and shapes are at their equilibrium distribution for a particular set of conditions and changes in solution composition shift the free energy minimum of the system. That the balance is delicate comes from the realization that small changes in surfactant solution composition sometimes has a dramatic effect on aggregate sizes and shapes without making or breaking any chemical bonds. Figure 12.1, from an old physics textbook, illustrates forces in balance. The orientation of the object is determined by the vector sum of the three forces on the object, which is zero at equilibrium. At equilibrium the object is also at its free energy minimum. If the orientation and/or strength of one of the forces is changed, the object will move to a new equilibrium position such that the vector sum of the forces is again zero. In aqueous solutions of ionic surfactants above the cmc in which surfactant monomer is in dynamic equilibrium with monomer in micelles, what are the forces responsible for the equilibrium structures of the micelles ?

12.2.2. The Driving Force, The Hydrophobic Effect

Scheme 12.1 illustrates the basic concept of the hydrophobic effect (and also ion- pairing, see below), where the ellipsoids are hydrocarbon, the circles are

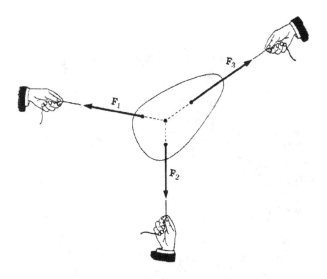

Fig. 12.1. An object is at rest because the vector sum of the three forces is zero. The free energy is also at a minimum.

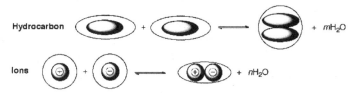

Scheme 12.1. Ion-pair / hydration model.

ions, and the solid lines indicate a layer water molecules interacting specifically with the hydrocarbon and ions. In this simple hypothetical example of Tanford's basic concept,[27] two hydrocarbon molecules in aqueous solution are in equilibrium with a hydrocarbon pair ($\Delta G_{pair} = \Delta H_{pair} - T\Delta S_{pair}$) in which the water/hydrocarbon surface contact is reduced and excess water molecules, mH_2O, are released into the surrounding aqueous phase. Because the intermolecular forces between the water and hydrocarbon molecules and between hydrocarbon molecules are small, the enthalpy of transfer for this reaction is small, ($\Delta H_{pair} \sim 0$). The position of equilibrium is determined by the amount of ordered water released into the bulk aqueous phase, i.e., $\Delta G_{pair} \approx T\Delta S_{pair}$. A similar argument is used for ion-pairing in Scheme 12.1.

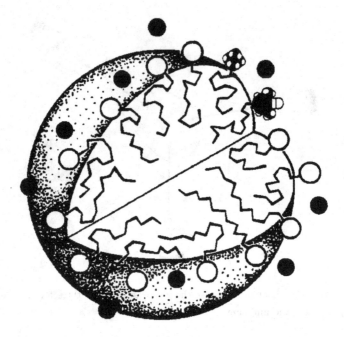

Fig. 12.2. Cartoon of an aqueous cationic micelle showing the cationic headgroups (o) and anionic counterions (•) surrounding a hydrocarbon core (stippled surface), and scaled space filling representations of one α-methylene-trimethylammonium surfactant headgroup and an arenediazonium group. Interfacial and bulk water molecules are not shown. Reproduced with permission of the American Chemical Society.[3]

12.2.3. *The Hydrophobic Effect Drives Micellar Aggregation*

Above the cmc, virtually all added surfactant spontaneously forms approximately spherical micelles, Figure 12.2, with aggregation numbers on the order 50-150 monomers per micelle, depending upon surfactant structure and other experimental conditions. The properties of micellar solutions are often interpreted using the pseudophase model, that is, the totality of the micelles in solution are treated as a separate phase distributed throughout the aqueous phase. Components in solution are associated with either the aqueous or micellar pseudophases. However, in its current form, this approach is limited because it cannot account for changes in micellar structure such as transitions from spherical to rod-like micelles. The pseudophase model is modified by the ion-pair/hydration model to allow for changes in aggregate structure with interfacial composition induced by changing the solution composition.

Measurements of the thermodynamics of micellization, expressed as the free energy of transfer of a surfactant monomer to a micelle, $\Delta G_{micelle} = \Delta H_{micelle} - T\Delta S_{micelle}$, shows that in the vicinity of room temperature, $\Delta G_{micelle}$ values obtained from measured cmcs, are generally large and negative.[28] However, measured ΔH values of micellization per monomer are generally small ($\Delta H_{micelle} \sim 0$), which means the difference in the strength of noncovalent intermolecular interactions on transfer of a surfactant tails from water to micelles is small.[28] Thus the large negative values of ΔG are determined primarily by large positive values of $T\Delta S_{micelle}$. The increase in $T\Delta S_{micelle}$ comes from the release of most of the water molecules that were organized around the surfactant tails in the bulk before the monomers aggregated into the largely dehydrated micellar core. Thus, ordering surfactants into micelles is driven by an increase in disorder of the water and a net increase in disorder of the system. Although the increase in organization of surfactant molecules into micelles is easy to observe experimentally, the increase in disorder of water molecules caused by the transfer of the more ordered water molecules in contact with the hydrocarbon tails of the surfactant monomers to the less ordered bulk aqueous solution has not been observed experimentally. The $\Delta H_{micelle}$ and $T\Delta S_{micelle}$ values for transferring surfactant headgroups and counterions from bulk solution to the micellar surface are usually not considered separately. Both are probably small because headgroups and counterions in the interfacial regions of spherical ionic micelles are essentially fully hydrated when the stoichiometric surfactant concentration is just above the cmc.[29] However, headgroups and counterions are not fully hydrated in rod-like micelles. If the hydrophobic effect drives micelle formation, what is/are the balancing force(s) that determines micelle size and shape?

12.2.4. The Balancing Force: the Traditional View

Above the cmc, hydrophobic tails of the surfactants aggregate and drag 50-150 charged headgroups together into the micellar interfacial region or "Stern Layer" around the core of approximately spherical micelles, Figure 12.2. The repulsive forces between the headgroups in the interfacial region of the micelles are screened by interfacial water and a fraction of the surfactants counterions, about 60-80%. The equilibrium balance of these two effects, hydrophobic effect versus headgroup repulsions, determines the cmc, aggregation number, and the fraction of "free" counterions in the surrounding electrical double layer extending out into the aqueous phase. A variety of experimental methods,[30] are used to estimate the fraction of free counterions, α, although different methods sometimes give significantly different values for the same surfactant. The bound ("not free") fraction of counterions, β, are estimated by difference from $\alpha + \beta = 1$.[31] There is no experimental method for measuring β independently. The dependence of the value of α on the experimental method used is unknown, both because different methods and investigators report different α values for the same surfactant in solution and because different methods may sense different fractions of "free" ions.[32] In addition, α is difficult to define theoretically because the boundary between "free" and "bound" counterions in the vicinity of the micellar interface cannot be defined unambiguously. Finally, the differences in α values for different headgroups and counterions and even surfactant chain lengths are not large and often less than the differences between experimental methods.[31] Ion specific effects are assumed to occur in the Stern layer, but the nature of these interactions are not understood.[33] The problem with the headgroup repulsion concept as the balancing force for the hydrophobic effect and determining aggregate shape is that, in its current form, it does not provide an explanation for why micelles change shape with increasing surfactant and salt concentration, i.e., why the sphere-to-rod and other structural transitions depend on headgroup and counterion type, counterion concentration, and changing temperature.

12.2.5. The Relationship of Surfactant and Aggregate Structure: The Packing Parameter

This visually appealing concept attempts to correlate the three-dimensional shape of a surfactant molecule to the three-dimensional shape of its aggregates,[20,34,35] but it has documented shortcomings.[36] The packing parameter is defined as: $p = v/la$, where v is the volume of the hydrocarbon chain, l is chain length of the surfactant, and a is the headgroup cross sectional area. Values of p provide a rationale for the relationship between surfactant shape and aggregate structure. For example, surfactants that are described as cone shaped such as single tailed anionic and cationic surfactants having headgroup cross sectional areas that are greater than the cross sectional area of their hydrocarbon tails tend to form spherical micelles. More cylindrically shaped surfactants tend to form rod-like aggregates and twin tail surfactants form nearly planar or planar surfaces found in vesicles and lamellar mesophases, respectively.[34,35,37] However, the packing parameter approach cannot account for several important properties connecting the shape of a particular surfactant and aggregate shape.[36] For example, the surfactant concentration-temperature phase diagrams for many surfactants show that increasing the concentrations of single chain surfactants (and decreasing the water concentration) give spherical, then rod-like then hexagonal closed packed, then lamellar, and finally reversed hexagonal closed packed structures, yet the headgroup cross section, a, and tail length, l, and tail volume, v, of the surfactant have, in principle, not changed.[20,36] Short chain, branched, twin tail surfactants such as AOT[20] and octyl to dodecyl phospholipids[38] form spherical micelles instead of reverse micelles or vesicles, respectively, which they are predicted to do based on the packing parameter. In fact, the tendency of single chain surfactants to form rod-like micelles depends on chain length, temperature, headgroup structure, and counterion type. Some examples are presented in Table 12.1 below. The cross sectional area, a, must depend not only on the diameter of the surfactant headgroup, but also on other components of the interfacial region, e.g., counterions and water.

12.2.6. The Balancing Effect: The Specific Ion-Pair / Hydration Model

The chemical trapping results summarized in the next section provide strong support for the ion-pairing/hydration model discussed below for sphere-to-rod transitions. The model provides a straightforward conceptual explanation for ion specific effects on structural transitions of micelles and is applicable to other association colloids. In this approach, the interfacial region of an ionic micelle, which is conceptually equivalent to the Stern Layer, is treated as separate region of unknown thickness composed primarily of headgroups, counterions, water of hydration around headgroups, counterions and some (unknown number) of methylenes in the surfactant tail (probably the first few methylenes extending from the surfactant headgroups), and free water. The interfacial region is a transitional layer between the non-polar hydrocarbon cores of the aggregates composed of surfactant tails and the polar aqueous solution, Figures 12.2 and 12.3. The properties of this region are assumed to be those of a concentrated salt solution with estimates of local interfacial headgroup and counterion concentrations on the order of 1-3 M (see below), although the interfacial region is adjacent to a hydrocarbon core (that is obviously not present in the reference aqueous solutions) that may (or may not) affect the ion-pairing properties of ions and molecules in the interfacial region.

In concentrated salt solutions, some of cations and anions form ion-pairs, although the presence of ion-pairs in water is sometimes difficult to demonstrate unambiguously.[39,40] The hydration demand of these neutral, but polar ion-pairs, is less than that of the separate free ions, and excess water of hydration is released into the surrounding aqueous phase, Scheme 12.1. In aqueous solution, the free energy of ion-pair formation, ΔG_{pair}, is composed an enthalpy term, ΔH_{pair}, which includes the specific interactions between the ions in the pair and the change in enthalpy of hydration of the ions on going from the hydrated free ions to the hydrated pair. The entropy term, $T\Delta S_{pair}$, includes the loss of entropy of ion translation, $T\Delta S_{trans}$ and the increase in $T\Delta S_{free}$ on release of excess water of hydration, Scheme 12.1.

In micellar solutions, the sphere-to-rod transition occurs when rod-like micelles become more stable than spherical micelles, i.e., when $\Delta G_{trans(s \to r)}$ is negative. To a first approximation, $\Delta H_{trans(s \to r)} \sim 0$, because the intermolecular interactions in the fluid hydrocarbon regions of spherical and rod-like micelles should be about the same. For example, the degree of order of deuterium labeled α, γ, and ω positions on the tail of sodium dodecyl sulfate are very similar in spherical and rod-like micelles indicating that degree of order in hydrocarbon cores of these two structures is about the same.[41] The size of the change in $\Delta H_{transinterface(s \to r)}$ for headgroups and counterions and their water of hydration is uncertain, but as in bulk solution, its value depends on differences in enthalpies of interaction of headgroups and counterions with water and the enthalpy of formation of the hydrated ion-pair. $T\Delta S_{transinterface(s \to r)}$ is the difference between the loss of translational entropy, $T\Delta S_{trans}$, which may be small because the headgroups and counterions are already at a high local concentration within the micellar interfacial region (on the order of 1-3 M in local interfacial counterion concentration), versus the increase of entropy that occurs because the ion-pair binds fewer waters of hydration than the free ions, $T\Delta S_{free}$. Water that is released on ion-pair formation leaves the interfacial region and enters the surrounding aqueous phase (see trapping evidence below), the interfacial region shrinks, and headgroup and counterion packing becomes

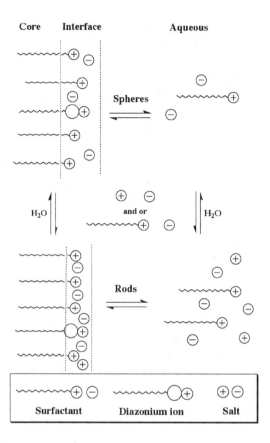

Fig. 12.3. Cartoon illustrating a small section of the interfacial regions of spherical (top) and rod-like (bottom) micelles. Added surfactant or salt increases the counterion concentration in the interfacial region and induces the sphere-to-rod transition. At the sphere-to-rod transition concentration, ion-pairs form, water is released, shrinking the interfacial region which permits tighter packing of the surfactant molecules, and the formation of rod-like structures. Water molecules are not shown. The locations of the dashed links marking the interfacial region are only illustrative.

tighter. Because this transformation is occurring under the same conditions for all the micelles in solution, the structures of all the micelles in solution change over the same concentration range from radially oriented, less tightly packed spherical micelles to more tightly packed cylindrical or rod-like micelles. The tendency to form ion-pairs follows the same basic order as the Hofmeister series. Addition of counterions as salt increases the interfacial counterion concentration, increases the extent of ion-pairing and, at sufficiently high counterion concentrations, induces the sphere-to-rod transition at a concentration that depends on surfactant tail length, headgroup structure, and counterion type.

The next three sections introduce: (a) the chemical trapping method; (b) some examples of ion specific effects on cmc and sphere-to-rod transition concentration values; and (c) a review published chemical trapping results on some of these surfactants that shows that the interfacial counterion concentrations increase and the interfacial water concentration decreases at the sphere-to-rod transitions for these surfactants, consistent with the ion-pair/hydration model.

12.3. The Chemical Trapping Method and Experimental Protocols-In Brief

12.3.1. *Method*

Probing interfacial compositions from measured product yields by the chemical trapping method is grounded in the well established heterolytic chemistry of arenediazonium ions[42,43] and in our published work[1-3,44-48] with our specifically designed probe: 4-alkyl-2,6-dimethylarenediazonium ion, z-ArN_2^+, prepared and stored as its stable BF_4^- salt, z-ArN_2BF_4, Scheme 12.2. In aqueous solutions, in the presence or absence of surfactant, below about pH 7, and in the absence of light, the probe reacts competitively with the weakly basic nucleophiles to give stable products that are analyzed quantitatively by HPLC. In aqueous solutions in the absence of surfactant, product yields from reaction with the water-soluble probe, 1-ArN_2^+, are proportional to their measured concentrations, Figure 12.4. In surfactant solutions, the long chain arenediazonium ion 16-ArN_2^+, which is itself a surfactant, is oriented with its reactive arenediazonium group in the interfacial region, Figures 12.2 and 12.3, and its product yields are proportional to the interfacial and not the stoichiometric nucleophile concentrations. Figure 12.4 graphically illustrates the application of the method for estimating interfacial molarities. The solid points are for product yields obtained from reaction of 16-ArN_2^+ in CTABr solutions from 0.05-0.2 M, where 16-ArOH is the phenol product from reaction with H_2O and 16-$ArBr$ is the bromo product from reaction with Br^-. The open points are for reaction of 1-ArN_2^+ in aqueous solutions of tetramethylammonium bromide (TMABr) to give 1-$ArOH$ and 1-$ArBr$ in aqueous solutions from 0 to 3.5

M. The 1-$ArBr$ data is a "standard curve" for estimating Br_m.[49] Interfacial Br^- molarities, Br_m, are obtained by assuming that when the yield of 16-ArBr is the same as yield of 1-$ArBr$ in aqueous solutions containing a known concentration of Br^-, then value of Br_m is the same as that of Br^- in the aqueous phase. In Figure 12.4, when the yields of 16-$ArBr$, and 1-$ArBr$ are both 35% (horizontal dashed line) the interfacial molarity of Br^- in 0.01M CTABr is 2.25M, the same as that in an aqueous solution 2.25M $TMABr$. The same procedure was used to determine Br_m at other CTABr concentrations and in the chemical trapping results described below. Interfacial water molarity, H_2O_m, is obtained from the 16-$ArOH$ and 16-$ArBr$ yields and the measured selectivity of the dediazoniation reaction toward Br^- versus H_2O in aqueous solution.[50]

In sum: *when the yield from reaction with a particular nucleophile is the same in the interfacial region and in a particular aqueous solution, then the nucleophile concentration in the interfacial region is the same as the stoichiometric nucleophile concentration in that aqueous solution. In brief:* **when the yields are the same, the concentrations are the same.**

We have used this approach repeatedly to estimate interfacial concentrations of a variety of weakly basic nucleophiles in a number of surfactant systems, including microemulsions and emulsions. The list of weakly basic nucleophiles that have or may be used includes many of the important functional groups in biological systems and surfactant based commercial products : water, alcohols (including the terminal OH groups of nonionic surfactants), urea and N substituted acetamides (both the O and N of the amide group), halide ions and alkyl carboxylates and sulfonates.[1,3] The method works in SDS micelles, but product yields are difficult to analyze because the alkyl aryl sulfate product from trapping by the weakly basic SDS headgroup hydrolyzes rapidly to an anionic aryl sulfate and dodecanol.[51] Strongly basic nucleophiles such as OH^-, CN^-, N_3^- and SO_3^{2-} react extremely rapidly at the terminal nitrogen[42] and their interfacial concentrations cannot be estimated by the chemical trapping method.

12.3.2. Experimental Protocols

The procedures for carrying out the chemical trapping experiments have been described in detail.[1,48] In brief, 16-ArN_2^+, a water insoluble surfactant itself and similar in size and shape to many surfactants, is used at low concentrations, 16-$ArN_2^+ = 10^{-4}-10^{-5}M$ with surfactant/16-ArN_2^+ ratios

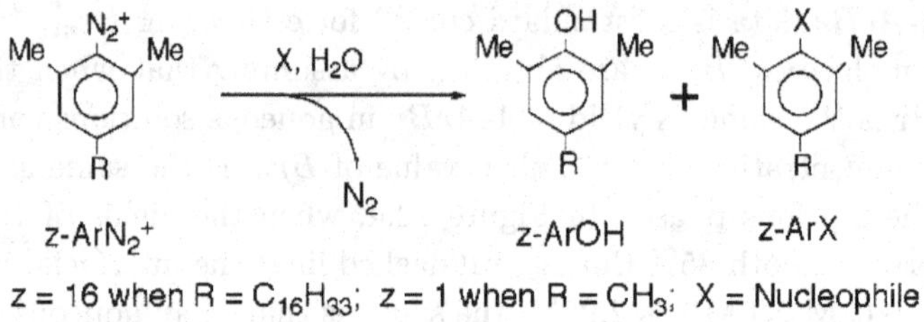

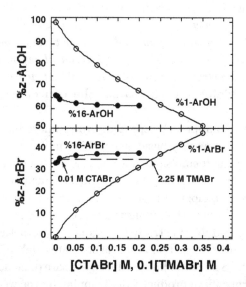

Scheme 12.2. Products formed from the chemical trapping reaction. X may be an anionic, e.g., Cl^-, or a neutral nucleophile, e.g., ROH, or acyl oxygen or nitrogen of a peptide bond.

Fig. 12.4. Dediazoniation product yields from reaction with H_2O (top) and Br^- (bottom) at $40°C$ from reaction of 16-ArN_2^+ in CTABr micelles (●) and 1-ArN_2^+ in TMABr salt solutions (o). To put the CTABr and TMABr data on the same scale, each stoichiometric TMABr concentration is multiplied by 0.1. Lines are drawn to aid the eye. Reproduced with permission from the American Chemical Society.[3]

$\geq$ 100, such that there is minimal perturbation of micelle structure by the probe. The reaction is run about 10 half-lives, typically 6-24 hours depending on temperature and products are analyzed quantitatively by HPLC on C-18 reverse phase columns with UV detection. Product yields are determined from peak areas by using calibration curves created with independently prepared samples of each product. The total yield of all products is generally within ± 5% of the amount of z-ArN_2^+ added and if they are less than 90%, the experiments are usually run again. The HPLC analyses are time consuming (ca. 30 min. per run) and all analyses are run in triplicate. Normalized yields are used to estimate interfacial concentrations and are reported as interfacial molarities, e.g., Br_m, H_2O_m, etc., moles per liter of interfacial volume.

Table 12.1. Effect of hydrocarbon chain length, headgroup structure, and counterion type on the cmc and sphere-to-rod transition concentrations of some cationic amphiphiles.

Row No.	Surfactant[a]	Tail # C's	Head Group, X	Cmc mM	Sphere-to-Rod, M	Ref.
1	CTABr	16	$-N(Me)_3^+ Br^-$	0.9	0.1	52
2	CTACl	16	$-N(Me)_3^+ Cl^-$	1	~ 1.0	53
3	DTABr	12	$-N(Me)_3^+ Br^-$	14.6	~ 1.8	54
4	DTACl	12	$-N(Me)_3^+ Cl^-$	19.4	none	55
5	DDABr	12	$-NH(Me)_2^+ Br^-$	12.4	~ 0.07	56
6	DDACl	12	$-NH(Me)_2^+ Cl^-$	14.9	~ 0.8	56
7	12-2-12 2Br	12× 2	$(Me)_2 N^+ (CH_2)_2 N^+(Me)_2 2Br^-$	1	~ 0.0042	57

[a] CTABr & CTACl cetyltrimethylammonium bromide and chloride; DTABr & DTACl, dodecyltrimethylammonium bromide and chloride; DDABr & DDACl, dodecyldimethylammonium bromide and chloride; and 12-2-12 2Br, bis(dodecyldimethyl)-α, ω-ethanediammonium dibromide.

12.4. Ion Specific Effects on cmcs and Sphere-to-Rod Transition Concentrations

Addition of surfactant or counterion as salt induces structural transitions from spheres to rods in many micellar solutions, but the concentrations at which the transitions occurs depends on a number of factors including, surfactant chain length, number of surfactant tails, headgroup structure, counterion type and concentration. Table 12.1 lists a few examples of cationic amphiphiles with monovalent headgroups and counterions. Note the change in the cmcs and sphere-to-rod transition concentrations when a methyl group is replaced by a proton (rows 3 and 5, 4 and 6), a Br^- is substituted for a Cl^- (1 and 2, 3 and 4, 5, 6), and the length of the alkyl chain increased (hydrophobicity) (rows 1 and 3, 2 and 4). Also, the gemini amphiphile with a dicationic headgroup (row 7) has a much lower cmc and sphere-to-rod transition than its single chain analog (row 3), even though the apparent hydrophilic-lypophilic balance (HLB)[20] of these two amphiphiles are essentially the same, i.e., the ratios of headgroup charge to number of carbons in each amphiphile are identical. Above the sphere-to-rod transition concentrations, the most stable equilibrium structures tend to be rod-like, thread-like, or worm-like cylindrical instead of spherical shapes (e.g., see references in Table 12.1).

12.5. Sphere-to-Rod Transitions: Chemical Trapping Results

This section reviews four sets of chemical trapping results in different surfactants that show that at the sphere-to-rod transition concentrations occur with significant increases in interfacial molarities of Br_m, Cl_m, dichlorobenzoate, OBz_m, counter-ions, and concomitant decreases in H_2O_m.

12.5.1. Gemini Micelles

The sphere-to-rod transitions for twin tail gemini surfactants, $12 - n - 12$ $2Br$ (spacer length, $n = 2 - 4$), are very sensitive to spacer length. Cryo-TEM results show that only 12-2-12 $2Br$ micelles grow below $5 mM$ added surfactant and 12-3-12 $2Br$ and 12-4-12$2Br$ remain spherical up to much higher concentrations.[57] Chemical trapping results show that only 12-2-12 $2Br$ exhibits a marked increase in Br_m (ca.2.5 to 3.6 M) and a concomitant decrease in H_2O_m, (ca. 33 M to 16 M) starting at about 2 mM 12-2-12 $2Br$ and going to a plateau above 3 mM, Figure 12.5. We used 1-n-12Br ($n = 2 - 4$) salts as the standard aqueous solutions to determine the interfacial molarities of these surfactants, matching the spacer lengths of the bolaform salts and the gemini surfactants.[58] Note that the interfacial molarities increase as the spacer length becomes shorter. As the spacer length decreases, the gemini headgroups pair more easily with Br^-. We determined the ion-pairing constants of the bolaform salts by chemical trapping and from Br^- linewidths and the ion-pairing constants and the results are in good agreement.[58] The other two surfactants show gradual increases in Br_m and gradual decreases in interfacial H_2O_m. The H_2O_m/Br_m ratio for 12-2-12 $2Br$ drops from ca. 14 to 5. Reported hydration numbers for Br^- in water vary, but are ~ 5.[59] These results are consistent with significant dehydration of the micellar interface when rods are formed. We estimated the fraction (dication •Br)$^+$ and (dication •$2Br$) pairs in the

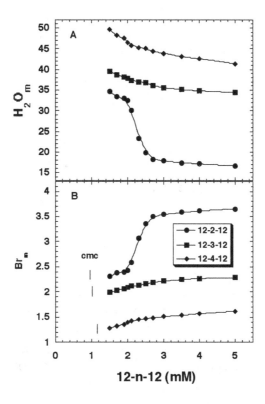

Fig. 12.5. Interfacial Br^-, Br_m, (A) and water, H_2O_m, (B) molarities in 12-n-12 $2Br (n = 2 - 4)$ gemini micelles at $25°C$ with $0.1m$ mol dm^{-3} HBr. Vertical lines are literature cmc values. Lines are drawn to aid the eye. Reproduced with permission of the American Chemical Society.[3]

interfacial regions of the three gemini surfactants.[47] 12-2-12 has 18% and 40% of the each pair, respectively, above 3 mM surfactant. Almost 60% of the headgroups are paired; a much higher fraction than for the other two surfactants. The paired headgroups should have a lower demand for water of hydration than the free ions and the release of excess water of hydration into the bulk aqueous phase permits tighter packing of headgroups and counterions in the interfacial region and formation of rod-like micelles.

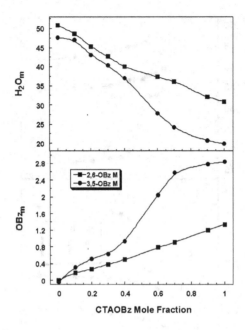

Fig. 12.6. Interfacial molarities of dichlorobenzoate ions, $2,6O\ Bz_m$ and $3,5\ OBz_m$, and water, H_2O_m, with increasing mole fraction (0 to 1) of CTAOBz and decreasing CTACl mole fraction (1 to 0) in 10 mM mixed micelles of CTAOBz/CTACl at 25°C. Lines are drawn to aid the eye.

12.5.2. Substituent Effects with Aromatic Counterions

Figure 12.6 shows some of the chemical trapping results in mixed micelles of CTACl and CTAOBz, where $OBz^- = 2,6$-dichlorobenzoate (2,6 OBz) and 3,5-dichlorobenzoate (3,5 OBz) ions.[47] A CTAOBz surfactant is mixed with CTACl such that the total surfactant concentration remained constant at 10 mM and the mole fraction of CTAOBz increases from 0 to 1. These experimental conditions mimic those of Magid et al.,[60] who determined the effect of changing the substituents on the aromatic counterion on the tendency to form rod-like micelles. Their cryo-TEM and NMR results showed that in mixed micelles of CTABr/3,5OBz (10 mM total amphiphile concentration) rods appear above CTA3, 5OBz mole fractions of ca. 0.2 CTA3, 5OBz, but mixed micelles of CTABr/CTA2, 6OBz remain spherical at all mole fractions of the two surfactants.[47] In Figure 12.6, the interfacial molar-

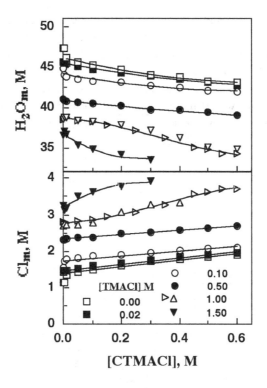

Fig. 12.7. Effect of increasing [CTACl] and added [TMACl] on Cl_m and H_2O_m in 0.1 M HCl at 40°C. Details on the calculations are in the reference.[50] Lines are drawn to aid the eye. Reproduced with permission of the American Chemical Society.

ity of CTA2, 6OBz increases essentially linearly with its mole fraction (and there is a concomitant decrease in H_2O_m). However, the interfacial molarity of CTA3, 5OBz shows a marked increase at a mole fraction of about 0.4 (with a concomitant decrease in H_2O_m). Below the sphere-to-rod transition for CTACl/CTA3, 5OBz, the H_2O_m/OBz_m molar ratios are those of relatively dilute salt solutions, i.e., H_2O_m is about 50 M. However, in 10 mM solutions of CTAOBz in the absence of $CTACl$, the H_2O_m/OBz_m molar ratios are those of concentrated salt solutions: 20 M/2.8 M = 7.1 (3,5OBz) and 32 M/1.2 M = 26.7 (2, 6 OBz). The interfacial regions of CTA3, 5OBz micelles are much less hydrated than CTA2, 6OBz micelles and therefore 3, 5 OBz ion-pairs with the quaternary ammonium headgroups much more than 2, 6 OBz. Such substituent effects on sphere-to-rod transitions are observed with other aromatic counterions[61-63] and are consistent with apparently small changes in counterion structure on the specific interactions in the interfacial region being amplified by the highly cooperative nature of the sphere-to-rod transition. In addition, organic ion-pairs, here quaternary ammonium and aromatic carboxylate pairs, are less polar than the free ions and a fraction of the pairs may partition more deeply into the interfacial region. This effect would increase the total fraction of paired ions in the interfacial region by shifting the overall equilibrium more toward paired ions. No evidence is available for this effect.

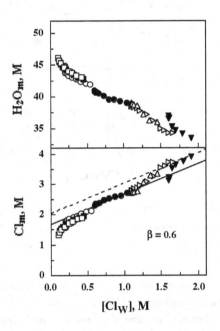

Fig. 12.8. Plots of Cl_m and H_2O_m versus $[Cl_w]$ at the optimal β value for CTACl/TMACl solutions. Data and symbols are the same as in Figure 12.7. The straight lines have slopes of 1 (one) and the intercepts were selected to give optimal contact of the data points with the solid and dashed lines. Note breaks from the smooth curve above ca. 1.0 M to the new line. Reproduced with permission of the American Chemical Society.[50]

12.5.3. Single Chain Surfactants: Cetyltrimethylammonium Chloride, CTACl

The sphere-to-rod transition for CTACl solutions containing added Cl^- as salt is reported to be at ca. 1.0 M total Cl^-, Table 12.1. Figure 12.7 shows values of Cl_m and H_2O_m obtained in CTACl solutions containing added tetramethylammonium chloride (TMACl) from 0 to 1.5 M using trapping results for $1\text{-}ArN_2^+$ in TMACl solutions (0.5 to 3.5 M) at 40°C as the standard curve.[50] Several observations are important here. First, Cl_m increases linearly with added CTACl at several [TMACl] (where square brackets here and throughout the text indicate stoichiometric solution molarity) and that added [TMACl] also increases Cl_m almost incrementally below 1.0 M [TMACl]. H_2O_m decreases concomitantly. At 1.0 and 1.5 M added TMACl, the Cl_m versus CTACl M profiles change shape and the H_2O_m values show larger decreases. The solutions at 1.0 and 1.5 M [TMACl] were visually more viscous indicating micellar growth. Note that the 1.0 M TMACl data is a superimposable set of two separate runs.

All the data in Figure 12.7 were replotted in Figure 12.8 as a function of the aqueous concentration of Cl^- in the solutions, $[Cl_w]$. The entire data sets for Cl_m and H_2O_m collapse into approximately continuous curves. Values for $[Cl_w]$ were obtained by assuming that the micelles contributed a constant fraction of their counterions to the aqueous phase, $\alpha = 1-\beta = 0.4$, that the $cmc \sim 0$ and HCl = 0.1 M, that virtually all the added TMACl is in the aqueous phase, and that the total excluded volume of the micellar cores, V, was estimated from the total volume of the hydrocarbon tails of CTACl, Eq. (1) (below).

The Cl_m^- $[Cl_w]$ and $H_2O_m - [Cl_w]$ profiles in Figure 12.8 are analyzed in three sections. Note that Cl_m M is always significantly larger than $[Cl_w]$ M. In the linear section from $[Cl_w] \approx 0.2$ to 1.2 M, the data fit a slope of ca. 1 (solid line), which means that each incremental increase in $[Cl_w]$ produces an equal increase in Cl_m, and a concomitant decrease in H_2O_m.

$$[Cl_w] = \frac{(1-\beta)\{[CTACl] - cmc\} + cmc + [HCl] + [TMACl]}{1 - V[CTACl]} \quad (1)$$

$$Cl_m = \frac{[Cl_m]}{\{[CTACl] - cmc\}V_m} + [Cl_w] = \frac{\beta}{V_m} + [Cl_w] \quad (2)$$

As the Cl^- and probably TMA^+ concentrations increase together in interfacial region, they displace water and H_2O_m decreases. Above ca. 1.2 M $[Cl_w]$, significant quantities of ion-pairs between CTA^+ headgroups and perhaps between TMA^+ and Cl^- ion begin forming, consistent with the release of more water into the aqueous phase. Cl_m increases to a new (dashed) line also with a slope of 1 and H_2O_m continues to decrease.

The linear regions between 0.2 to 1.2 M Cl_w and above 1.5 M are consistent with equation 2 (above), which is composed of two terms, one that relates Cl_m to its stoichiometric concentration, $[Cl_m]$, and one to $[Cl_w]$. If ratio of $[Cl_m]/\{[CTACl] - cmc\} = \beta/V_m$, where V_m is the molar volume of the interfacial region, is constant, then the increase in Cl_m is caused by the increase in $[Cl_w]$ with a slope of 1 (one). When rods are present above 1.5 M Cl_w, Cl_m continues to increase with the same slope. The most difficult region to interpret is between ca. 0.1 to 0.2 M $[Cl_w]$. If our assumption that $\beta = 0.6$ is constant and independent of [TMACl] is correct, then the initial rise in Cl_m to the solid line must be caused by a decrease in V_m (and the interfacial volume), which could be a consequence of a salt induced decrease in water/hydrocarbon contact in the interfacial region with increasing [TMACl] and [CTACl]. When $[Cl_m]/\{[CTACl] - cmc\} >> [Cl_w]$ in equation 2 (e.g., Figure 12.8, $[Cl_w] = 0.2$ M, $Cl_m = 1.5$ M), Cl_m is given by β/V_m. If both β and V_m were constant, then Cl_m should be constant, which is not the case, supporting the idea that added salt helps "dehydrate" the micellar interface.

12.5.4. *Single Chain Surfactants: Cetyltrimethylammonium Bromide (CTABr) and Cetyltri-n-propylammonium Bromide (CTPABr)*

The results in Figures 12.9 and 12.10[50] show a similar pattern to those for CTACl in Figures 12.7 and 12.8, but with some distinct differences. The amount of added CTABr, CTPABr, and tetramethylammonium bromide (TMABr) is less because CTABr undergoes the sphere-to-rod transition at much lower concentrations, ca. 0.1 M $[Br_w]$, Table 12.1. The CTABr profiles at different [TMABr]s show similar changes in shape to that of CTACl, but at 0.1 and 0.2 M added TMABr instead of at 1.0 and 1.5 M TMACl. CTPABr, however, shows only a series of parallel lines up to 0.5 M TMABr and the CTPABr solutions remained fluid unlike the CTABr solutions, which showed visual increases in viscosity at 0.1 M and 0.2 M [TMABr], consistent with the formation of rod-like micelles. The results in Figure 10 were obtained by the same procedure as those for CTACl in Figure 8, except that the values of β used were 0.7 and 0.75 for CTPABr and CTABr, respectively, Eq. (1). For both CTPABr and CTABr there are

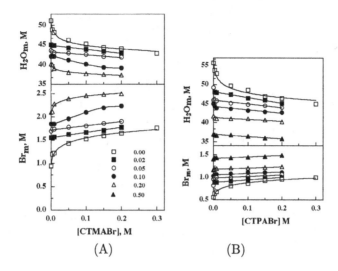

Fig. 12.9. Change in Br_m and H_2O_m at 40°C and 0.001 M HBr in CTABr and TMABr (A) and CTPABr and TMABr (B). Lines drawn to aid the eye. Reproduced with permission of the American Chemical Society.[3]

initial rises in Br_m that we attribute to the dehydration of hydrocarbon in the interfacial region. After the initital rise, most of the data at elevated salt concentrations for CTPABr fall on a line with the slope of 1.0. The steady increase in Br_m is caused by Br^-, and probably TMA^+, entering the interfacial region and displacing water. However, for CTABr, the initial rise is followed by a jump in Br_m at about 0.1 M $[Br_w]$ and a concomitant decrease in H_2O_m without a plateau region. Pairing of CTA^+ headgroups and Br^- releases water into the aqueous pseudophase and rod-like micelles are formed.

Although parts of the above interpretation in CTACl, CTABr and CTPABr micellar solutions are tentative, the most important conclusions from these results are three: (a) the interfacial Cl^- and Br^- concentrations as measured by chemical trapping, Cl_m and Br_m are essentially continuous functions of $[Cl_w]$ and $[Br_w]$, which makes sense because the interfacial concentrations of these ions should be proportional to their concentrations in the aqueous phase whether added as surfactant or salt; (b) the values of Cl_m and Br_m increase and H_2O_m decreases at the reported sphere-to-rod transition concentrations for these two ions; and (c) these changes are consistent with the ion-pair/dehydration model for the sphere-to-rod transition.

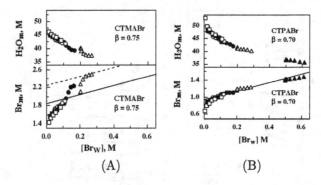

Fig. 12.10. Change in Br_m and H_2O_m verus aqueous Br^- at $40°C$ and 0.001 M HBr in CTABr and TMABr (A) and CTPABr and TMABr (B). $[Br_W]$ values were obtained by using Eq. (1). The solid lines and dashed line have a slope of 1 (one). Reproduced with permission of the American Chemical Society.[3]

12.6. Discussion

The primary focus of this review has been to show that the ion-pair/hydration model provides a natural and consistent explanation for specific ion effects on the properties of ionic micelles and by implication on the properties of other ionic association colloids such as microemulsions and vesicles. The basic components of the model are:

- the interfacial regions of ionic micelles have the same properties as concentrated salt solutions containing the same ions (or model ions, e.g., TMA^+ for the trimethylammonium headgroup of cationic surfactants) at the same molarity;
- ion-pair formation results in the release of water of hydration and an increase in entropy of the system;
- the tendency to form ion-pairs increases with ion size and depends on the free energies of hydration of the free and paired ions and the specific interactions between cation and anion in the pair; and
- the sphere-to-rod transition is induced by addition of surfactant and/or counterion (as salt) to a solution of spherical micelles and shifts the free energy minimum for the most stable structures from spheres to rods.

The implications of these results for structural transitions of micelles and other association colloids are multiple:
- increasing the temperature should shift the ion-pair equilibrium to-

ward that of free ions and should lead to a rod-to-sphere transition with increasing temperature as has been observed for alkali-metal salts of alkyl sulfates;[64]
- paired organic counterions with headgroups (*e.g.*, aromatic tosylates, carboxylates, *etc.*) are less hydrated (more hydrophobic) than the free organic counterions and headgroups and the average location of the free ions may be more toward the micellar core than the free ions. This effect should shift the equilibrium in favor of the paired ions; an effect that cannot occur with the model compounds in water.
- increasing the surfactant chain length increases the hydrophobic effect and, for surfactants of the same headgroup and counterion, the free energy minimum of the aggregates will shift toward larger micelles (higher aggregation numbers), with a concomitant increase in ion-pairing and decrease interfacial hydration; and
- the same logic should hold for other structural transitions, e.g., rods to lamellae or vesicles because these transitions occur at higher stoichiometric surfactant concentrations and lower stoichiometric water concentrations and the interfacial region should be less hydrated and the surfactant packing in the aggregates will be tighter.

12.7. Conclusions

These results illustrate the surprising richness of information that comes from the study of micellar solutions. Because a delicate balance-of-forces controls micelle formation, changes in micellar sizes and shapes can be induced by modest changes in headgroup structure, counterion type and concentration. Conversely, the interfacial regions of micelles are a "laboratory" for studying weak intermolecular and inter ionic interactions. Small changes in free energies of interaction per surfactant monomer produce dramatic and easily observed changes in aggregate structure because the highly cooperative structural transitions in each micelle involve hundreds to thousands of molecules.

More generally, the ion-pair/hydration model suggests that the net curvature of the surfaces of aggregates is determined both by the nature of the headgroups and counterions and their interactions between each other and with water molecules in the interfacial region. That is, short-range forces govern the driving force for aggregation, the collective hydrocarbon molecule-water molecule interactions of the hydrophobic effect, and also the collective balancing forces in the interfacial region that govern aggregate size and shape.

The ion-pair/hydration model provides a reasonable explanation for aggregate growth, but cannot explain the equilibrium shapes of the aggregates, why rods and not disks, why vesicles and not lamellar or cubic mesophases. These differences might be caused, for example, by multiple weak crystal-like orientation interactions between headgroups and counterions in the interfacial region.[65]

Future work on the ion-pair/hydration model by chemical trapping will be directed at the understanding the spontaneous transition from micelles to vesicles on dilution by catanionic micelles, a deeper understanding of surfactant chain length and ion specific effects controlling the sphere-to-rod transition, and the temperature dependence of ion specific effects on the sphere-to-rod transitions.

Acknowledgments

I am grateful (yet again) to Clifford A. Bunton for his extremely helpful discussions about earlier versions of this manuscript, to Kim Collins for suggesting that I create a thermodynamic description of the ion-pair/hydration model, to Krishnan Gunaseelan and Yan Cai for careful proof reading, and to the Molecular Dynamics Program (CHE-0411990) and International Programs of NSF for financial support.

References

1. L. S. Romsted, in : *Reactions and Synthesis in Surfactant Systems*, Ed. J. Texter, (Marcel Dekker: New York, 2001), p. 265.
2. L. S. Romsted, in : *Adsorption and Aggregation of Surfactants in Solution*, Eds. K. L. Mittal; D. O. Shah, (Marcel Dekker: New York, 2002), p. 149.
3. L. S. Romsted, *Langmuir* **23**, 414 (2007).
4. F. Hofmeister, *Naunyn-Schmiedebergs Arch. Exp. Pathol. Parmakol. (Leipzig)* **24**, 247 (1888).
5. W. Kunz, P. Lo Nostro and B. W. Ninham, *Curr. Opin. Colloid Interface Sci.*, **9**, 1 (2004).
6. K. D. Collins and M. W. Washabaugh, *Q. Rev. Biophys.*, **18**, 323 (1985).
7. S. Berr, R. R. M. Jones and J. S. J. Johnson, *J. Phys. Chem.* **96**, 5611 (1992).
8. K. D. Collins, *Proc. Natl. Acad. Sci. U. S. A.* **92**, 5553 (1995).
9. M. G. Cacace, E. M. Landau and J. J. Ramsden, *Q. Rev. Biophys.*, **30**, 241 (1997).
10. B. W. Ninham and V. Yaminsky, *Langmuir*, **13**, 2097 (1997).
11. M. T. J. Record, W. Zhang and C. F. Anderson, *Adv. Protein Chem.*, **51**, 281 (1998).

12. B. W. Ninham, *Adv. Colloid Interface Sci.*, **83**, 1 (1999).
13. M. Bostrom, D. R. M. Williams and B. W. Ninham, *J. Phys. Chem. B* **106**, 7908 (2002).
14. M. Bostrom, D. R. M. Williams, P. R. Stewart and B. W. Ninham, *Phys. Rev. E*, **68**, 0419021 (2003).
15. D. T. Bowron and J. L. Finney, *J. Chem. Phys.*, **118**, 8357 (2003).
16. M. C. Guarau, S.-M. Lim, E.T. Castellana, F. Albertorio, S. Kataoka and P.S. Cremer, *J. Am. Chem. Soc.*m **126**, 10522 (2004).
17. M. Bostrom, W. Kunz and B. W. Ninham, *Langmuir*, **21**, 2619 (2005).
18. B. Lonetti, P. Lo Nostro, B. W. Ninham and P. Baglioni, *Langmuir*, **21**, 2242 (2005).
19. K.D. Collins and G.W. Neilson, J. E. Enderby, *Biophys. Chem.*, **128**, 95 (2007).
20. B. Jonsson; B. Lindman; K. Holmberg; B. Kronberg, *Surfactants and Polymers in Aqueous Solution*, (John Wiley & Sons, Chichester, 1998).
21. G. Savelli; R. Germani; L. Brinchi, in : *Reactions and Synthesis in Surfactant Systems*, Ed. J. Texter, (Marcel Dekker: New York, 2001), **100**, p. 175.
22. L. S. Romsted, in : *Surfactants in Solution*, Ed. K. L. Mittal; B. Lindman, (Plenum Press: New York, 1984), **2**, p. 1015.
23. A. W. Adamson; A. P. Gast, *Physical Chemistry of Surfaces*. 6th ed., (J. Wiley & Sons: New York, 1997).
24. D. F. Evans; H. Wennerstrom, *The Colloidal Domain: Where Physics*, Chemistry, Biology and Technology *Meet*. (VCH Publishers: New York, 1994), p. 515.
25. M. Bostrom, D. R. M. Williams and B. W. Ninham, *Langmuir*, **18**, 6010 (2002).
26. C. J. Cramer and D. G. Truhlar, *Chem. Rev.*, **99**, 2161 (1999).
27. C. Tanford, *The Hydrophobic Effect: Formation of Micelles and Biological Membranes*. 2nd ed., (Wiley: New York, 1980).
28. M. N. Jones; D. Chapman, *Micelles, Monolayers, and Biomembranes*, (Wiley-Liss: New York, 1995).
29. B. Lindman, H. Wennerstrom, H. Gustavsson, N. Kamenka and B. Brun, *Pure Appl. Chem.*, **52**, 1307 (1980).
30. M. J. Rosen, *Surfactants and Interfacial Phenomena*. 3rd ed., (John Wiley & Sons: New York, 2004), p. 464.
31. L. S. Romsted, *Rate Enhancements in Micellar Systems*. Ph.D., Indiana University, 1975.
32. G. Gunnarsson, B. Jonsson and H. Wennerstrom, *J. Phys. Chem.*, **84**, 3114 (1980).
33. N. J. Buurma, P. Serena, M. J. Blandamer and J.B.F.N. Engberts, *J. Org. Chem.*, **69**, 3899 (2004).
34. J. N. Israelachvili, D. J. Mitchell and B. W. Ninham, *J. Chem. Soc. Faraday Trans 2*, **72**, 1525 (1976).
35. J. Israelachvili, *Intermolecular and Surface Forces*. 2nd ed., (Academic Press: London, 1991).
36. S. Svenson, *J. Dispersion Sci. Tech.* 25, 101 (2004).
37. K. Holmberg, *Curr. Opin. Colloid Interface Sci.* **8**, 187 (2003).

38. W. J. Paugh and L. Saunders, *J. Pharm. Pharmacol.* **26**, 286 (1974).
39. Y. Marcus, *J. Phys. Chem. B* **109**, 18541 (2005).
40. Y. Marcus and G. Heftler, *Chem. Rev.* **106**, 4585 (2006).
41. O. Soderman, G. Carlstrom, U. Olsson and T. C. Wong, *Journal of the Chemical Society Faraday Transactions I*, **84**, 475 (1988).
42. A. F. Hegarty, in : *The Chemistry of the Diazonium and Diazo Groups*, Part 2, Ed. S. Patai, (John Wiley & Sons: New York, 1978), p. 511.
43. H. Zollinger, *Diazo Chemistry I : Aromatic and Heteroaromatic Compounds*, (VCH Publishers, Inc.: Weinheim, 1994), p. 1.
44. J. Keiper; L. S. Romsted; J. Yao; V. Soldi, *Colloids abd Surfs. A*; **176**, 53 (2001).
45. C. A. Bunton; L. S. Romsted, in : *Handbook of Microemulsion Science and Technology*, Eds., P. Kumar; K. L. Mittal, (Marcel Dekker: New York, 1999), p. 457.
46. Y. Geng, L. S. Romsted and F. M. Menger, *J. Am. Chem. Soc.* **128**, 492 (2006).
47. Y. Geng, L. S. Romsted, S. Froehner, D. Zanette, L. Magid, I. M. Cuccovia and H. Chaimovich, *Langmuir*, **21**, 562 (2005).
48. L. S. Romsted, J. Zhang, I. M. Cuccovia, M. J. Politi and H. Chaimovich, *Langmuir*, **19**, 9179 (2003).
49. A. Chaudhuri, J. A. Loughlin, L. S. Romsted and J. Yao, *J. Am. Chem. Soc.*, **115**, 8351 (1993).
50. V. Soldi, J. Keiper, L. S. Romsted, I. M. Cuccovia and H. Chaimovich, *Langmuir*, **16**, 59 (2000).
51. I. M. Cuccovia, M. A. da Silva, H. M. C. Ferraz, J. R. Pliego, J. M. Riveros and H. Chaimovich, *J. Chem. Soc., Perkin Trans.*, **2**, 1896 (2000).
52. T. Imae, R. Kamiya and S. Ikeda, *J. Colloid Interface Sci.*, **108**, 215 (1985).
53. T. Imae and S. Ikeda, *Colloid Polym. Sci.*, **265**, 1090 (1987).
54. S. Ozeki and S. Ikeda, *J. Colloid Interface Sci.*, **87**, 424 (1982).
55. S. Ozeki and S. Ikeda, *Bull. Chem. Soc. Jpn.*, **54**, 552 (1981).
56. S. Ikeda, in : *Surfactants in Solution*, Eds. K. L. Mittal, B. Lindman, (Plenum: New York, 1984), **2**, p. 825.
57. A. Bernheim-Groswasser, R. Zana and Y. Talmon, *J. Phys. Chem. B.*, **104**, 4005 (2000).
58. Y. Geng and L. S. Romsted, *J. Phys. Chem. B.*, **109**, 23629 (2005).
59. Y. Marcus, *Ion Solvation* (John Wiley & Sons: Chichester, U.K., 1985).
60. P. J. Kreke, L. J. Magid and J. C. Gee, *Langmuir*, **12**, 699 (1996).
61. L. J. Magid, Z. Han, G. G. Warr, M. A. Cassidy, P. D. Butler and W. A. Hamilton, *J. Phys. Chem.*, **101**, 7919 (1997).
62. S. J. Bachofer and U. Simonis, *Langmuir*, **12**, 1744 (1996).
63. M. Vermathen, P. Stiles, S. J. Bachofer and U. Simonis, *Langmuir*, **18** 1030 (2002).
64. P. J. Missel, N. A. Mazer, M. C. Carey and G. B. Benedek, *Phys. Chem.*, **93**, 8354 (1989).
65. B. Regler, J. T. Emge, J. J. Elliot, R. R. Sauers, J. Potenza and L. S. Romsted, *J. Phys. Chem. B.*, **111**, 13668 (2007).

Chapter 13

Biocatalytic Studies in Microemulsions and Related Systems

Aristotelis Xenakis

Institute of Biological Research and Biotechnology
The National Hellenic Research Foundation
48, Vassileos Constantinou Ave.
11635 Athens, Greece

Biocatalytic studies of lipases immobilized in water in oil microemulsions and related systems are presented in this review article. Microemulsion-based organogels which are produced using lecithin, AOT or surfactant-less microemulsions were used as media for lipase catalyzed esterifications. Systems consisting of hexane, a short chain alcohol and water were also used as media for such biotransformations. The factors that affect lipase activity were determined by following the esterification reactions of 1-propanol with lauric acid or some phenolic acids, which show low solubility in organic solvents and water.

Microemulsions containing lecithin or AOT as well as surfactantless microemulsions, were used for producing gels of biopolymers such as hydroxypropylmethyl cellulose. These organogels constitute a suitable matrix for lipase immobilization and can be used for the catalytic synthesis of fatty acid esters as they are stable in various organic solvents. The immobilized lipases presented an enhanced activity in comparison with the same lipases immobilized in microemulsions, while they also retained their catalytic ability for longer time periods.

In addition, the external organic solvent used for the esterification reactions that are catalyzed in the organogels, was replaced by supercritical CO_2. Lipase retained catalytic activity whereas the kinetic study confirmed the above mechanism. Lipase stability and biocatalyst's reusability were determined.

13.1. Introduction

Microemulsions are systems that can be successfully used as media for hosting various biocatalytic reactions.[1-3] Enzymes can be encapsulated in the water core of reverse micelles catalyzing reactions of both hydrophilic and hydrophobic substrates.[4,5] However, their drawback is the difficulty in biocatalyst separation and product isolation, which is due to the presence of surfactants. The optimization of such systems can lead to many applications within the field of Industrial Biotechnology. The present chapter presents the use of microemulsion based systems as effective biocatalysts for performing synthetic reactions leading to high added value products. The use of alternative systems without surfactants, the transformation of the biocatalytic medium to gel-like solid systems and finally the use of solvent free systems are presented.

13.2. Enzymes in Microemulsions

Microemulsions are homogeneous systems of water and oil stabilized by surfactants.[6] Depending on the relative quantities of the two major components, water and oil, they can be distinguished in oil in water microemulsions or in water in oil ones.[7] In the later case the water phase is dispersed in the oil in form of tiny reverse micelles formulated by a monolayer of surfactant molecules (Fig. 13.1).

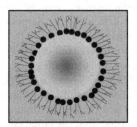

Fig. 13.1. Model of a reverse micelle with water droplets of about 10 - 100 nm

The main features of these colloidal dispersions are the following:

- They are thermodynamically stable
- They are optically isotropic allowing thus spectroscopical studies
- They form spontaneously without any needs for energy
- They offer the possibility to solubilize both hydrophilic and lipophilic substances

A major consequence of the latter property is that microemulsions represent a unique micro-heterogeneous medium of simultaneous coexistence of various compounds with different solubilities. They can, thus, act as microreactors between hydrophilic and lipophilic substrates.

Of particular interest are the water in oil microemulsions where water droplets are dispersed in an oil continuous medium. Within the water pools hydrophilic biomolecules, such as enzymes, can be hosted, whereas lipophilic substrates can be solubilized in the external continuous phase. The enzymes can catalyze the bioconversion of the substrates through the surfactant monolayer, which being of dynamic nature is easily permeable. This is illustrated in Figure 13.2, where the substrate diffuses through the membrane into the reverse micelle forming the intermediate complex with the enzyme and finally releasing the product.

Fig. 13.2. Model presentation of an enzymatically catalyzed reaction of lipophilic substrates in a w/o microemulsion

A typical example is the case of lipases.[5] The role of these enzymes in nature is to cleave the triglycerides of the fats and oils to release fatty acids and glycerol. When a lipase is in a medium with very low water content, as the reverse micellar system, it can catalyze the inverse reaction towards the synthesis of esters as shown in Scheme 1.

Esterification : $R_1COOH + R_2OH$ → $R_1COOR_2 + H_2O$
                      Fatty acid + Alcohol   →   Fatty ester
Transesterification : $R_1COOR_2 + R_3OH$ → $R_1COOR_3 + R_2OH$
                      Fatty ester 1 + Alcohol → Fatty ester 2.

Scheme 13.1. Synthetic reactions catalysed by lipase

The important point of these synthetic reactions is the fact that they cannot occur in homogeneous aqueous environments because of the low solubility of the fatty substrates in water and mainly because of a particularity of the lipase molecule: the active site of this enzyme is blocked by a "lid" which can get in the open conformation only in presence of a hydrophobic medium.[8] This phenomenon is called interfacial activation and is favored in systems owing important interfaces as in the case of water in oil microemulsions.

Many synthetic reactions have been carried out in various microemulsions formed with different surfactants such the anionic AOT, the cationic CTAB, the nonionic $C_{12}EO_4$ and the zwitterionic lecithin.[5] The latter is of particular interest because of its biocompatibility as it is a natural product.

Nevertheless a serious drawback of the microemulsions to be considered as ideal media for performing biocatalytic conversions at an industrial scale is the difficulty to isolate the product keeping intact the biocatalyst. Many procedures have been suggested to overcome this serious handicap but in most cases the activity of the enzyme suffered important losses. One alternative could be the case of the so-called microemulsion-based organogels (MBGs) presented in the next paragraph.

13.3. Catalytic Studies in Microemulsion-based Organogels (MBGs)

By mixing an aqueous solution of a biopolymer, such as gelatin, agar or cellulose derivatives, with a water in oil microemulsion a gel - like solid is obtained.[9] This microemulsion-based organogel can be used as a handy enzyme immobilizing matrix to perform biocatalytic conversions.[10,11] The substrates are transferred with an adequate solvent to the organogel, the bioconversion is carried out after the diffusion of the substrates to the encapsulated enzymes and finally the product is easily isolated whereas the MBG with the biocatalyst can be reused (Fig. 13.3).

Fig. 13.3. Enzymatic bioconversion in a microemulsion based gel (MBG)

The system is based on a matrix of the biopolymer with entrapped droplets of the water in oil microemulsion in the core of which resides the enzyme.

The solvent to be used for carrying the substrates to the enzyme should not affect the composition of the MBG neither alter its texture. From a series of solvents tested the more adequate were found to be the more apolar ones such isooctane. Interestingly, some alcohols with relatively long aliphatic chains can be applied with the advantage to be considered also as a substrate, avoiding thus the use of an additional solvent.

The model reaction of esterification was shown to proceed in the MBG - lipase systems in a quite comparable pattern as in the case of the relative simple microemulsions. The major advantage of these systems was the remarkable stability of the biocayalyst and the possibility to recycle the system.

Regarding the enzymatic reaction itself in the MBGs, it was shown that it depended linearly on the lipase concentration proving that the rate limiting step is the catalytic reaction and not other parameters such as the substrates' diffusion.

As the system allows the controlling of the substrate concentrations a complete kinetic analysis was carried out leading to the determination of the kinetic constants $V_{\max}$, K_m and K_I as the reaction obeyed to the Ping Pong Bi Bi mechanism encountered in almost all cases of lipase catalyzed synthetic reactions.[12] Furthermore, the stability of the immobilized lipases as well as the reusability of the biocatalyst was studied, since they are both factors that determine the practical usability of the system. The study showed that the immobilized lipases present an enhanced activity in comparison with the same lipases immobilized in microemulsions, while they also retain their catalytic ability for longer time periods.

13.4. Catalytic Studies in Microemulsion-like Surfactant-free Systems

Another type of media that can be considered for biocatalytic applications is the ternary systems of oil, short-chained alcohol and water, also called as surfactantless microemulsions.[13] These systems are homogeneous solutions of oil and water just as microemulsions but in absence of a typical surfactant. Some structural studies on these systems applying the fluorescence energy transfer revealed the existence of distinct confined water domains.[14] Enzymes can be solubilized in such systems and retain their catalytic ability.[15]

Scheme 13.2. Typical esterification reaction of a phenolic acid with an alcohol.
$R_1 = R_2 = H$: cinnamic acid;
$R_1 = OH, R_2 = H$: p-coumaric;
$R_1 = OCH_3$ and $R_2 = OH$: ferulic acid

A typical example is the use of lipases to catalyze esterification reactions. The system of hexane/1-propanol/water was shown to be adequate for such reactions in a similar way to the classic microemulsions formulated with surfactants such as AOT, CTAB or lecithin.[16] The advantage of the surfactant-free systems is the simplification of the product recovery procedure which can be accomplished by simple extractions.

Furthermore, these systems were shown to be well suited for carrying out esterification reactions of phenolic acids. It is well known that phenolic acids are natural antioxidants but their relatively low solubility both in water and in oils inhibits their use as additives in various food or cosmetic formulations. By esterifying the phenolic acids with medium chained alcohols their hydrophobicity is increased making possible their incorporation in oil rich products (margarines, creams etc.) (Scheme 13.2).

Ternary surfactant-free systems were also used to form MBGs using HPMC or agar as gelling agents in a similar way as described above. As the systems are quite rich in alcohol, the same alcohol was considered as a substrate. Alternatively, the non-reactive t-butanol was used to form the systems, allowing longer alcohols such as octanol, to be used as substrate and external solvent simultaneously.

13.5. Catalytic Studies in MBGs in scCO$_2$

Finally, the use of supercritical CO_2 was examined as a medium to carry out lipase catalyzed reactions. CO_2 can be at its supercritical state when the temperature is higher than $31.5°C$ and the pressure over 90 bars, conditions that can be easily attained using pressure cells. Supercritical CO_2 is a fluid with remarkable solubilizing capacities and has been shown to be an adequate medium for enzymatically catalyzed conversions.[17,18]

In the present case MBGs formulated with either AOT or lecithin as surfactant with encapsulated lipase were used as biocatalysts for carrying out esterification reactions of fatty acids and alcohols. The kinetic study of the same model reaction of lauric acid esterification by propanol, confirmed the above mechanism. In order to determine the physicochemical parameters that affect the esterification synthesis in organogels in $scCO_2$, the effect of pressure was studied. In addition, lipase stability and biocatalyst's reusability were determined. The determination of the same parameters in isooctane as an external solvent was used for comparison reasons.[19] Table 13.1 shows the comparative results obtained for the esterification of exanoic acid with 1-octanol in $scCO_2$ and isooctane.

Table 13.1. Conversion yields of lipase (from *C. antarctica* and *R. miehei*) catalyzed esterification of exanoic acid with 1-octanol in $scCO_2$ and isooctane.

Solvent	Conversation %	
	CaL	RmL
sc CO_2	47.95	4.23
Isooctane	57.89	5.59

13.6. Conclusions and Outlook

This article has presented the various approaches that can be undertaken to achieve biocatalytic conversions of hydrophobic substrates using lipases immobilized in water in oil microemulsion based systems. The common basis of all applied systems is the encapsulation of the biomolecule in a water restricted medium in the vicinity of a hydrophobic environment. This situation allows the lipase molecule to be in its active conformation and catalyze the reverse reactions towards synthesis. The criteria for adopting a medium for such a bioconversion comprise the effectiveness of the system, its stability against enzyme denaturation, its resistance to external solvent application, its reusability, and the ease of product recovering.

References

1. P. L. Luisi and L. Magid, *CRC Crit. Rev. Biochem.* **20**, 409 (1986).

2. K. Martinek, A. V. Levashov, N. L. Klyachko, Y. L. Khmelnitsky and I. V. Berezin, *Eur. J. Biochem.* **155**, 453 (1986).
3. M.A. Biasutti, E.B. Abuin, J.J. Silber, M. Correa, E.A. Lissi, *Adv. Colloid Interface Sci.* **136**,1 (2008).
4. B. Orlich and R. Schomäcker, in *History and Trends in Bioprocessing and Biotransformations, Adv. Biochem. Engin. Biotechnol.* Ed. T. Scheper, Springer, Berlin **75**, 185 (2002).
5. H. Stamatis, A. Xenakis and F. N. Kolisis, *Biotechnology Advances* **17**, 293 (1999).
6. L. Danielsson and B. Lindman, *Colloids and Surfaces*, **3**, 81, (1982).
7. S. P. Mulik and B. K. Paul, *Adv. Colloids Interface Sci.*, **78**, 99, (1998).
8. L. Brady, A. M. Brzozowski, Z. S. Derewenda, E. Dodson, S. Tolley, J. P. Turkenburg, L. Christiansen, B. Huge-Jensen, L. Norskov, L. Thim, U. Menge, *Nature* **343**, 767 (1990).
9. G. D. Rees, M. G. Nascimento, T. R. J. Jenta and B. H. Robinson, *Biochim. Biophys. Acta* **1073**, 493 (1991).
10. H. Stamatis and A. Xenakis, *J. Mol. Catalysis B*, **6**, 399 (1999).
11. A. Pastou, H. Stamatis and A. Xenakis, *Progr. Colloid Polym. Sci.*, **115**, 192 (2000).
12. C. Delimitsou, M. Zoumpanioti, A. Xenakis and H. Stamatis, *Biocatal. Biotranf.*, **20**, 319 (2002).
13. Y. L. Khmelnitsky, A. vanHoek, C. Veeger and A. J. W. G. Visser, *J. Phys. Chem.*, **93**, 872 (1989).
14. M. Zoumpanioti, H. Stamatis, V. Papadimitriou and A. Xenakis, *Colloids and Surfaces B: Biointerfaces*, **47**, 1 (2006).
15. Y. L. Khmelmtsky, R. Hilhorst, C. Veeger, *Eur. J. Biochem.*, **176**, 265 (1988).
16. M. Zoumpanioti, M. Karali, H. Stamatis and A. Xenakis, *Enzyme Microb. Technol.* **39**, 531(2006).
17. A. Ballesteros, U. Bornscheuer, A. Capewell, D. Combes, J. S. Condoret, K. Koenig, F.N. Kolisis, A. Marty, U. Menge, T. Scheper, H. Stamatis and A. Xenakis *Biocatal. Biotranf.*, **13**, 1 (1995).
18. A. Marty, W. Chulalaksananukul, J. S. Condoret, R. M. Willemot and G. Durand, *Biotechnol. Bioeng.*, **39**, 273 (1992).
19. C. Blattner, M. Zoumpanioti, J. Kroner, G. Schmeer, A. Xenakis and W. Kunz, *J. Supercritical Fluids* **36**, 182 (2006).

Chapter 14

Colloidal Dispersions for Drug Delivery

Syamasri Gupta

Centre for Surface Science
Department of Chemistry
Jadavpur University
Kolkata 700032, India

Colloidal dispersions are recently considered as potential vehicles for drug and gene delivery. These dispersed systems can accommodate a variety of molecules due to their intrinsically compartmentalized nature with polar and non-polar regions separated by definite interfaces. Some drugs perform better when administered *in vivo* with drug delivery systems. Liposome, noisome, nanoparticle, microemulsion, organogel, hydrogel and polymeric micelle are some such systems which have shown good potentiality for drug delivery through improved solubilization, enhanced uptake, altered pharmakokinetics and biodistribution and reduction in toxic side effects of the drugs as well. The present review briefly describes the studies undertaken on these systems as drug delivery vehicles and their applicability with reference to drugs and their specificities.The prospects of these dispersed systems for pharmaceutical uses have been addressed. Futuristic aspects of these delivery systems have also been indicated.

14.1. Introduction

Twentieth century has witnessed significant growth in drug development. The new drugs, mostly synthetic organic compounds, are effective in curing many deadly diseases, but the harmful side effects of the drugs often lead to major damage to vital organs of the body,[1] which are sometimes irreversible. The lipophilic drugs when administered *in vivo* in the form of tablets and suspensions cause toxicity due to incomplete absorption and improper metabolism and biodistribution.

Encapsulation and delivery of the lipophilic drugs with aqueous based drug delivery systems could ensure improvement in absorption and decrease the toxic side effects through enhanced bio-availability and reduced dose frequency.

Over the past few decades colloidal systems have been explored as potential drug delivery systems because of their structured hydrophobic domains (cavities) where the drug molecules can snugly reside. The dispersed systems studied for the delivery of different kinds of drugs are liposome, niosome, microsperes, nanoparticles microemulsions, hydrogels, organogels, lipid emulsions, etc. These systems often lead to improvement in the efficacy of the lipophilic drugs through increased solubilization and modification of their pharmacokinetic profiles. They alter the plasma clearance kinetics, tissue distribution, metabolism and cellular interaction of the drugs.[2-12] For pharmaceutical use, tolerance towards additives, stability over a wide range of temperature range, low viscosity, small size, biodegradability and easy elimination from the body are some of the essential criteria. For intravenous administraion, the size of the delivery system needs to be controlled to avoid capillary blockage and hence sub-microen sized entities are preferred.[13] The exipients that could be used are, cholesterol, phospholipids, biocompatible amphiphiles,esters and ethers of fatty acids, bile salts, vegetable oils and biodegradable polymers.[14] The dispersed systems are mostly structured aggregates, formed when amphiphilic molecules (Fig. 14.1A) self aggregate in aqueous medium due to hydrophobic effect. Depending on the chemical composition, length of hydrophobic tail, polarity of the headgroup, nature(polarity) of the dispersion medium and the interplay of different forces, these aggregates can assume a wide variety of structures

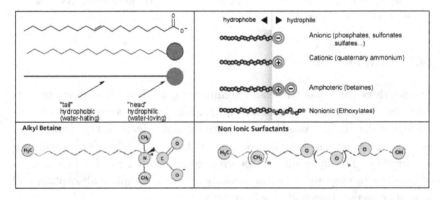

Fig. 14.1A. Amphiphiles are the building blocks of the organized systems.

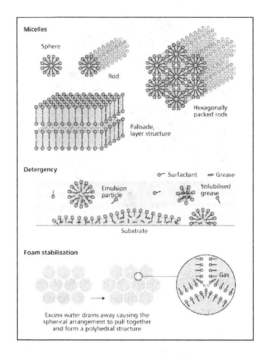

Fig. 14.1B. Examples of different organized structures.

(Fig. 14.1B); viz., spherical, oblate, bilayer, multilayer, lamellar, vesicular, rod shaped, liquid crystalline, etc. These systems are compartmentalized with well-defined hydrophilic and hydrophobic regions where both nonpolar and polar compounds can be accommodated and are useful for encapsulation and delivery of drugs.

The present article gives an overview of the commonly studied colloidal dispersions for drug delivery with subsections on liposome, niosome, microsphere, nanoparticle, microemulsion, hyrogel and organogel.

14.1.1. *Liposomes*

Liposomes are lipid vehicles consisting of one or more concentric phospholipid bilayers with alternating aqueous and lipophilic compartments. They are vesicles of lecithin (phospholipid) and cholesterol, composed of naturally-derived phospholipids with mixed lipid chains (like eggphosphatidylethanolamine), or of pure surfactant components like dioleoyl phosphatidylethanolamine (DOPE) (Fig. 14.2A).[15,16]

Fig. 14.2A. Liposome, with sectional view of the interior.

Liposomes are classified according to structural features as large multilammelar vesicles (LMVs), small unilamellar vesicles (SUVs) and large unilamellar vesicles (LUVs) depending on their size and the number of concentric lipid bilayers. These vesicles are capable of entrapping polar compounds within the aqueous domains or non polar compounds in the lipid bilayers. They have good bioacceptability, poor shelf life, and their size lies in the range of 100-1000nm. Due to the presence of ester bonds, phospholipids are easily hydrolysed.[17] This can lead to phosphoryl migration at low pH. Another type of degradation is the peroxidation of unsaturated phospholipids.[18] Unreliable reproducibility arising from the use of lecithins in liposomes sometimes leads to problems. Liposomes have been tested for encapsulation and delivery of a variety of drug molecules through intravenous, intraperitonial, intramuscular and subcutaneous routes.[19-21]

While liposomes administered through intravenous route are rapidly cleared from the circulation by cells of the mononuclear phagocyte systems (MPS) in liver, lung and spleen, those administered by the other modes remain in circulation for longer period of time. More recently other techniques are in use so as to by pass recognition by the MPS by coating liposomal surface with polyethtylene glycols. The pegylated liposomes represent a new class of delivery system to significantly improve the efficacy of anticancer drugs.[22,23] These PEG coated liposomes, the "Stealth liposomes" are thus designed to improve efficacy and reduce toxicity of drugs with site specific targeting.[24-26] Independent of the route of administration, the *in vivo* fate of the liposomes and the phrmacokinetics of the encapsulated drug are mainly determined by liposome size, lipid composition (surface hydrophobicity/hydrophilicity and net charge) and lipid bilayer fluidity.[27]

"Long circulating liposomes" are more useful in extravasating tumour tissues. These modified liposomes can penetrate cancer tissues in mice more efficiently.[28] Reduction of cancer mass and increased survival have been reported for Dauxorubicin encapsulated in PEG coated liposome in mouse mammary carcinoma model,[29] and for Epirubicin in mouse colon tumour *in vivo*.[30] Conventional Amphotericin B is the drug of choice for treating systemic fungal infection. Tolerance of high doses of Amphotericin B was achieved by infusion of a liposomal formulation. This new form of administration was developed in order to lower the acute side effects and to offer the possibility of administering high doses of Amphotericin B.[31] PEGlycated liposomal Doxorubicin has shown substantial efficacy in breast cancer treatment both as monotherapy and in combination with other chemotherapeutics.[32] Treatment with the drug, often leads to congestive heart failure and death. Liposomal encapsulation of the drug brings a beneficial change in Doxorubicin biodistribution which provides local accumulation of the drug to the tumor tissues without any loss in the therapeutic activity.[33] Tumor specific drug delivery has become increasingly important in cancer therapy where receptor-mediated liposomes are developed with suitable "anchor" to bind to a partcular type of cancer cell.[34–39] Multifunctional stimuli-sensitive pharmaceutical nanocarriers were developed which were capable of responding to certain local stimuli, such as decreased pH values in tumors or infarcts. These were the targeted pharmaceutical carriers, the PEGylated liposomes and PEG-phosphatidylethanolamine (PEG-PE)-based micelles.[40] Recent advances with liposomes as pharmaceutical carriers has been explicitly discussed by Torchillin,[41,42] Banerjee[43] and Basu.[44]

14.1.2. *Niosome*

Niosomes are vesicles mainly consisting of nonionic surfactants (Fig. 14.2B).[45]

These synthetic surfactants possess greater chemical stability than the phospholipids. Niosomes are prepared from several classes of synthetic nonionic surfactants polyglycerol alkyl ethers,[46] glucosyl dialkyl ethers, crown ethers,[47] polyoxyethylene ethers, and esters.[48] Often a charged surfactant is intercalated in the bilayers in order to introduce electrostatic repulsion in the vesicles, thus increasing their stability. Niosomes have poor bio-

Fig. 14.2B. A cross sectional view of Niosome.

compatibility and their size lie in the range of 100-1000nm. Apart from conventional spherical shapes, vesicles of different shapes such as tubular, polyhedral can be prepared by changing the composition. These vesicles show different release profile of the encapsulated drug from the spherical ones.[49,50] Anticancer therapy can be made more effective by targeting the delivery of the drugs to the tumor site more quantitatively. Attempts have been made to activate and use macrophages to deliver niosomal Bleomycin more quantitatively to tumour site using noisome encapsulated immunomodulator muramyl dipeptide. The mean survival time of the animals increased significantly after macrophage activation. Accumulation of higher Bleomycin levels after macrophage activation exerted increased antitumour effect.[51] Niosomal encapsulated Plumbagin and Withraferin also showed macrophage activation, improved efficacy and reduced toxicity.[52,53] The effects of PEG chain length and particle sizes on the niosome surface properties, in vitro drug release, phagocytic uptake, in vivo pharmacokinetics and antitumor activity is reported by Shi et al.[54] The model antitumor drug Hydroxycamptothecin (HCPT) was encapsulated in a series of stealth niosomes prepared with the amphiphilic co-polymer of poly (methoxy-polyethylene glycol cyanoacrylate-co-n-hexadecyl cyanoacrylate) (PEG-PHDCA), PEG(Mw varying between 2,000-10,000). The tumor inhibition rate of PEG 5000-PHDCA niosomes (92.5nm) at a dose of 2mg/kg was five times that of HCPT injection at 4mg/kg. The influence of drug thermodynamic activity and nisome composition, size, lamellarity and charge on the (trans)dermal delivery of Tretionin (TRA) showed that drug delivery is strongly affected by vesicle composition and thermodynamic activity of the drug. Particularly, small negtively charged niosomal formulations, saturated with the drug show higher cutaneous drug retention than both liposomal and commercial formulation.[55] Carotinoids, with preventive effects on cancer and atherosclorosis are difficult to solubilize due to its hydrophobicity and are unstable when exposed to sunlight or oxygen. But when β-carotine is solubilized and stabilized in niosomal formulation of non-ionic surfactant and cholesterol and delivered to cultured cells at concentrations spanning in the range of physiological levels, the carotinoid is extremely stable in the culture medium up to 96 hours. It was easily taken up by both immortalized and transformed cells at carotinoid concentrations 0.1 - 2μM.[56] A novel noisome, the Bola - niosomes made of α, ω- hexadecyl-bis-(1-aza-18- crown -6) (Bola surfactant)- span 80-cholesterol(2:3:1) molar ratio was studied for the topical delivery of the natural product Ammonium glycyrrhizinate with anti- inflammatory activity. The confocal laser scanning microscopy showed that the encapsulation of the drug in the Bola-niosomes significantly improved its intracellular uptake than the free drug.[57] Improved selectivity and specificity of the immuno niosomes, composed of Sorbitan monostearate (Span60), polyoxyethylene sorbitan monostearate (Tween 61), cholesterol, and dicetyl phosphate conjugated with a purified monoclonal antibody CD44 (IM7) through a cyanuric chloride (CC) linkage on the head group of the Tween 61 molecule was observed over the control when incubated with synovial lining cells expressing CD44. The findings indicate that immunoliposomes offer an effective method for targeted drug delivery.[58] Highly stable niosomes, prepared with Span 80, PEG 400 and water is reported by Hua and Liu.[59] The niosomes were stable over one year and was of diameter 100-180nm. PEGylated cationic liposomes, composed of Cholesterol, PEG2000-DSPE and the non-ionic surfactant - SPAN (300nm) improves the stability and cellular delivery of oligo nucleotides. PEG modification significantly decreases the binding of serum protein and prevents particle aggregation in serum. Compared to cationic liposomes, the PEGylated niosomes showed a higher efficiency of oligonucleotide uptake in serum.[60]

14.1.3. *Microspheres and Nanoparticles*

Nanoparticles consist of either a polymeric matrix (nanospheres) or of a reservoir system in which an oily or aqueous core is surrounded by a thin polymeric shell (nanocapsules) These nanospheres are polymeric aggregates of poly (methyl acrylate), poly (alkyl cyanoacrylate), poly (methylidene malonate) and polyesters such as poly (lactic acid), poly (glycolic acid) and their copolymers. The drug can be firmly adsorbed at their surface, entrapped or dissolved in the matrix.

Fig. 14.2C. A cluster of Nanoparticles.

These delivery systems have good shelf-life, poor biocompatibility and their size lie in the range of 10-1000nm. Depending on the process used for the preparation of nanoparticles, nanospheres or nanocapsules can be obtained (Fig. 14.2C).

Microspheres are suited for the delivery of a wide variety of compounds into the human body and they have been used successfully for fast and effective delivery. The advantages of nanoparticles used as drug carriers are high stability, high carrier capacity, choice of various routes of administration, including oral application and inhalation. They can be designed to allow controlled (sustained) drug release from the matrix.[61]

Improved intracellular delivery of a lipophilic derivative of a muramyl dipeptide analog could be achieved by means of nancapsules of poly (DL Lactide).[62] The nanocapsule form was more effective than the free drug in activating rat alveolar macrophages for a cystostatic effect towards syngeinic tumor cells. The increased efficacy is due to improved intracellular delivery by phagocytosis of nanocapsules. Nanoparticles of a highly soluble macromolecular drug, Heparin were formulated with two biodegradable polymers and two non-biodegradable polymers by double emulsion and solvent evaporation method. The encapsulation efficiency and Heparin release profiles were studied as functions of the type of polymers employed (alone or in combination) and the concentration of Heparin. High drug entrapment was observed in the non-biodegradable polymers compared with the biodegradable ones. Combination of two types of polymers enhanced the efficiency further, however, the in vitro drug release was not modified and remained low.[63] Encapsulation of insulin was possible in poly (ethyl-2-cyanoacrylate) nanocapsules. These nanocapsules were prepared of medium chain triglycerides and a mixture of polysorbate -80 and sorbitan mono-oleate by interfacial polymerization and charactertized by physicochemical methods. Entrapment and release of insulin from the nanocapsules were determined. The method is an useful one for entrapment of bioactive peptides.[64] Microspheres made from biodegradable polymers such as poly(glycolic acid) (PGA), poly (lactic acid) (PLA), and their co-polymers are particularly attractive for drug delivery because they degrade via hydrolysis into non-toxic byproducts, lactic and glycolic acids. Microspheres made from PGA, like most biodegradable microspheres are more attractive in drug delivery than lipid vesicles because of their great flexibility in design which allows high control over the rate of drug release. Administration of anti inflammatory drugs using these microspheres are very useful to avoid the gastrointestinal side effects and transport barriers commonly encountered with administration of anti-inflammatory drugs for the treatment of chronic diseases which require prolonged treatment with drugs.[65] Circulation half life of the drug Methotrexate (a drug for cancer and autoimmune diseases) in lipid nano particles, (lipoprotein-mimicking biovectorized systems LMBVs) was enhanced and the encapsulated drug exhibited capability to accumulate in tissues for longer periods and thus could be useful for targeting and systemic controlled release required for treatment of various types of cancers.[66] Diclofenac sodium is a non-steroidal anti-inflammatory drug with pronounced analgesic and antipyretic properties. Its half-life in plasma is only 1−2 hrs and it also causes adverse side effects such as bleeding, ulceration or perforation in the intestinal wall. Due to the short biological half life and the side effects, it is an ideal drug for controlled drug release formulations. The in *vitro* release kinetics of the drug from chitosan (a non toxic and bioadsorbable natural poly-aminosaccharide having structural similarities with glycosaminoglycans) beads are slower in comparison to the microgranules. The percent and amount of drug release were much higher in acidic solution than in basic solution (probably due to the swelling properties of the matrix in acidic pH), suggesting possibility of modifying the formulations suitable for controlled drug delivery systems through the gastrointestinal tract where the drug is subjected to changing environments of acidity.[67] Chitosan based particulate delivery systems could be useful for colon targeted delivery, mucosal delivery, cancer therapy, gene delivery, topical delivery and ocular delivery.[68] Sustained release behavior of the anti-inflammatory drug Indomethacin loaded in amphiphilic diblock co-polymeric nanospheres composed of methoxypoly (ethyleneglycol) and glycolide was observed without any burst effect. These nanospheres could remarkably reduce cell damage compared with unloaded free drug.[69] Polyethylecyanoacrylate (PECA) nanoparticles coated with polyethylene glycol (PEG) are good injectable colloidal systems for the drug Amoxycillin and can evade recognition by mononuclear phagocyte systems and remain in circulation for long period of time. Phagocytosis studies showed significant difference between nanoparticles prepared in the presence and absence of PEG and demonstrated that PEG coating reduces the macrophage uptake.[70] Better immune responses have been observed after a single intranasal or intramascular administration of the protein Bovine serum albumin (BSA) from polymeric microparticles made of biodegradable polymer than free BSA. The microencapsulation has been beneficial in inducing high and long lasting systemic immune responses after a single dose by both parenteral and mucosal delivery.[71] The cytotoxicity profile of poly (isobutylcyanoacrylate) nanoparticles is affected by the design of their surface. The *in vivo* tolerance of these nanoparticles could be improved 100-fold by coating their surface with polysaccarides and haemoglobin.[72] Biodegradable nanoparticles formulated from poly (D, L-lactide-co-glycolide)(PLGA) are effective for sustained and targeted/localized delivery of drugs and gene,plasmid DNA, proteins, peptides and low molecular weight compounds to cells and tissues.[73] To increase the local concentration of Tamoxifen in estrogen receptor (ER) positive breast cancer, Chawla and Amiji[74] developed and characterized nanpoparticle formulations using poly (ϵ-caprolactone) (PCL). The maximum Tamoxifen loading efficiency was 64%. A large fraction of the administered nanoparticle dose was taken up by MCF-7 cells through non specific endocytosis. The results suggest, nanoparticle formulations of selective ER modulators like Tamoxifen would provide increased therapeutic benefit by delivering the drug in the vicinity of ER. Nanoparticle based drug delivery systems have considerable potential for treatment of tuberculosis[75] as they offer feasibility of variable route of administration including oral administration and inhalation. Nanoparticles provide sustained release of the anti-tuberculosis drugs and considerably enhance their efficacy after oral administration. Three frontline drugs Rifampin (RMP), isoniazid (INH) and pyrazinamide (PZA) co-encapsulated in poly(lactide-co-glycolide) (PLG) nanoparticles could be detected in the circulation for four to nine days. In contrast, free drugs were cleared from the plasma within 12-24 hours of administration.[76] The usefulness of biodegradable polymeric nanoparticles for drug delivery has been discussed by Soppimath et al.[77]

14.1.4. Microemulsion

Microemulsions are spontaneously forming single phase colloidal dispersions of either oil or water stabilized by an interfacial film of surfactant (s) and co-surfactant (s) (optional). (Fig. 14.2D).

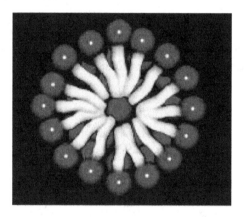

Fig. 14.2D. A schematic diagram of Microemulsion.

Systems devoid of cosurfactants are the ternary systems and those requiring cosurfactants are the "pseudoternay" systems (where the surfactant and cosurfactant are together taken as a single component). The surfactants are amphphilic molecules with a polar head and a non-polar (hydrophobic tail) and the cosurfactants can be short chain alcohols, amines and similar substances. The dispersions are formed when oil, water and surfactant/cosurfactant are mixed in appropriate proportions. Development, characterization and biological studies on "biocompatible microemulsion" as potential vehicles for drug delivery constitute a thrust area of research as they satisfy most of the required criteria of drug delivery systems.[78-99] Depending on composition and type of amphiphile, there may be dispersion of oil droplets in water continuum or vice versa. In some systems, on increasing the water volume fraction, a transition from a water-in-oil (w/o) to oil-in-water (o/w) microemulsion can occur through a bicontinuous state wherein the swollen micelles and swollen reverse micelles constantly interchange between themselves forming interspersed regions of oil and water with undefined structures. Because of their internal heterogeneity, the microemulsion systems have dual solubilizing capability of both polar and non polar entities. These self-assembled systems have low viscosity, ultralow interfacial tension, enormous interfacial area, good shelf-life (stability with time), high solubilizing capacity, macroscopic homogeneity and microscopic heterogeneity. The encapsulated drugs are protected from enzymatic degradation by the interfacial layer[100] and their membrane permeability is facilitated due to the presence of surfactants and cosurfactants.[101] Oil-in- water microemulsions have been proposed as aqueous based vehicles for delivery of a range of drugs.[102-109] Microemulsions used as vehicles for topical, dermal and transdermal admistration are relatively well studied than those used for oral, parenteral and other modes of delivery.[14]

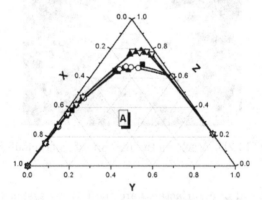

Fig. 14.3A. Pseudo-ternary for different systems at 303K. (■) corn oil/Brij - 92 + iPrOH (1:1 v/v)/water; (o) cottonseed oil / Brij-92+iPrOH (1:1 v/v) / water; (▲) corn oil / Brij-30+iPrOH (1:1v/v) / water; (△) cottonseed oil / Brij-30+iPrOH (1:1 v/v) / water. In the diagrams, the scale magnitude is $1/100^{th}$ of the actual. The axis representations are, X = oil; Y = water and Z = (surfactant + cosurfactant).

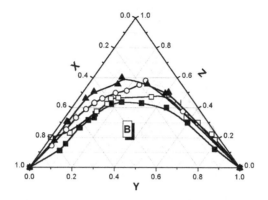

Fig. 14.3B. Ternary and pseudo ternary phase diagrams for different systems at 303K. (■) clove oil/ Tween-20/ water; (□) orange oil/ Tween-20/ water; (○) clove oil / Tween-20 / EtOH (1:1 v/v)/water; (△) orange oil /Tween-20 + EtOH (2:1v/v) / water. Scale magnitude and axis representation as in Fig. 14.3A.

Since the drug delivery system should be biocompatible, the choice of excipients is restricted. Plant and vegetable oils (corn oil, cottonseed oil, orange oil, clove oil, peppermint oil, eucalyptol oil and coconut oil) (Fig. 14.3) triglycerides and esters of fatty acids, isopropyl myristate, ethyl oleate etc have been used as the oil components for preparation of biocompatible microemulsions.[86] The commonly used surfactants are aerosol OT, polysorbates, alkyl poly ethers and sorbitan mono esters[94–96] there the use of short chain alcohols as co-surfactants is generally considered undesirable[14] Phospholipids are probable replacements for synthetic surfactants.[89,110–115]

A number of novel plant oil derived microemulsion systems as potential drug delivery vehicles have been reported by Gupta et al.[86] In them, corn oil, cottonseed oil, clove oil, orange oil and peppermint oil were dispersed in water continuum with Tween-20, Brij-30, and Brij-92 as surfactants and ethanol (EtOH) and isopropyl alcohol (i-PrOH) as co-surfactants. The ternary systems produced larger microemulsion forming zones than the pseudoternary systems. The prepared systems had a shelf life of one year, and they could withstand temperature variations in the range of 4-40°C.

Colloidal drug delivery systems accumulate and are taken up non-specifically by the reticuloendothelial systems (RES).[4] So the drugs to be targeted to the macrophages of the RES are expected to yield good results when they are encapsulated in micro emulsions and administered intra-

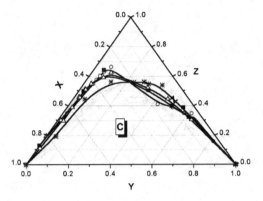

Fig. 14.3C. Ternary and pseudo-ternary phase diagrams for different systems at 303 K. (★) peppermint oil+IPM (1:1v/v)/ iPrOH/water; (o) peppermint oil/Tween-20/water; (■) peppermint oil / Tween-20 + iPrOH (1:1v/v) / water; (△) peppermint oil / Tween-20 + EtOH (2:1 v/v) / water; Scale magnitude and axis representation as in Fig. 14.3A.

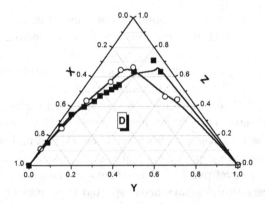

Fig. 14.3D. Pseudo-ternary phase diagrams for he ternary systems at 303 K. (■) IPM/ iPrOH / water; (o) peppermint oil / iPrOH/water. Scale magnitude and axis representation as in Fig. 14.3A.

venously. Experimental Leishmaniasis was used as a model disease where the amastigotes of the causative parasite *leishmania donovani* resides and proliferates within the macrophages of the RES. The novel microemulsion system of clove oil and Tween-20 was simple and very stable (temperature and salt resistant) system with good potentiality for drug delivery. Natural product quercetine[91] (with antileishmanial property) was solubilized and encapsulated in clove oil/Tween 20 (5/30/65 wt% composition) with reasonably low droplet dimensions (10-20nm). The efficacy of the drug in the vehicle against leishmaniasis in hamster models showed considerable improvement compared to maceration in DMSO. Dissolution of a plant derived natural product Bassic acid[4] (unsaturated tri terpene acid with antileishmanial properties) in the same clove oil-based microemulsion composition was effective both *in vitro* and *in vivo*. The *in vivo* activity of the bioactive compound was evaluated and compared in hamster models of visceral leishmaniasis both in free form as well as incorporated in two different delivery systems, viz. microemulsions and poly-lactide nanoparticles. A model drug diospyrin[93] (a highly lipophilic bioactive compound derived from an indigenous plant with potential chemotherapeutic properties) was successfully encapsulated and stabilized in the same composition as well as in two other modifications(prepared by addition of dipalmitoyl phosphatidyl choline, DPPC, and a mixture of DPPC and cholesterol, respectively). The stability of the vehicles, before and after encapsulation, was assessed under varying conditions of time and temperature. The microemulsion compositions for drug delivery before and after encapsulation of diospyrin were found to be physically stable for one year and over a temperature range of 4-40°C. The *in vivo* single-dose acute toxicity (survival study and enzyme assay in Swiss Albino mice) showed good tolerance for intraperitonoal application at a particular dose. Several microemulsion systems has been studied for adminstration by oral route. The solubility of Ibuprofen-euginol ester (with anti-inflammatory activity) and Biphenyl Dimethyl Dicarboxylate, a drug for treating liver disease improved significantly after encapsulation in microemulsion and the area under the curve (AUC) showed a remarkable increase compared to oral Ibuprofen suspension.[116–118] The intraduodenal absorption of a water soluble RGD peptide SK&F106760 was enhanced after microemulsification and the pharmacological activity was not compromised due to encapsulation.[119,120] The low solubility of Amphotericin *B* leads to poor bioavailability through oral route, which showed remarkable improvement when administered through intraperitonial route by encapsulation in microemulsion.[121,122] Longer circulation time, enhanced drug uptake at specific organs and tissues were observed for Norcantheridine, Flurbiprofen, Clonixic acid etc.[123–125]

The cytostatic activity of the anticancer drug Carmustine was preserved after encapsulation while its toxicity was reduced.[126] The anti cancer activity of the drug all trans-retinoic acid, the drug specially used for acute promyelocytic leukemia was not impaired by loading in phospholipid-based microemulsion.[127]

Use of microemulsion for nasal routes are also reported for anti-epileptic and anti migrain drugs.[128,129] Microemulsion-based Dexamethasone eye drop showed greater penetration efficacy along with good physiological tolerance[130,131] and encapsulation of Chloramphenicol in microemulsion improved the stability of the drug.[132–133]

Self microemulsifying drug delivery systems (SMEDDS) are a promosing technology for drug delivery through oral route. They are comprised of mixture of drug, oil, surfactant(s) that form w/o or o/w microemulsion upon dilution with aqueous medium or after *in vivo* administration. SMEDDS enhance the bioavailability of poorly water-soluble drugs through solubilization in the excipient matrix or at the interface, and dispersion in the gastrointestinal tract.[134–136]

Cholesterol-rich microemulsion (LDE) are prospective delivery systems for drugs directed against neoplastic cells. These microemulsions bind to low density lipoproteins (LDL) receptors and have the ability to concentrate in acute myloid leukemia cells and in ovarian and breast carcinomas. A cholesterol rich microemulsion that binds to LDL receptors is selectively taken up by malignant cells that overexpresses those receptors and may be used as vehicles for anti-neoplastic agents.[126,137]

14.1.5. *Organogel, Hydrogel and other Colloidal Dispersions*

The hydrogel systems are three- dimensional polymer networks characterized by a remarkable ability to absorb water. The hydrogels are able to release drug molecules in a prolonged and/or controlled way and sometimes in site-specific way thus preventing drug accumulation in non-target tissues and increasing their bio-availability. Preparation of hydrogels for combination delivery of antineoplastic agents was done by Bouhadir *et al.*[138] Organogels are formed by the addition of geletin in the aqueous phase of water-in-oil microemulsion. At low geletin concentration, the system retains microemulsion-like macroscopic properties such as low viscosity and conductivity. The addition of geletin at sufficiently high concentrations, results in the formation of a semi-rigid three dimensional gel matrix, cross-linked via gelatin.[139]

The passage of the encapsulated drugs Fluconazol and Diclofenac in lecithin microemulsion based organogels through a membrane formulated with cellulose derivatives such as hydroxypropyl methyl cellulose and hydroxypropyl cellulose studied with Franz diffusion cell (determined spectroscopically) show that the large amount of the drugs could be transported from the organogels through the membrane in the receptor solution depending on the type of organogel used.[140] Nicardipine hydrochloride (NC-HCl) was encapsulated in a novel lecithin organogel (microemulsion-based gel) composed of soybean lecithin, polypropylene glycol, oleic acid, dimethyl isosorbide, and isopropyl myristate. In vitro precutaneous penetration studies revealed that the organogel system has skin penetrating potential and could be used as a promising matrix for the transdermal delivery of nicardipine. Higher permeation rates were observed when nicardipine free base was incorporated into gel matrix instead of hydrochloride salt.[141] Alginate was chemically modified into low molecular weight oligomers and cross linked with a biodegradable spacer (adipic dihydrazide) to form biodegradable hydrogels. Three model neoplastic agents, Methotrexate, Dauxorubicin and Mitoxantrone were loaded in the cross linked oxidized alginate hydrogels by different mechanisms based on their structures and they showed three different release profiles (diffusion controlled, covalent bond degradation and ionic dissociation controlled mechanisms). The system could be used for the controlled delivery of a variety of anticancer drugs sequentially or simultaneously as recent studies indicate that sequential and simultaneous delivery of drugs can reduce the side effects associated with systemic delivery of anticancer drugs. Also several drugs have been found to amplify the anticancer activity of other drugs.[142]

Ionizable polymer networks, prepared from oligo (ethylene glycol) (OEG) multiacrylate and acrylic acid (AA) exhibited a complex dependence on the network structure, and composition when tested for controlled release devices. The release kinetics of model solute Proxyphyllin exhibited a strong dependence on the polymer network structure and the pH of the release medium.[143] A comparative study on the drug release capacity of four water swellable polymeric systems were carried out by differential scanning calorimetry by Castelli and co-workers.[144] The release kinetics of the drug from hydrogels were followed at different temperatures to evidence the influence of temperature on the drug release and on the successive transfer to biological membrane model. The total amount of drug transferred and the release rates are affected by the polymer crosslinking degree as well as by the cross linking agents.

Amphiphilic branched polymers self assemble into nanoscopic supramolecular structures with hydrophobic core and hydrophilic shell.

Polymeric micelles are considerably more stable than surfactant micelles and their size ragnes to 10-100nm. Polymeric micelles have good circulation time *in vivo*, evade host defences and gradually release the drugs. They are preferentially accumulated in solid tumor tissues. Poorly water soluble anticancer drugs Doxorubicin, Propofol, Amphotericin B, Paclitaxel could be solubilized in polymeric micelles.[145-150]

14.2. Future Perspective

Different colloidal drug delivery systems have been developed using different excipients for different routes of administration. The studies on their physicochemical properties have been focused more compared to their applications in vivo. Though among the encapsulating systems, nanomaterials, microemulsions, lipid dispersions, hydrogels, organgels and polymeric micelles are more recent developments than liposomes, studies on these systems have gathered a lot of momentum in the recent past. These materials have advantages over liposomes and have shown good biocompatibility, stability and effectivity. Their formulation, physicochemistry and uses for drug encapsulation and in vivo delivery require more exploration in future under the topic of colloidal drug delivery systems.

Acknowledgment

The author acknowledges with thanks the financial support from the Department of Science and Technology, India.

References

1. A. Klein-Szanto, *J. Prog Clin. Bio* Res. **74**, 356, (1992).
2. J. Kreuter, *Colloidal Drug Delivery Systems*, Ed. J. Kreuter, Marcel Dekker, New York, (1994).
3. G. Gregoriadis, Ed. *Liposome Technology*, 3 Vols. 2nd. edn., CRC Press, Boca Raton, Florida, (1993).
4. S. Lala, S. Gupta, N.P. Sahu, D. Mandal, N. Mondal, S.P. Moulik and M.K. Basu, *J. Drug Target.*, **14**, 171, (2006).
5. B.K. Paul, and S. P. Moulik *J. Disp Sci. Technol.*, **78**, 301 (1997).
6. S.P. Moulik and B. K. Paul, *Adv. Colloid Interface Sci.* **78**, 99 (1998).
7. S. Tenjarala. *Crit. Rev. Ther. Drug Carrier Systems*, **16**, 461 (1999).
8. S. Kantaria, G. D.Rees and M. J. Lawrence, *J. Contr Del.* **60**, 355 (1999).
9. S.S. Watnasirichaikul, N. M. Davis, T.Rades and I.G. Tucker, *Pharm Res.*, **17**, 684 (2002).
10. M.J. Lawrence, *Chem. Soc. Rev.* **23**, 417 (1994).

11. R.Cortessi and C.Nastruzzi, *Pharm. Sci. and Tech. Today.* **2**,288 (1999).
12. B. McCormack and G.Gregoriadis, *Int. J. Pharm.* **162**, 59 (1998).
13. J.A.C, Daan and H. Schreier, in *Colloidal Drug. Delivery Systtem*, Ed.J. Kreuter, Marcel Dekker,. New York, (1994).
14. S.Gupta and S. P. Moulik, *J. Pharm. Sci.* **97**, 22 (2008).
15. F. Sozoka and D. Papahadjopoulos, *Proc. Natl. Acad. Sci. 308*, 250 (1978).
16. T. Lian, and R. J. Y Ho, European *J. Pharmaceutics and Biopharmaceutics*, **90**, 667 (2001).
17. J.M.A. Kemp and D.J.A. Crommelin., *Pharm Week Bull.* **123**, 355 (1998).
18. J.M.A. Kemp and D.J.A. Crommelin, *Pharm. Week. Bull.* **23**, 457 (1988).
19. T. Ohsawa, Y. Matsukawa, Y. Takakura, M.Hashida and H. Sezaki., *Chem. Pharm. Bull.* **33**, 5013 (1985).
20. H. Shreier, M. Levy and P. Mihalco, *J. Contr. Rel.* **5**, 187 (1987).
21. F.Kadir, W.M. C. Eling, D.J.A. Crommelin and J. Zuidema, *J. Contr. Rel.* **17**, 277 (1991).
22. A.A. Gabizon, *Cancer Invest.* **19**, 424 (2001).
23. A.Gabizon,H. Shmeeda and Y. Barenholz, *Clin Pharmacokinet.* **42**, 419 (2003).
24. D.D. Lasic and D. Needham, *Chem. Rev.* **95**, 2601 (1995).
25. L. Cattel, M. Ceruti, F.Dosio, *Tumori.* **89**, 237 (2003).
26. S. M. Moghimi and J. Szebeni, *Progr. Lipid. Res.* **42**, 463 (2003).
27. D. J. A Crommelin and H. Schreier, *Lipsomes*, in *Colloidal Drug Delivery Systems.* Ed. J. Kreuter, Marcel Dekker, New York (1994).
28. T. M.Allen, A. Gabizon, E. Mayhew, K. Matthey, S.K Huang, K.Lee, K.M.C. Woodle, D. Lasic, C. Redemann and F.J. Martin, *Proc. Natl. Acad. Sci. U.S.A*, **88**, 11460 (1991).
29. J.Vaage, E. Mayhew, D. Lasic, F. Martin, *Int. J. Cancer.* **51**, 942 (1992).
30. E.G. Mayhew, D. Lasic, S. Babbar and F. Martin, *Int. J. Cancer.* **51**, 202 (1992).
31. W. Emminger, W. Ganinger, W. Emminger-Schmidmeir, A. Zoiubek,.K. Pillwein, M. Susani and H. Gadner, *Ann. Hematol.* **68**, 27 (1994).
32. J.W. Park, *Breast. Cancer. Res.* **4**, 942 (2002).
33. S.A. Abraham, D.N. Waterhouse, L.D. Mayer, P.R. Cullis, T.D. Madden and M.B. Bally, *Meth. In Enzymol.* **391**, 71 (2005).
34. W. Guo, T. Lee, J. Sudimac and R.J. Lee, *J. Liposome Res.* **10**, 179 (2000).
35. K.M. Hege, D.L. Daleke, T.A. Waldman and K.K. Matthey, *Blood. J.* **74** 2043 (1989).
36. L.D. Leserman, J. N. Weinstein,R. Blumenthal and W. D. Terry, *Proc. Natl Acad. Sci.* **77**, 4089 (1980).
37. J. Lu, E. Jeon, B. S. Lee, H. Onyuksel and Z. J. Wang, *J. Control Rel.* **110**, 505 (2006).
38. T.L. Andersen, S. S. Jensen and K. Jorgensen, *Progr. in Lipid Res.* **44**, 68 (2005).
39. C. Mamot,D. C. Drummond, U. Greiser, K. Hong,D. B. Kirpotin,J. D. Marks and J. W. Park, *Cancer Res.* **63**, 3154 (2003).

40. R.M. Sawant, J. P. Hurley, S. Salmaso, A. Kale, E. Tolcheva, T. S. Levchenko and V. Torchillin, *Bioconjug. Chem.* **17**, 943 (2006).
41. V.P. Torichilin, *Nat. Rev. Drug. Discov.* **4**, 145 (2005).
42. V.P. Torchillin, Ed. *Liposomes : A Practical Approach*, Oxford University Press, USA, (2003).
43. R. Banerjee, *J. Biomat Appl.* **16**, 3 (2001).
44. M.K. Basu, *Biotechnol and Genet. Engg Rev.* **12**, 364 (1994).
45. R.M. Handjani-Vila, A. Riber, A. Rondot and G.Vanlerverge., *Int. J. Cosm. Sci.* **1**, 303 (1979).
46. A.J.Baillie,A. T Florence, L. R. Hume,G. T Muirhead and A.Rogerson, *J. Pharm. Pharmacol.* **37**, 863 (1985).
47. L.E. Echoyen, J.C. Hernandez, G. Akaifer. and W. Egokel, *J. Chem Soc. Chem. Commn.* **8**, 836 (1988).
48. H.E. Jhofland, J.A. Boustra, M. Ponec, H.E. Bodde, F. Spies, J.C. Verhoef and H.E. Junginger, *Control. Rel.* **16**, 155 (1991).
49. P.Arunothyanun, I.F.Uchgbu, D. Q. Mgcraig, J. A.Turton and A. T. Florence, *J. Pharm Pharmacol* (supplement) **50**, 169, (1998).
50. M.N. Azmin, A.T. Florence, R.M. Handjani-Vila, J.F. Stewart, G. Vanlerberghe and J.S. Whittmaker, *Pharm. Pharmacol.* **37**, 237 (1985).
51. R.A. Raja Naresh, N. Udupa and P. Uma Devi, *Ind. J. Pharacol.* **28**, 175 (1996).
52. R.A. Raja Naresh, N. Udupa and P. Uma Devi, *J. Pharm. Pharmacol.* **48**, 1128 (1996).
53. E. Oomen, S.B. Tiwari, N. Udupa, R. Kamath and P. Uma Devi, *Ind. J. Pharmacol.* **31**, 279 (1999).
54. B. Shi, C. Fang and Y Pei, *J. Pharm. Sci.* **95**, 1873 (2006).
55. M. Manconi, C. Sinico, D. Valenti, F. Lai and A.M. Fadda, *Int. J. Pharm.* **311**, 11 (2006).
56. P. Palozza, R. Muzzalupo, S. Trombino, A. Vladinni and N. Picci, *Chemistry and Physics of Lipids.* **139**, 32 (2006).
57. D. Paolino,R. Muzzalipo,A. Ricciardi,C. Celia,N. Picci and M. Fresta, *Biomed. Microdev* **9**, 421 (2007).
58. E. Hood,M. Gonzales, A. Plaas, J. Strom and M. Van Auker, *Int. J. Pharm.* **339**, 222 (2007).
59. W. Hua and T. Liu, *Colloids and Surfaces (A)*, **302**, 377 (2007).
60. Y. Huang, J. Chen, X. Chen, J. Gao and W. Liang, *J. Material Sci : Materials in Med.* **19**, 607 (2007).
61. V.P. Torchillin Ed. *Nanoparticles as Drug Carriers*, Imperial College Press U. K., (2007).
62. G. Barrat, C. Morin, H. Fessi, J. Devissaguet and F. Puisieux *Int. J. Immuno Pharmacol.* **16**, 451 (1994).
63. Y.Y. Jiao, N. Ubrich, M. Marchand-Arvier, C. Vigneron, M. Hoffman, T. Lecpmpte and P.Maincent, *Circulation.* **105**, 230 (2002).
64. S. Wanasirichaikul, N.M. Davies, T Rades and I.G. Tucker, *Pharm. Res.* **17**, 684 (2000).
65. A.O. Eniola, S. Rodgers and D A. Hammer, *Bioat.* **23**, 2167 (2002).

66. S. Utreja, A.J. Khopade, N. K Jain, *Pharm Acta Helv.* **73**, 275 (1999).
67. K.C. Gupta and M.N. V. Ravi Kumar, *Biomat* **21**, 1115 (2000).
68. S. A. Agnihotri,N. N. Mallikarjuna and T. M. Aminabhavi, *J. Contr. Rel.* **100**, 5 (2004).
69. S. Kim., I. G. Yshin and Y.M. Lee, *Biomat.* **20**, 1033 (1999).
70. G Fontana, M. Licciardi, S. Mansueto, D. Schillaci. and G. Giammona, *Biomat.* **22**, 2857 (2001).
71. I. Spiers, J. E. Deyles, S.W.J. Baillie, E.D. Williamson and H.O. Alper, *J. Pharm Pharmacol.* **52**, 1195 (2000).
72. C. Chauvierre, L Leclerc,D. Labarre,M. Appel,M. C. Marden,P. Couvreur and C. Vauthier, *Int. J. Pharm.* **338**, 327 (2007).
73. J. Panyam and V. Labhasetwar, *Adv. Drug. Del. Rev.* **55**, 329 (2003).
74. J.S. Chawla and M. M. Amiji, *Int. J. Pharm.* **249**, 127 (2002).
75. S. Gelperina,K. Kisich,D. Iseman and L. Heifets, *Am. J. Resp and Crit Care Med.* **172**, 1487 (2005).
76. R. Pandey, A. Zahoor, S. Sharma and G.K. Khuller. *Tuberculosis.* (Edin). **83**, 373 (2003).
77. K.S. Soppimath, T.M. Aminabhavi, A.R. Kulkarni and W.E. Rudzinki, *J. Contr. Rel.* **70**, 1 (2001).
78. J. Danielsson and B. Lindmann, *Coll. and Surf.* **3**, 391 (1988).
79. T. P.Hoar and J. H Schulman, *Nature.* **152**, 102 (1943).
80. M. Kahlweit, G.Busse and B.Faulhaber, *Langmuir.* **12**, 861 (1993).
81. M. Trotta, E. Ugazio, M.R Gasco, *J. Pharm Pharm.* **47**, 451 (1995).
82. M.L. Das, P.K. Bhattacharya and S.P. Moulik. *Indian J. Biochem. Biophys.* **26**, 24 (1989).
83. L. Mukhopadhyay, N. Mitra, P.K. Bhattacharya and S.P. Moulik *J. Colloid Interface Sci.* **186**, 1 (1997).
84. A, Acharya, S.P. Moulik, S.K. Sanyal, B.K. Mishra and P.M Puri., *J. Colloid Interface Sci.* **245**, 163 (2002).
85. P. Majhi, and S.P.Moulik, *J. Disp Sci. Tech.* **20**, 1407 (1999).
86. S. Gupta, S.K. Sanyal, S. Datta and S.P. Moulik *Ind. J. Biochem Biophys.* **43**, 254 (2006).
87. A. Shukla, M. Janiach, K. Jahn and R.H Neubert, *J Pharm. Sci.* **92**, 730 (2001).
88. M. Kreilgaard, E.J. Pederson and J.W Jarozewski, *J. Contr. Rel.* **69**, 421 (2000).
89. C. Malcomson and M. J. Lawrence, *J. Pharm Pharmacol.* **45**, 141 (1993).
90. P.P. Constantinides, *Pharm Res.* **12**, 1561 (1995).
91. S. Gupta, S.P. Moulik, S. Lala, M.K. Basu, S.K. Sanyal and S. Datta, *Drug Del.* **12**, 267 (2005).
92. A. Kagan and N. Garti, *Adv. Colloid Interface Sci.* **123**, 369 (2006).
93. S. Gupta, S. P. Moulik, B. Hazra, R. Ghosh, S. K. Sanyal and S. Datta, *Drug Del* **13**, 193 (2006).
94. C. Von Corswant, S. Engstrom and O. Soderman, *Langmuir.* **13**, 5061 (1997).
95. M. Trotta, M.R. Gasco and S. Morel, *J. Contr. Rel.* **10**, 237 (1989).

96. C. Von Corswant, P. Thorne and S. Engstrom, *J. Pharm Sci.* **87**, 200 (1998).
97. M.A. Moreno, P. Frutos and M.P. Ballesteros, *J. I. Lastres. Pharm Res.* **18**, 344 (2001).
98. K.M. Park and C. K. Kim, *Int. J. Pharm.* **181**, 173 (1999).
99. M.J. Sarciaux, L. Alan and P. A. Sado. *In. J. Pharm.* **12**, 127 (1995).
100. E.C. Swenson and W. J.Curatolo, *Adv. Drug. Del. Res.* **8**, 39 (1992).
101. S. Sarker, S. Mandal, J.Sinha, S. Mukhopadhyay, N. Das, M.K Basu. *J. Drug Target.* **10**, 573 (2002).
102. B. Baroli, M.A Lopez -Quintela, A.N. M. Delago Charro- M, Frades, *J. BLanc-Mendez, J Contr. Rel.* **69**, 209 (2000).
103. I. L. Lianly, I. Nandi and K.H. Kim, *Int. J. Pharm.* **37**, 77 (2002).
104. C. Malcomson, C. Satra, S. Kantaria, A. Sidhu and M.J Lawrence, *J. Pharm Sci.* **87**, 109 (1998).
105. M. Changez and M. Varshney, *Drug. Dev. Ind. Pharm.* **26**, 507 (2000).
106. M. Trotta, S. Morel and M.R. Gasco, *Pharmazie*, **52**, 50 (1997).
107. M.J. Lawrence, *J. Pharm. Pharmacol.* **45**, 141 (1998).
108. P.J. Lee, R. Langer, and V.P. Shastri V P. *Pharm. Res.* **20**, 264 (2003).
109. M.J. Alvarez- Figueroa and J. Blanco -Mendez, *J. Int J. Pharm.* **21**, 57 (2001).
110. B. Hazra, S. Gupta, R. Sarkar and S.P. Moulik, *J. Pharm. Pharmacol.*, (suppl) **50**, 191, (1998).
111. P.P. Constantinides, *Pharm. Res.* **12**, 1561 (1995).
112. A.T. Florence and D. Attwood Eds. *Surfactant Systems. Their chemistry, Pharmacy and Biology*, Chapmann and Hall, London and New York, (1983).
113. A.T. Florence and D.Attwood Eds 1998. *Physicochemical Principles of Pharmacy*, Macmillan Press, London, (1998).
114. M.E. Leser, W.C. Vanevert and W.G. MAgterof, *Colloid and Surfaces (A)*, **116**, 293 (1996).
115. R. Aboofazeli, N. Patel, M. Thomas, M.J Lawrence, *Int. J. Pharm.* **125**, 107 (1995).
116. X.L. Zhao, D.W. Chen P. Gao. Y.F. Luo and K.X. Li, *Pharmazie.* **60**, 883 (2005).
117. X.L. Zhao, D.W. Chen P. Gao. Y.F. Luo and K.X. Li, *Chem. Pharm Bull.* **10**, 1246 (2005).
118. C.K. Kim, Y.J.Cho and Z.G. Gao, *J.Contr. Rel.* **70**, 149 (2001).
119. P.P. Constantinides, H. Welzel,P.L. Ellens, S.Smith, S. H. Strurgisb and A.B.Y. Owen, *Pharm Res.* **13**, 210, (1996).
120. P.P. Constantinides, J.P. Scarlet, C. Lancaster, J. Marcello, G. Marks, H. Ellens and P.L. Smith. *J. Pharm Res.* **11**, 1385 (1994).
121. M.A. Moreno, M.P. Ballesteros and P. Frutos, *J. Pharm. Sci.* **92**, 1428 (2003).
122. B. Brime, M.A. Moreno, G. Frutos, M.P. Ballesteros and P. Frutos, *J. Pharm. Sci.* **91**, 1178 (2002).
123. J.M. Lee, K.M. Park., S. J. Lim, M.K. Lee and C.K. Kim, *J. Pharm Pharmacol.* **54**, 43 (2002).
124. L. Zhang, X. Sun, Z. R and Zang. *Drug Del.* **12**, 289 (2005).

125. K.M. Park, M.K. Lee, K.J. Hwang and C.K. Kim, *Int J. Pharmacol.* **183**, 145 (1999).
126. C.J. Valduga, D.C. Fernandes, A.C. Lo Prete, C.H. Azevedo, D.G. Rodrigues and R.C. Maranhoe, *Int. J. Phrmacol.* **54**, 1615 (2002).
127. S.R. Hwang, S.J. Lim, J.S. Park and C.K. Kim, *Int. J. Pharm.* **276**, 175 (2002).
128. I.L. Lianly, I. Nandi and K.H. Kim, *Int J. Pharm.* **237**, 77 (2002).
129. T.K. Vyas, A.K. Babbar, R.K. Sharma, S. Singhand A. Misra, *J. Pharm Sci.* **95**, 570 (2006).
130. S.L. Fialho and A. Da Silva-Cunha. *Clin. Expt Opthal.* **32**, 626 (2004).
131. A. Radomska and R. Dobrucki, *Int. J. Pharm.* **196**, 131 (2000).
132. FF LV, L.Q. Zheng and C.H. Tung, *Int. J. Pharm.* **301**, 237 (2005).
133. FF LV, L.Q. Zheng, and C.H. Tung, *Eur. J. Pharm. Biopharm.* **62**, 288 (2006).
134. N. Subramanian, S. Ray, S.K Ghosal, R. Bhadra and S. P. Moulik, *Biol. Pharm Bull.* **27**, 1993 (2004).
135. W. Wu, Y. Wang and L. Que, *Eur. J. Pharm. Bopharm.* **3**, 288 (2006).
136. T. Andrysek, *Mol. Immuno.* **39**, 1061 (2003).
137. V.T Hungaria, M.C. Latrilha, D.G. Rodrigues, S.P. Bydlowsky, C.S. Chiattone and R.C Maranhao, *Cancer Chemother Pharmacol*, **53**, 51 (2004).
138. K.H. Bouhadir, E. Alsberg and D.J. Mooney, *Biomat*, **22**, 2625 (2001).
139. L.S. Guo, *Adv. Drug. De. l Rev.* **47**, 149 (2001).
140. M. Zoumpanioti, E. Karavas, C. Skopelitis, H. Stamatis and S. Xenakis, *Progr. Coll. Polym. Sci.* **123**, 199 (2004).
141. R. Aboofazli, Z. Zia and T.E. Needham, *Drug. Del.* **9**, 239 (2002).
142. C A. Presant, W. Wolf, V. Walush, C.L Wiseman, I. Weitz and J. Shani *J. Clin. Oncol.* **18**, 255 (2000).
143. R.A. Scott and N. A. Peppas, *Biomat*, **20**, 1371 (1999).
144. F. Castelli, G. Pitaressi and G. Giammona, *Biomat*, **21**, 821 (2000).
145. M.C Jones, and J.C. Leroux, *Eur. J. Pharm. Biopharm.* **48**, 101 (1999).
146. N. Nasongkla, X. Shuai, A. Hua, B.D. Weinberg, J. Pink, D.A. Boothman and J. Gao, *Angew Chem. Int. Ed.* **43**, 6323 (2004).
147. G.S. Kwon, *Adv. Drug Del Rev.* **54**, 1145 (2002).
148. D. Sutton, N. Nasongkla, E. Blanco and J. Gao, *Pharm Res.* **24**, 1029 (2007).
149. S. C. Lee, K. M. Huh, J. Lee, Y.W. Cho, R. E. Gallinsky and K. Park, *Biomacromol.* **8**, 202, (2007).
150. S. R Croy and G. S. Kwon, *Curr. Pharm. Design*, **12**, 4669 (2006).

Chapter 15

Nanoscale Self-Organization of Polyampholytes

H.B. Bohidar

Polymer and Biophysics Laboratory,
School of Physical Sciences,
Jawaharlal Nehru University,
New Delhi-1100067, India

Amarnath Gupta

Faculty of Science,
Department of Physics,
Science Drive 3
National University of Singapore,
Singapore 117546

Gelatin, a biopolymer, is a low charge density polypeptide that exhibits polyampholyte behaviour in aqueous solutions. Thus, morphological changes in this system could be easily induced by changing the solution pH and ionic strength. The self-assembly of gelatin molecules and aggregates was studied in both two and three dimensions. When spread on a $2-D$ hydrophilic substrate self-organized nano-structures arranged as self-similar objects, giving a mass fractal dimension, $d_f = 1.67$ for solutions made with KCl salt as estimated from atomic force microscopic studies, were observed. The dehydration driven self-organization of particles changed the d_f values to 1.78 after 24 hours which further changed to 1.83 after a time lapse of 10-days The dynamics of formation of these structures are modeled through spin-exchange kinetics in the non-equilibrium steady state regime in order to understand their complex behaviour. Kawasaki spin exchange dynamics has been applied to a diffusion limited aggregation (DLA) type fractal object, and the growth of the domains was observed by minimizing the free-energy. The fractal dimension of such a system changed from 1.70 to 1.82 which inferred the loss of fractal behaviour and the generation of a more compact object. A brief discussion on nano-scale self-assembly of gelatin molecules and clusters spread on hydrophilic surface and its modeling is presented here.

15.1. Introduction

Biopolymers can be found in nature either as polyelectrolytes (examples: polysaccharides, polynucleic acids *etc.*) or molecules carrying ionizable acid and base groups. Under appropriate conditions, such as in aqueous solutions, these groups dissociate, leaving ions on chains, and counterions in solution. The net charge of a polyampholyte in aqueous solution can be changed by varying the *pH*. At a particular pH called the isoelectric *pH(pI)*, there are equal numbers of positive and negative charges on the polyion, giving the polymer chain a minimum net charge, and a collapsed conformation. The conductivity and viscosity of the weakly hydrophobic polyampholyte solutions show a minimum at this pH. On the other hand strongly hydrophobic polyampholytes will undergo precipitation when the solution pH approaches *pI* value.[1,2] In the vicinity of the isoelectric *pH*, the polymers are nearly charge-balanced and exhibit the unusual properties associated with polyampholytes. Far above or far below the isoelectric *pH*, these polymers demonstrate polyelectrolyte-like behavior arising from asymmetric charge distribution. The net polymer charge, in turn, is an artifact of quenched and annealed charges located on the contour of the chain. The charge distribution present on the polymer chain affects the persistence length of the chain considerably, which in turn governs binding.

The polyampholyte (*PA*) to polyelectrolyte (*PE*) crossover mechanism in real and model systems has generated considerable interest in the past. Higgs and Joanny[3] assumed that equilibrium charge distribution on a long *PE* to be alike that of an ionic solution. This model estimates the electrostatic interactions through Debye-Hckel formalism. The same problem, when approached through random phase approximation yielded identical results. However, these models did not distinguish between quenched and annealed charges, which was the major drawback associated with these approaches. This issue was addressed later and a refined proposition was made. In the new model conformation of PA was shown to be very sensitively dependent on the overall charge, Q of the polymer. It was proposed that for a chain having N_0 number of total charges with each charge unit being $\pm q_0$, one can write,[4]

$$\mathbf{Q} \approx q_0 N_0^\alpha; \quad \alpha = \begin{cases} < 0.5; & Anti-correlation \\ > 0.5; & Positive-correlation \\ 0.5; & no-correlation \\ 1; & PE-behaviour \end{cases} \quad (1)$$

This quantified the origin of interactions prevailing inside the solution environment rather explicitly. Thus, the charge distribution on the polymer chain assumes enormous importance. The binding mechanism can be significantly different depending on whether or not the charge is symmetrically distributed on the PA molecule.

Free-energy driven self-organization of molecules and clusters is observed in a variety of systems of interest in biology, immunology, polymer science, colloidal chemistry, metallurgy *etc.* For example, the self-assembly dynamics underlying the protein shell formation in spherical viruses is an interesting biological problem. Normally, self-organization and aggregation processes have been described through diffusion limited cluster aggregation (DLA),[5] reaction limited cluster aggregation (RLA),[6] random sequential adsorption (RSA)[7,8] *etc.* Adhesion of macromolecules onto surfaces, polymer chain reactions, association of non-crystalline granular materials and arrangement of nano-structures, quantum memory devices, biosensors on substrates, car parking etc. fall in to the category of random sequential adsorption process (RSA). These processes are generated with sequential addition of particles at a constant rate to a substrate where overlapping and leaving the cluster once a part is forbidden. Brilliantov *et al.*[9] have shown that geometric and kinetic features of the RSA clusters for a system showing polydispersity in size are governed by the smallest size particles.

15.2. Experimental Observations

The biopolymer, Gelatin (type-B, 300 Bloom, microbiology grade devoid of $E.$ coli, $pI = 4.9$) was obtained from $E.$ Merck (India) which was dissolved in Millipore deionized water. The gelatin solutions (1% w/v) were prepared by dispersing gelatin in 0.1 $mol\ dm^{-3}$ KCl solution at $60°C$. The macromolecules were allowed to hydrate completely; this took about 1hr. The gelation concentration of gelatin in water is 2% (w/v), the gelatin concentration chosen in these experiments was deliberately kept lower than this to avoid formation of gels.

The zeta potential measurements were performed using an electrophoresis instrument (Model: ZC-2000, Microtec, Japan). The sample solution was much diluted with deionized water to isolate all individual particles from the aggregates in order to know the distribution of charges on the surfaces of streaming particles. In order to minimize the influence of electrolysis to the measurements, molybdenum (+) and platinum (−) were used for electrodes. Also, during the measurements, the cell chamber tap on molybdenum electrode was kept open to release the air bubbles, for the purpose of reducing their effects on the particle movements. During the measurements, the molybdenum anode was cleaned each time as it turned from a metallic color to blue-black. If one uses the zeta potential (ζ) as an approximation of the surface potential ϕ of a uniformly charged sphere, the theory gives[10]

$$\zeta \simeq \phi - 4\pi(\sigma/\in \kappa), \qquad (2)$$

where σ is the surface charge density of the particle, and ϵ and κ are the dielectrical constant and Debye-Hückel parameter of the solution, respectively. The relationship between the mobility (μ) and the zeta potential (ζ) is $\zeta = 4\pi(\mu\eta/\epsilon)$. Then, μ can be written as $\mu = \sigma/\eta\kappa$ where η is the viscosity of the solution. Since the polyelectrolytes are in random coil conformations, the quantitative application of Eq. (2) is not expected. However, it has been shown experimentally that the mobility of a polyelectrolyte is identical to the mobility of the polymer unit, as given by[10]

$$\mu = q/f_0, \qquad (3)$$

where q and f_0 are the charge and friction of the polymer unit, respectively. The measured electrophoretic mobility (μ) values could be converted to equivalent zeta potential (ζ) values through the relation $\mu = \epsilon\epsilon_0\zeta/\eta$, where the solvent viscosity and dielectric constant values are given by η and ϵ respectively. The ϵ_0 parameter stands for permittivity of free space.

The particle sizing was done using a Photocor (Photocor Instruments, USA) multi-angle laser light scattering instrument operating in multi-τ mode. The hydrodynamic radius (R) data was evaluated from the measured intensity correlation function using Stoke-Einstein relation. These measurements also permitted the evaluation of particle size distribution shown in Figure 15.1 which follows a power-law distribution. More details are available in ref.[11] The zeta-potential data is presented in Figure 15.2. The iso-electric point thus established was ≈ 5 invariant of gelatin concentration, and in consistence with literature data. This established the robustness and reliability of our measurement procedure; the effect of gelatin surface charge on intermolecular interactions is discussed in ref.[12]

A drop of the solution was removed from the beaker and allowed to spread out uniformly on a quartz plate over a period of 30 minutes. Atomic Force Microscope (AFM) pictures were taken using an Autoprobe CP Research AFM system, model AP-2001 (Thermo- microscopes, USA) using a 90 mm scanner and in tapping mode. All experiments were carried out at

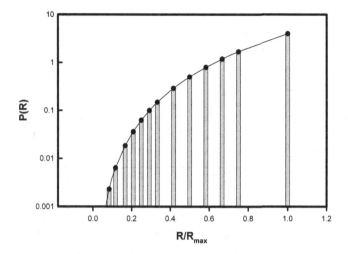

Fig. 15.1. Particle size distribution of gelatin molecules and clusters in solution. The R_{max} has been fixed at 600nm. The distribution follows a power-law behaviour (taken from ref.[11])

20°C. The evolution of AFM micrographs was recorded for 10 days which is displayed in Figure 15.3 for samples made with KCl salt. These self-assemblies are made of gelatin nano-particles and clusters that are generated through intra and intermolecular interactions.[11–13] Such interactions produce partial charge neutralization of gelatin molecules, and release counterions previously bound to the charged segments into the solution thereby increasing the solution entropy which facilitates more interactions.[13] The self-similarity and scaling in the fractal structures observed in Figure 15.1 was ascertained from the fractal dimension, d_f. This is defined as,[14]

$$d_f(s) = \lim_{\epsilon \to 0} \frac{lnM(\epsilon)}{ln(1/\epsilon)}, \qquad (4)$$

where s is a subset of N-dimensional space occupied by the fractal object and M is the number of N-dimensional cubes required to cover the subset s. Our analysis yielded $d_f = 1.67 \pm 0.08$, 1.78 ± 0.09 and 1.83 ± 0.08 for Figures 15.3 (left to right) respectively which corresponds to time 0, 1 day and 10 days respectively (see ref 11 for details on determination of d_f). This clearly indicated that the dehydration driven self-organization was driving the system towards the formation of more compact structures.

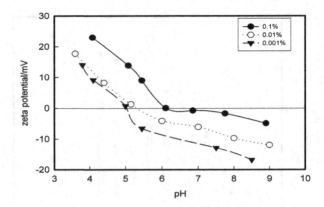

Fig. 15.2. Determination of iso-electric pH (pI) of alkali processed type-B gelatin in aqueous solutions, pI was found to be 5.5±0.5. All measurements were done at 20°C.

Fig. 15.3. Temporal evolution of self-organized nano-structures of gelatin nano spread on quartz substrate. The mass fractal dimensions of figure (left) was 1.67 ± 0.08 that increased to 1.78 ± 0.09 after 24 hours (middle). This further increased to 1.83 ± 0.08 after a lapse of 10 days (right). Figures are $1\mu \times 1\mu$ scans.

15.3. Simulation Studies

The retraction pattern of liquid-vapour interface observed was modeled[15]. We have taken a square lattice size (128 × 128) in which we fixed the origin as initial site for inorganic low molecular weight salt which acts as a nucleation centre for a local fractal structure to grow via the DLA model. We have taken 5000 particles that perform random walk (diffusion) within the lattice until hit is registered with the nucleation site or a growing cluster. Following this the particle sticks to the site or cluster leading to the growth of a microscopic fractal structure (defined as $t = 0$). We allow local spin variable to interchange which is appropriate for kinetics of phase separation studies in binary mixtures. The Kawasaki spin exchange model adequately mimics phase segregation via diffusion. There are systems where the bulk mobility diminishes drastically due to physical processes, where one component may undergo gelation transition.

We mapped the water and ampholytic molecules (gelatin) by Ising spin variables with spin $S = \pm 1/2$. Monte-Carlo simulation was performed on a fractal employing Kawasaki spin exchange dynamics. In order to incorporate the dehydration mechanism into the model the following protocol was used. The rate of creation of new vacancies was considered as directly proportional to evaporation of solvent molecules. As the solvent evaporates it causes cooling to the system which develops a temperature gradient in the system. The probability of evaporation of solvent from the outer surface of the film is more. We select a gelatin molecule at random and flip the same with the nearest solvent molecule, and then calculate the total energy for the new system, if the new energy is lower than the previous then accept the change otherwise discard that change. This process gives rise to a domain type of growth mechanism. Thus, clearly the system prefers to be in a phase-separated state at the lower temperature.

The standard model for binary mixtures is the Ising model, and the Hamiltonian is given as,[16]

$$H = -J \sum_{i \to 1, j \to 1}^{N, N_1} \sigma_i \sigma_j, \quad (5)$$

where the spins $\{\sigma_i\}$ with $i = 1 \to N$ are located on a discrete lattice and J is the strength of the exchange interaction between spins. The states $\sigma_i = +1/2$ or $-1/2$ denote the presence of a gelatin particles or solvent molecules at site i, respectively. The concentration of gelatin particle is less in comparison to the solvent. In a Monte Carlo simulation of this model, a pair of nearest-neighbor sites i and j is randomly selected, and the spins σ_i and σ_j are exchanged. The probability that this exchange is accepted as,

$$P = \min[1, \exp(-\beta \Delta H)], \quad (6)$$

$$\Delta H = J(\sigma_i - \sigma_j) \left[\sum_{L, i \neq j} \sigma_{I,j} - \sum_{I, j \neq i} \sigma_{I,j} \right], \qquad (7)$$

here, ΔH is the energy change due to the proposed spin exchange, and $\beta = 1/k_B T$, is the inverse temperature, with k_B denoting the Boltzmann constant. L_i denotes the nearest neighbors of 'i' on the lattice. We have performed the simulations at 22°C. A single Monte Carlo step (MCS) corresponds to N such attempted exchanges.

The phase separation kinetics follows a simple diffusion equation,[17]

$$\frac{\partial u(r,t)}{\partial t} = \eta \nabla^2 u(r,t), \qquad (8)$$

where $u(r,t)$ is density and η is diffusion constant, so future profile of $u(r,t)$ can be predicted by solving the ordinary differential equation. The evaporation of solvent induces an outward flow, which brings the solute towards the perimeter of the polymer film. We have taken snaps in the interval of 10×10^6 Monte Carlo steps. The first snap matches with the initial stage of grown fractal (Figure 15.4), and then we allow the system to evolve with Monte Carlo Steps. Finally, the snap (6th) taken after 50×10^6 MCS can be compared with the corresponding AFM data [Figure 15.3 (right)]. We are comparing the Monte Carlo Steps with the real time of experiments only at the end points. It must be realized that one to one correspondence in the intermediate steps is difficult to establish. Regardless, a qualitative matching between the experimental data (Figure 15.3) and simulation pictures (Figure 15.4) could be clearly seen.

15.4. Results and Discussions

In a real system, the gelatin particles are charged and strong screened electrostatic interactions prevail between these entities. When we make a thin film on quartz plate the particles diffuse and adhere to each other to form fractal structures through diffusion-limited aggregation process. The dehydration driven growth process owes its origin to the presence of hydrodynamic instability caused due to the asymmetric drying of the film. There is a positive osmotic pressure gradient at the tip of the fractal structures with respect that of the backbone which sets up a solvent flux. Recall that free energy of hydration is negative which forces the structural changes. The diffusion between microscopic zones is possible only if a percolating solvent channel connects these regions. The death of branches as they are

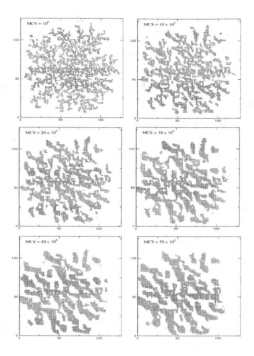

Fig. 15.4. Simulated evolution of the initial DLA fractal structure (top left) $d_f = 1.7$ shown in various Monte Carlo time steps. The fractal dimension undergoes continuous change. Notice that the final states (bottom right) are not too different from the AFM pictures shown on panels Figure 15.3. Final structure shown in panel has $d_f = 1.82$.

screened by their neighbors is balanced by the creation of new branches via microscopic tip splitting process. Temperature influences a number of parameters which play a key role in the drying process; the interfacial viscosity and surface tension of liquid are of paramount interest here. But, these are not amenable to experimental measurements. As the water evaporates the salt present in the system crystallizes and the crystal generates a stress on the fractal structure leading to its disintegration. In our case, the gelatin molecules are highly miscible in water so these form finger-like structure initially, but as the water molecules evaporate the local concentration of the polymer gets affected which results in the loss of the specific structure. Ultimately, the disintegrated structure appears as a polydisperse assembly of clusters scattered over the quartz plate randomly.

The process described in the present study pertains to the retraction pattern of a viscous liquid phase at the liquid-vapor interface. The highest rate of evaporation occurs at the edges whereas the bulk of the solvent is concentrated closer to the center of the film. In order to replenish the liquid removed by evaporation at the edge, a solvent flux from the inner to the outer regions must exist inside the film. Such a flux will drive the solute molecules towards the center in order to minimize the free energy of hydration, thus leading to more compact structures and the concomitant loss to the fractal features. The effective concentration of the gelatin molecules is increasing during the evaporation of solvent which will exceed the gelation concentration ($\approx$ 2% w/v) when 50% of the solvent is lost due to evaporation. Thus, there is a definite possibility of formation of thin gelatin gel films on the substrate which will hinder further solute movement. Notice that the experimental temperature is less than the gelation temperature of gelatin which is $\approx 28^\circ C$.[18]

Here it is worthwhile to compare the present experiments to another similar observation.[11] An appropriate composition of a binary solvent, water-ethanol, provided the necessary thermodynamic environment for gelatin molecules to form self-assembled nano-clusters having fractal dimension $d_f \approx 2.6$ in the bulk $(3 - D)$ and 1.7 on surfaces $(2 - D)$. The aggregation in the bulk of the solution appeared to be an anomalous process and could be explained through Smoluchowski aggregation model. It gave a DLA (diffusion limited aggregation) type fractal dimension to the cluster, but showed extremely low polydispersity, which in fact is a signature of a slowly growing RCA (reaction controlled aggregation) process. The temporal growth in hydrodynamic radii (R) of these structures in the bulk followed: $R(t) \sim t^z$ and the corresponding growth in scattered intensity $I(t)$ followed, $I(t) \sim t^z$ with $z = 1/d_f$. Experimental results obtained from light scattering, rheology and atomic force microscopy experiments enabled the kinetics of such growth processes to be probed. These nano-structures formed on surfaces followed a sequence of morphological time dependent changes, driven by selective but slow evaporation of ethanol, to yield several types of self-organized nano-assemblies. It must be stated here that the MC simulations did not account for gelatin polydispersity and its polyampholyte character. It is not easy to estimate how these parameters will effect the structure morphology and their fractal dimension.

15.5. Conclusions

The process described in the present study pertains to the retraction pattern of a viscous liquid phase at the liquid-vapor interface. The highest rate of evaporation occurs at the edges whereas the bulk of the solvent is concentrated closer to the center of the film. In order to replenish the liquid removed by evaporation at the edge, a solvent flux from the inner to the outer regions must exist inside the film. Such a flux will drive the solute molecules towards the center in order to minimize the free energy of hydration, thus leading to more compact structures and the concomitant loss to the fractal features. The effective concentration of the gelatin molecules is increasing during the evaporation of solvent which will exceed the gelation concentration ($\approx 2\%$ w/v) when 50% of the solvent is lost due to evaporation. Thus, there is a definite possibility of formation of thin gelatin gel films on the substrate which will hinder further solute movement. The contents of this article relies heavily on our previously published works where various kinetic aspects of self-organization has been discussed at length.[11-13,15,18]

Acknowledgments:
AG is thankful to Council of Scientific and Industrial Research, Government of India for a senior research fellowship.

References

1. A.V. Dobrynin, R.H. Colby and M. Rubinstein, *J. Polym. Sci. Part B : Polym. Phys.*, **42**, 3513 (2004).
2. R. Everaers, A. Johner and J. F. Joanny, *Europhys. Lett.*, **37**, 275 (1997).
3. P.G. Higgs and J.F. Joanny, *J. Chem. Phys.*, **94**, 1543 (1991).
4. A.M. Gutin and E.I. Shakhnovich, *Phys. Rev. E.*, **50**, R3322 (1994).
5. T.A. Witten, Jr., and L.M. Sander, *Phys. Rev. Lett.*, **47**, 1400 (1981) ibid *Phys. Rev. B* **27**, 5686 (1983).
6. J. Feder, T. Jossang and E. Rosenqvist, *Phys. Rev. Lett.*, **53**, 1403 (1984).
7. J.W. Evans, *Rev. Mod. Phys.*, **65**, 1281 (1993).
8. A. Di Biasio, G. Bolle, C. Cametti, P. Codestefano, F. Scortino, and P. Tartagalia, *Phys. Rev. E*, **50**, 1649 (1994).
9. N.V. Brilliantov, Y.A. Andrienko, P.L. Krapivsky and J. Kurths, *Phys. Rev. Lett.*, **76**, 4058 (1996).
10. J. Xia, P. L. Dubin, Y. Kim, B.B. Muhoberac and V. J. Klimkowski, *J. Phys. Chem.*, **97**, 4528 (1993).
11. B. Mohanty and H. B. Bohidar, *Phys. Rev. E.*, **69**, 021902 (2004).
12. Amarnath Gupta, Reena and H. B. Bohidar, *J. Chem. Phys.*, **125**, 054904 (2006).

13. B. Mohanty and H. B. Bohidar, *Biomacromolecules*, **4**, 1080 (2003).
14. K.S. Birdi, *Fractals in Chemistry, Biochemistry and Biophysics*, (Plenum Press, New York, 1991).
15. Amarnath Gupta and H. B. Bohidar, *Phys. Rev. E.*, **76**, 051912 (2007).
16. Han Zhu and Zian-Yang Zhu, *Phys. Rev. E.*, **66**, 017102 (2002).
17. G. Reiter, R. Khanna and A. Sharma, *J. Phys. Condens. Matter*, **15**, 331 (2003).
18. H.B. Bohidar and S.S. Jena, *J. Chem. Phys.*, **98**, 8970 (1993) and **100**, 6888 (1994).

Chapter 16

Polymer-Modified Microemulsions as a New Type of Template for the Nanoparticle Formation

Joachim Koetz, Carine Note, Jennifa Baier and Stefanie Lutter

Institut für Chemie, Universität Potsdam,
Karl-Liebknecht-Strasse 24-25,
Haus 25, 14467 Potsdam, Germany

Polymer-modified microemulsions can be obtained by adding polymers to a mixture consisting of water, oil, and a cosurfactant. Depending on the type of polymer used quite different effects can be observed. For example water soluble polymers can be incorporated into the individual water droplets of a water-in-oil microemulsion, can induce a cluster formation, or the formation of a sponge phase, that means a bicontinuous microemulsion.

It is shown that the cationic polyelectrolyte, i.e. poly (diallyldimethylammonium chloride) (PDADMAC), of low molar mass can be incorporated up to a polymer concentration of 20% into individual inverse microemulsion droplets consisting of water, heptanol, and a surfactant with a sulfobetaine head group (SB). These PDADMAC-modified microemulsions, well characterized by means of conductometry, rheology, calorimetry, ^1H NMR self-diffusion, ultrasound relaxation measurements, and electron microscopy, can be successfully used as a template for the formation of ultrafine spherical $BaSO_4$ nanoparticles.

By adding nonionic polymers like poly (N-vinyl-2-pyrrolidone) or poly (ethyleneglycol) to the quasiternary system water/ toluene-pentanol/ sodium dodecyl sulfate (SDS), polyampholytes or polycations, one can induce the formation of a single phase channel between the water-in-oil and the oil-in-water microemulsion. The resulting sponge phase can be used as a template for producing $BaSO_4$ nanorods.

16.1. Introduction

Microemulsions, i.e. thermodynamically stable, transparent mixtures of two immiscible liquids, have been known for a long time. One interesting field of application for microemulsions is their use as a template for making ultrafine particles[1]. For example Ag^2, Ag_2S^3, $CaCO_3^4$, Pd^5, or TiO_2^6 nanoparticles with diameter smaller than 10 nm can be successfully produced inside the water-in-oil droplets.

Ideally, the particle diameter will be given by the droplet size. However, often the surfactant film stability and bending elasticity is not strong enough to stop the particle growth, and in consequence the particles become polydisperse. To solve this problem the surfactant film stability has to be enhanced. One possibility is to use polymerizable surfactants,[7] and another one is to add polymers, which can interact with the surfactant head groups. However, when the interaction becomes too strong polymer-surfactant complexes are formed, and a macroscopic phase separation is observed. Therefore, moderate interactions between the surfactant head group and the polymer are required. Already, different authors have shown that water-soluble polymers, ionic as well as nonionic ones, can be successfully incorporated into w/o microemulsions.[8-10] In addition polymers, especially polyelectrolytes, can show a size regulating effect during the process of nanoparticle formation, and an additional stabilizing effect during the nanoparticle recovering process.[11] Based on this knowledge, we tried to incorporate non-charged and charged polymers (i.e. polyelectrolytes) in different microemulsion systems with surfactants having cationic (CTAB),[12] anionic (SDS)[13-15] or zwitterionic (SB) head groups.[11,16]

Surprisingly, we were able to show that an oppositely charged cationic polyelectrolyte, i.e. poly(diallyldimethylammonium chloride) (PDADMAC), can be incorporated into a SDS-based microemulsion up to a polymer concentration of 20%.[17] The reason for this "unusual" behaviour is the high ionic strength inside the water droplets given by the counterions. Therefore, the shielding effect is strong enough to realize moderate electrostatic interactions. Recently, we have shown that also poly(ethyleneimine) can be incorporated into the SDS-based microemulsion without problems.[13] Newly synthesized polyampholytes with quaternary amino groups, namely poly(N,N-diallyl-N,N-dimethylammonium-alt-maleamic carboxylate) (PalH), and hydrophobically modified derivatives with a butyl (PalBu) or octyl (PalOc) side chain, can be incorporated into the SDS-based microemulsion as well.[18-19]

By using the oppositely charged cationic surfactant cetyltrimethylammonium bromide (CTAB), it is possible to incorporate polyanions, i.e. hydrophobically modified polyacrylates.[12]

When a zwitterionic surfactant N,N-dimethyldodecylammonio propanesulfonate (SB) is used in the system pentanol/toluene/water the area of the inverse microemulsion phase can be significantly enlarged by adding the polycation PDADMAC.[20] However, the type of the oil phase used is of importance, too. In the ternary system SB/water/long chain alcohol we were able to show that the L2 phase region can be tuned by varying the chain length of the alcohol, and the PDADMAC concentration, respectively.[16]

In this paper we focus our interest on a brief overview of our work about polymer-modified microemulsions, and their use as template for the formation of nanoparticles. Therefore, the influence of the added cationic polyelectrolyte PDADMAC on the physicochemical behaviour of the inverse microemulsion in dependence on the type of surfactant used, i.e. SB or SDS, is discussed in more detail. In addition it is shown, that non-ionic, amphoteric and weak cationic water soluble polymers can induce the formation of a bicontinuous microemulsion. Finally, the use of these polymer-modified microemulsions as a template for the formation of barium sulfate ($BaSO_4$) nanoparticles is discussed.

16.2. Experimental

16.2.1. *Materials*

Poly(ethyleneimine) (PEI) a highly branched, commercial product of the BASF (Polymin P) with $M_n =$ 35,000 g/mol, and $M_w =$ 600,000 g/mol was used. Poly(diallyldimethylammonium chloride) samples of different molar mass were purified by ultrafiltration. The molecular weight of the samples used was 7000, 31000, and 246000 g/mol.

The average molecular weight of the Poly(ethylene glycol) (PEG) given by Fluka was 3000 g/mol, and 20000 g/mol. The poly(N-vinyl-2-pyrrolidone) (PVP) samples K30 and K90 were purchased from Fluka with a molecular weight of 40000, and 3600000 g/mol. The amphoteric poly(N,N'-diallyl-N,N'-dimethylammonium-alt-N maleamic carboxylate) (PalH) with a molecular weight of *ca.* 22000 g/mol (determined from viscometric data) was synthesized in comparison to the butyl-modified derivative (PalBu) with adequate molar mass.[18]

Pentanol (99+%), toluene (99+%), sodium dodecylsulfate (SDS), obtained from Fluka were used without further purification. Barium chloride ($BaCl_2$) and sodium sulfate (Na_2SO_4) were purchased from Merck, and heptanol (99+%), and 3 (N, N dimethyldodecylammonio) propanesulphonate (SB > 97%) from Fluka. Water was purified with the Modulab Pure One water purification system (Continental).

16.2.2. Methods

The phase diagrams were determined optically by titrating the alcohol/surfactant mixture with water or the corresponding aqueous polymer solution at room temperature (22°C), 35°C and 45°C. Each phase diagram was constructed on the basis of at least 20 measuring points.

The droplet structure of the microemulsions was investigated by means of ultra-high resolution cryo scanning electron microscopy (Cryo-SEM; S-4800, Hitachi). Phenomena of percolation were determined by conductometric measurements by using a micro-processor conductivity meter (LF 2000, WTW) at different temperatures.

Calorimetric measurements were carried out with a Micro-DSC III (Setaram) in a temperature region between −20°C and +80°C. The heating and cooling rate was fixed to 0.3 K/min. After cooling, the sample was kept frozen at −20°C for 3 hours, before the heating curve starts again. The DSC curves were repeated several times. The different types of frozen water were discussed in more detail.

The shear viscosity of the microemulsions was determined in dependence on the amount of added polymer and water content by using a low stress rheometer (LS 100, Physica).

^{1}H NMR self-diffusion measurements were realized using a Bruker AMX-300 spectrometer operating at a proton resonance frequency of 300.13 MHz using the LED pulse sequence with bipolar pulse pairs (bpp-LED) at 25°C.[20]

16.2.3. Synthesis of the Nanoparticles

The nanoparticle formation process was initiated spontaneously by mixing at least two microemulsions containing the reactants $BaCl_2$, and Na_2SO_4, respectively.

16.2.4. Characterisation of the Nanoparticles

The particle diameters were determined by dynamic light scattering, i.e. by using a Nano Zetasizer (Malvern) equipped with a He-Ne laser (4 mW) and a digital autocorrelator. For determining the average size of the main particle fraction, an automatic peak analysis was used.

The shape and size of the nanoparticles was determined in addition by transmission electron microscopy (EM 902, Zeiss). Samples were prepared by dropping a small amount of the nanoparticle dispersion on the copper grids, dried and examined in the electron microscope at an acceleration voltage of 80 KV.

16.3. Results

16.3.1. Characterisation of the Polymer-modified Microemulsion Template Phases

16.3.1.1. SB/water/heptanol system

The partial phase diagram, given in Figure 16.1, shows the influence of the molar mass of the cationic polyelectrolyte PDADMAC on the area of the water-in-oil (L2) microemulsion at a constant polymer concentration of 1%.

One can see that the L2 phase becomes significantly smaller, when the molar mass exceeds a critical value. That means, when the molar mass is low (< 10.000 g/mol) the polymer coil can be incorporated without any problems into individual water droplets. At higher molar mass the polymer can be solubilized only into droplet clusters,[21] and therefore often a partial phase separation results.

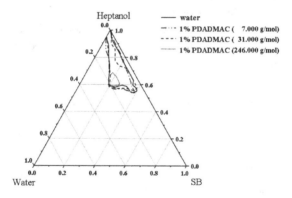

Fig. 16.1. L2 phase of the SB/heptanol/water system in absence and presence of 1% PDADMAC of different molar mass

However, by using the PDADMAC sample of low molar mass it is possible to increase the polymer concentration. Figure 16.2 shows a concentration dependent shift of the L2 phase in direction to the water corner, and to a higher surfactant concentration.

This means that in presence of 10% PDADMAC the microemulsion

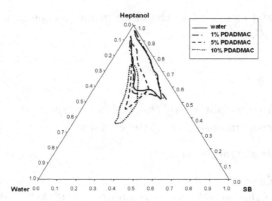

Fig. 16.2. L2 phase of the SB/heptanol/water system in dependence on the PDADMAC concentration (molar mass: 7,000 g/mol)

droplets can solubilize more water. However, the increase of the water incorporation capacity can be related to a more stable surfactant film due to polyelectrolyte-surfactant head group interactions. Freeze fracture TEM micrographs show individual water droplets. ^{1}H NMR diffusion measurements demonstrate that the water diffusion is strongly restricted as indicated by the very low reduced self-diffusion coefficient (Figure 16.3). From these results one can conclude that the water is confined to the inside of inverse water-in-oil droplets.

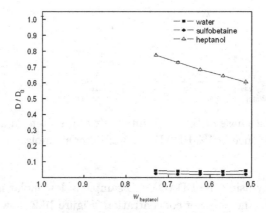

Fig. 16.3. Reduced self-diffusion coefficients of the components in the L2 phase of the SB/heptanol/water system in presence of 10% PDADMAC

16.3.1.2. SDS/water/xylene-pentanol system

It is well known, that in the quasi ternary system SDS/water/xylene-pentanol two separated optically clear phase regions exist as to be seen in Figure 4. In the water corner an oil-in-water (L1) microemulsion, and in the oil corner a water-in-oil (L2) microemulsion is formed.

Surprisingly, the addition of the nonionic polymers PVP[14] or PEG[15] leads to the formation of a bicontinuous phase channel between the L1 and L2 phase (Figure 16.4) already at room temperature.

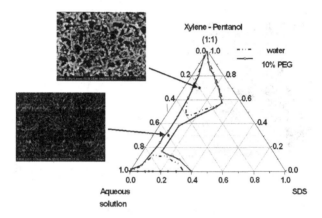

Fig. 16.4. Partial phase diagram of the system SDS / xylene-pentanol / water in absence and presence of 10% PEG, including a Cryo-SEM micrograph of the L2- and the bicontinuous phase

When the cationic polyelectrolyte PEI is added at a polymer concentration of 10% no significant effect is observed. Only if the system is heated up to 30°C, a "boostering" effect becomes visible and the isotropic phase is expanded in direction to the water corner. A further increase of the PEI concentration to 20% or 30% leads to a similar result, and a phase channel between the L1 and L2 phase is formed.

Conductometric investigations have shown that the percolation boundary of the non-modified water-in-oil microemulsion can be shifted to lower water concentrations by adding the non-ionic polymer PVP, but to higher water contents by adding cationic polyelectrolytes, e.g. PDADMAC or PEI. It has to be mentioned here that the percolation can be completely suppressed, when the PEI concentration is increased up to 30%. Therefore, one can conclude that the electrostatic interactions between the cationic functional groups of the poly(ethyleneimine) and the SDS head groups lead to a more stable surfactant film.

In general, the shift to lower water concentrations can be related to more attractive, and the shift to higher water contents to more repulsive droplet-droplet interactions in the L2 phase.

The formation of an isotropic phase channel between the L1 and L2 phase can also be induced by adding hydrophobic polyampholytes, *e.g.* PalBu, and heating up the system to 40°C, as to be seen in Figure 16.5.

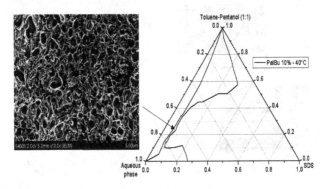

Fig. 16.5. Partial phase diagram of the system SDS / toluene - pentanol / water in presence of 10% PalBu, including a Cryo-SEM micrograph of the bicontinuous phase

Cryo SEM investigations show isolated water-in-oil droplets in the non-percolated range of the L2 phase (compare Figure 4), and droplet clusters in the percolated phase range. In the polymer-induced phase channel a typical sponge structure can be visualized by means of cryo-SEM (compare Figures 16.4 and 16.5).

Rheological measurements indicate the transition from the L2 phase to the bicontinuous phase by reaching a plateau level in the shear viscosity, and the transition to the L1 phase by reaching a shear viscosity maximum.[15] The disappearence of an interphasal water peak in the DSC heating curves is an additional hint for the transition from the L2 phase to the bicontinuous microemulsion phase.[15]

Taking these results into account, one can conclude that interactions between the polymer and the surfactant are predominantly responsible for the formation of a sponge phase. Due to the incorporation of the partly hydrophobic, non-ionic polymer into the interfacial region the surfactant film is destabilized, and in consequence the droplet-droplet interactions in the L2 phase are reinforced. In addition the spontaneous curvature is changed, and a bicontinuous phase is formed already at room temperature.

When the polymer-surfactant head group interactions are enhanced by adding polyampholytes or oppositely charged polyelectrolytes, the stability of the surfactant film is increased, and repulsive droplet-droplet interactions are induced in the L2 phase. The transition to a sponge phase is observed only at higher temperatures or higher polymer concentration, this means by boosting the surfactant head group polyelectrolyte interactions, and changing the spontaneous curvature of the surfactant film.

16.3.2. $BaSO_4$ Nanoparticle Formation in the Microemulsion Template Phase

16.3.2.1. *Nanoparticle Formation in the L2 Phase*

When two adequate microemulsions, one containing $BaCl_2$ and the other one Na_2SO_4, were mixed together, due to the spontaneous collision of the water droplets $BaSO_4$ nanoparticles are formed. The resulting nanoparticles can be characterized by means of dynamic light scattering and transmission electron microscopy.

In the SDS-based L2 phase in presence of PEG we were able to produce $BaSO_4$ nanoparticles of about 5 nm. In presence of PalBu 3.2 nm and in presence of PalH 2.0 nm sized particles were formed inside the water droplets. These results show a strong regulating effect of the polymers in the water-in-oil microemulsion.

16.3.2.2. *Nanoparticle Formation in the Bicontinuous Phase*

When the particle formation is realized in the bicontinuous phase channel in presence of PEG also small nanoparticles of about 5–10 nm can be formed, when the polymer concentration of the polymer is high enough and the molar mass is low (M_w = 3000 g/mol). Recently, we have shown that the molar mass of the polymer can control the size of the bicontinous phase channels, and therefore the size of the nanoparticles formed.[22]

16.3.3. Redispersed $BaSO_4$ Nanoparticles

16.3.3.1. Nanoparticles Redispersed from the L2 Phase

After a complete solvent evaporation realized in a vacuum oven at 30°C for 3 days, the crystalline powder can be redispersed in water. When these experiments were made in absence of polymer only particles with an average particle size $\geq$ 300 nm can be redispersed.

When the same experiments were made in the PDADMAC-modified SB-based microemulsion template phase[11] nanoparticles < 10 nm can be redispersed again.

In Figure 16.6 one can see $BaSO_4$ nanoparticles (7 nm in size) redispersed from the SB-based microemulsion in presence of PDADMAC. By means of electrophoretic measurements a positive zeta potential of $+29 \pm 5$ mV can be detected. From these data one can conclude that the polycation is adsorbed at the particle surface, and can stabilize the particles during the process of solvent evaporation and redispersion.

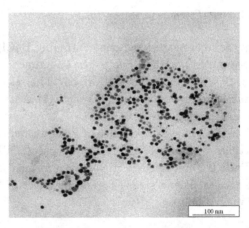

Fig. 16.6. TEM micrograph of redispersed BaSO4 particles formed in the L2 phase of the SB/heptanol/water system in presence of 10% PDADMAC

In the L2 phase of the SDS-based microemulsion in presence of the polyampholytes PalBu and PalH quite smaller particles of about 3.4–3.9 nm can be successfully redispersed, but not yet in presence of the nonionic PEG. This means that only polyelectrolytes can adequately stabilize nanoparticles during the process of solvent evaporation and redispersion.

16.3.3.2. Nanoparticles Redispersed from the Bicontinuous Phase

When the nanoparticles formed in the bicontinuous microemulsion in presence of PEG of low molar mass, are redispersed after solvent evaporation, fine particles of about 5 nm can also be redispersed. However, the nanoparticles tend to grow to larger aggregates and $BaSO_4$ nanorods and triangular plates are formed (Figure 16.7), especially when the molar mass of the PEG is increased. This means the PEG is not able to suppress the growth of the preliminary formed spherical nanoparticles to plate-like or rod-like aggregates. It has to be mentioned here that similar formed $BaSO_4$ aggregates, *i.e.* filaments, can be produced for example in AOT-based microemulsions by changing the molar ratio between the reactants.[23] The formation of cubic $BaSO_4$ aggregates has been shown already by other groups.[7,24]

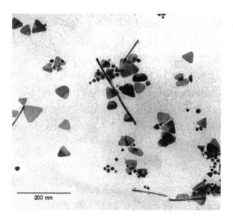

Fig. 16.7. TEM micrograph of redispersed BaSO4 particles formed in the bicontinuous phase of the SDS/xylene-pentanol/water system in presence of 10% PEG

16.4. Conclusions

Our investigations have shown that water soluble polymers can influence the physicochemical behaviour of the L2 phase in a characteristic way. Hydrophobic polymer-surfactant interactions lead to a destabilization of the surfactant film, and therefore more attractive droplet-droplet interactions can be observed. Electrostatic interactions between the surfactant head groups and polyelectrolytes increase the film stability, and more repulsive droplet-droplet interactions have to be taken into account. In addition both types of interactions can induce the formation of a bicontinuous microemulsion due to a change of the spontaneous curvature of the surfactant film.

The polymer-modified microemulsions can be used successfully as a template phase for producing nanoparticles. Ultrafine spherical nanoparticles with particle dimensions below 10 nm can be produced predominantly in the reverse microemulsion droplets due to the regulating effect of the polymers. However, also in the sponge phase small sized nanoparticles of about 5 nm can be formed, when the molar mass of the PEG is low and the concentration high enough. In the following solvent evaporation, and redispersion process the polyelectrolytes can stabilize the nanoparticles formed, and in consequence ultrafine spherical nanoparticles with the same diameter, already observed in the microemulsion, can be redispersed. In contrast to the polyelectrolytes used, the "stabilizing power" of the non-ionic polymers is often not strong enough to hinder aggregation phenomena during the recovery process. Nevertheless, these systems are of special interest for producing nanorods or multi-angular platelets.

When ultrafine nanoparticles are required, polyelectrolytes are much better stabilizing components, but the final particle size mainly depends on the type of polyelectrolyte used. Polyampholytes show the best effects and nanoparticles smaller than 4 nm can be redispersed.

References

1. Capek, *Adv. Colloid Interface Sci.* **110**, 49 (2004).
2. W.Z. Zang, X.L. Qiao, J.G. Chen, *Colloids Surf. A: Physicochem. Eng. Aspects* **299**, 22 (2007).
3. M.P. Pileni, L. Motte, F. Billoudet, C. Petit, *Surface Review and Letters* **3**, 1215 (1996).
4. F. Rauscher, P. Veit, K. Sundmacher, *Colloids Surf. A : Physicochem. Eng. Aspects* **254**, 183 (2005).
5. M. Lade, H. Mays, J. Schmidt, R. Willumeit, R. Schomäcker, *Colloids Surf. A : Physicochem. Eng. Aspects* **163**, 3 (2000).
6. T. Hirai, H. Sato, I. Komasawa, *Industrial & Engineering Chemistry Research* **32**, 3014 (1993).
7. M. Summers, J. Eastoe, S. Davis, *Langmuir* **18**, 5023 (2002).
8. H. Bagger-Jorgensen, U. Olsson, I. Iliopoulos, K. Mortensen, *Langmuir* **13**, 5820 (1997).
9. P. Lianos, *J. Phys. Chem.* **100**, 5155(1996).
10. P. Plucinski, J. Reitmeir, *Colloids Surf. A : Physicochem. Eng. Aspects* **122**, 75 (1997).

11. J. Koetz, J. Bahnemann, G. Lucas, B. Tiersch, S. Kosmella, *Colloids Surf. A : Physicochem. Eng. Aspects* **250**, 423 (2004).
12. C. Note, J. Koetz, S. Kosmella, *Colloids Surf. A : Physicochem. Eng. Aspects* **288** 1-3, 158 (2006).
13. C. Note, S. Kosmella, J. Koetz, *J. Colloid Interface Sci.* **302**, 662 (2006).
14. T. Beitz, J. Koetz, G. Wolf, E. Kleinpeter, S.E. Friberg, *J. Colloid Interface Sci.* **240**, 581 (2001).
15. J. Koetz, S. Andres, S. Kosmella, B. Tiersch, *Composite Interfaces* **4-6**, 461 (2006).
16. J. Koetz, J. Bahnemann, S. Kosmella, *J. Polymer Sci., Chem. Ed.* **42**, 742 (2004).
17. T. Beitz, J. Koetz, S.E. Friberg, *Prog. Colloid Polym. Sci.* **111**, 100 (1998).
18. C. Note, J. Ruffin, J. Koetz, *J. Dispersion Sci. and Technology* **28** (1), 155 (2007).
19. C. Note, J. Koetz, L. Wattebled, A. Laschewsky, *J. Colloid Interface Sci.* **308**, 162 (2007).
20. J. Koetz, C. Gnther, S. Kosmella, E. Kleinpeter, G. Wolf, *Prog. Colloid Polym. Sci.* **122**, 27 (2003).
21. J. Koetz, J. Baier, S. Kosmella, *Colloid Polym. Sci.* **285**, 1719 (2007).
22. S. Lutter, J. Koetz, B. Tiersch, S. Kosmella, *J. Dispersion Sci. Technol.* **30**, 745 (2009).
23. M. Li, S. Mann, *Langmuir* **16**, 7088 (2000).
24. L.M. Qi, J.M. Ma, H.M. Cheng, Z.G. Zhao, *Colloids Surf. A : Physicochem. Eng. Aspects* **108**, 117 (1996).

Chapter 17

Maximizing the Uptake of Nickel Oxide Nanoparticles by AOT (W/O) Microemulsions

Nashaat N. Nassar and Maen M. Husein

Department of Chemical & Petroleum Engineering,
University of Calgary, Calgary,
Alberta, Canada T2N 1N4

Ultradispersed colloidal nanoparticles in organic media are applied as catalysts to promote organic reactions and sorbents for gaseous and other organic species. Their catalytic activity and sorption capacity are greatly influenced by their surface area/volume of the organic medium which, in turn, is dictated by their size and stable colloidal concentration. In this study, the highest possible concentration of stable colloidal nickel oxide nanoparticles, nanoparticle uptake, having an average particle diameter of 5 nm were obtained in AOT/water/isooctane microemulsions. Variables affecting nanoparticle size, stability and uptake were investigated and conditions leading to maximum uptake were identified. Nanoparticle uptake increased as AOT and precursor salt concentrations increased. Water to AOT mole ratio, R, resulted in a maximum uptake around $R = 3.0$. The surface area of nanoparticles/volume of the microemulsion followed the same trend as the nanoparticle uptake. The particle size increased as R, AOT concentration and precursor salt concentration increased. The stability of the nanoparticles and their size were determined using inductively coupled plasma, dynamic light scattering, and UV-Vis absorption spectroscopy.

17.1. Introduction

Colloidal metal oxide nanoparticles have been the subject of interest due to their attractive physical and chemical properties. Metal oxide colloidal nanoparticles have promising applications in the fields of ultradispersed catalysts and sorbents for *in-situ* heavy oil upgrading,[1-4] drug delivery,[5-7] dip coating,[8,9] *etc.* These applications can be better served when the highest possible concentration of stable colloidal nanoparticles is maintained.

Water-in-oil (w/o) microemulsions, or reverse micelles, have been used to prepare stable colloidal nanoparticles, however, they suffer from low reactant solubilization and product stabilization capacities.[10,11] Nonetheless, for applications such as ultradispersed catalysis and sorption within oil phases, microemulsions serve as an ideal media for the preparation and dispersion of colloidal nanoparticles.[1,2] Moreover, (w/o) microemulsions are capable of preparing the ultradispersed catalyts and sorbents *in-situ*, which limits particle aggregation associated with particle storage and transportation. Maintaining the highest concentration of stable colloidal nanocatalysts is essential for achieving high rates of reaction. For given values of microemulsion and operating variables, the time-independent maximum possible concentration of colloidal nanoparticles in the microemulsions is referred to as nanoparticle uptake.[12,13]

The current work demonstrates the use of AOT microemulsions for *in-situ* preparation of high concentration of nickel oxide nanoparticles. In addition, this work aims at maximizing the surface area of colloidal nanoparticles/volume of the microemulsion system, and studies the effect of some microemulsion and operation variables that favor high nanoparticle uptake of nickel oxide. Maintaining high surface area/volume of the microemulsion systems requires maintaining smallest particle size with the highest possible uptake.

17.2. Experimental Section

17.2.1. *Chemicals*

AOT, dioctylsulfosuccinate sodium salt (98%, Sigma-Aldrich Fine Chemical, Toronto, ON, Canada) was used as the surfactant, and isooctane (HPLC grade, Fisher scientific, Toronto, ON, Canada) was used as the oil phase in all the experiments. Nickel (II) nitrate hexahydrate, $Ni(NO_3)_2.6H_2O$, (99.99%, Alfa Aesar, Toronto, ON, Canada) was used as the precursor salt and sodium hydroxide (5.0 N, Alfa Aesar, Toronto, ON, Canada) was used as the precipitating agent. All chemicals were used without further purification.

17.2.2. *Nanoparticle Synthesis*

The colloidal nickel oxide nanoparticles were prepared *in-situ* by exposing AOT/water/isooctane microemulsions to $Ni(NO_3)_2.6H_2O$ precursor prior to the addition of aqueous NaOH.[12,13] Following the addition of $NaOH_{(aq)}$, nickel oxide/hydroxide nanoparticles stabilized in the water pools of the microemulsions formed in addition to a bulk precipitate of nickel oxide/hydroxide. All the experiments were performed by mixing 20 mL volume of AOT microemulsions with a specified amount of solid $Ni(NO_3)_2.6H_2O$ until it completely dissolves, unless otherwise noted. The concentration of $Ni(NO_3)_2$ was considered on the basis of the volume of the microemulsions. A stoichiometric amount of NaOH was then added to the microemulsions and the system was left to mix for 1.5 h at 300 rpm and 25°C.

Stock AOT microemulsions were prepared by adding AOT and distilled water to a volumetric flask and completing the volume to the mark with isooctane. Aqueous NaOH solution was accounted for the calculation of the mole ratio of water to AOT, R. A change in R was achieved by adding distilled water to the microemulsions before introducing any reactant. All microemulsions were freshly prepared and optically clear before introducing $Ni(NO_3)_2.6H_2O$ powder. A complete clear greenish solution was obtained after dissolving the $Ni(NO_3)_2.6H_2O$. When NaOH was added to the system, the solution became cloudy and, then, turned clear greenish after shaking for 1 min. The system was left to mix for 1.5 h after the addition of NaOH, and bulk greenish/blackish precipitate of nickel oxide/hydroxide formed at the bottom of the jar in all the samples. Volumes of 5 ml were collected from the experimental samples after mixing for 1.5 h, and filtered out using Fisher Brand 0.45 μm microfilters (Cole Parmer, Toronto, ON, Canada). Clear samples were obtained after filtration. The presence of colloidal nickel oxide/hydroxide nanoparticles was confirmed using the UV-Vis spectroscopy (Nicolet Evolution 100, ThermoInstruments Canada Inc., Mississauga, ON, Canada). Microemulsions identical to those used in the experimental samples were employed as blank. A range of wavelength between 250 and 450 nm was covered. The nanoparticle size was characterized using dynamic light scattering (DLS) (Malvern Instruments Ltd., UK). The nanoparticle uptake of the colloidal nickel oxide was determined using inductively coupled plasma (ICP) (IRIS Intrepid II XDL, ThermoInstruments Canada Inc., Mississauga, ON, Canada). A standard containing 5 ppm was prepared by diluting 1000 ppm Ni stock standard in the form of nickel colloidal particles with PermiSolv^{TM} ICP solvent (CONOSTAN Oil Analysis Standards, Ponca City, OK). During the standard preparation no bulk precipitate formed. Volumes of the filtered experimental samples were diluted with PremiSolv^{TM} to concentrations between 2 to 8 ppm. Three replicates were prepared for some samples, and the 95% confidence intervals were calculated and presented in the figures.

17.3. Results and Discussion

17.3.1. Stability of the Colloidal Nanoparticles

The stability of the colloidal nanoparticles was examined by leaving samples, prepared after mixing for 1.5 h, standing for more than a week without mixing. No significant difference in the nanoparticle uptake was observed. The difference between the sample prepared after 1.5 h and the one left standing for more than a week was 0.12 mM which falls within ± 0.13 of 95% confidence interval. This indicates that the microemulsions provide excellent stability for nickel oxide/hydroxide nanoparticles at the condition considered in this study. In an attempt to maximize the nanoparticle uptake the following variables were examined, AOT concentration, water to AOT mole ratio (R), and the concentration of nickel nitrate precursor, [$Ni(NO_3)_2$].

17.3.2. Effect of AOT Concentration

Recent studies involving reactive[14–16] and non-reactive surfactants[12,13] showed that nanoparticle uptake can be increased by increasing the surfactant concentration at constant water to surfactant mole ratio. Higher surfactant concentration corresponds to higher population of reverse micelles, nano-reactors, in the continuous oil phase[17] and, potentially higher nanoparticle uptake. On the other hand, the rate of collision among nanoparticle-populated reverse micelles increases with the increase of the surfactant concentration, which increases the probability of particle collision and aggregation. In this study, the effect of increasing surfactant concentration was investigated at $R = 3.0$ and constant [$Ni(NO_3)_2$] of 11 mM, and the stoichiometric amount of NaOH. Fig. 17.1 shows the nanoparticle uptake as a function of AOT. Fig. 17.2 depicts the UV-Vis absorption spectra of the colloidal nanoparticles for 120, 150, and 200 mM AOT. The particle size distribution histograms of the same samples are shown in Fig. 17.3.

Figure 17.1 indicates that, at $R = 3.0$ the nanoparticle uptake increased linearly with the AOT concentration, in agreement with our previous findings.[12,13] Consistently, Fig. 17.2 shows an increase in the UV-Vis absorption peaks as AOT concentration increased. Husein et al.[14–16] reported an increase in the peak maximum as the nanoparticle uptake increased. However, the maximum could not be used as a reproducible measurement of particle uptake; since in addition to particle size,[18,19] it varies with the excess lattice ion, surfactant species and surfactant concentration.[20] A shift towards a bimodal distribution is seen in Fig. 17.3 at high AOT concentration suggesting particle aggregation. The increase in the nanoparticle uptake is attributed to the increase in the population of the nano-reactors. Also, high population of the nano-reactors increased the rate of collisions between nanoparticle-populated reverse micelles and led to particle aggregation as discussed earlier.[21,22]

Table 17.1. Mean N_i^{2+} ion occupancy number and surface area of the stabilized nanoparticles.

Before reaction		After reaction			
Mean occupancy no. of Ni^{2+} ions[a]		Nanoparticle uptake (mM)	Average particle diameter (nm)	Surface area/mass (m^2/g)[b]	Surface area / volume of microemulsions (m^2/L)[c]
Effect of AOT concentration (mM)					
100	0.018	2.0			
120	0.042	2.6	5.1	150	36
150	0.067	3.0			
200	0.083	4.3	5.6	123	50
250	0.085	5.3	6.3	116	57
Effect of R					
2	0.016	3.0	4.6	157	43
2.5	0.043	3.5			
3	0.083	4.3	5.6	123	50
4	0.200	1.3	6.2	119	14
5	0.369	0.8			
6.5	0.718	1.0			
Effect of [Ni(NO$_3$)$_2$] (mM)					
3.5	0.058	1.7	4.4	153	24
5.4	0.074	2.2			
7	0.082	3.3	5.0	127	39
9	0.085	3.5			
11	0.083	4.3	5.6	123	50
12.5	0.076	4.8			

[a] Calculated according to reference[23] with R on the basis of the volume of distilled water added before introducing NaOH aqueous solution.

(b) Calculated on the basis of DLS measurements.
(c) Calculated on the basis of DLS measurements and nanoparticle uptake in the given volume of the microemulsions.
(d) For no bulk precipitate mean occupancy no. of Ni^{2+} ions(a) and nanoparticle uptake (nM) are 0.031 and 1.5 respectively. Sample prepared at $R = 3.0$, [AOT]= 200 mM, and [Ni(NO$_3$)$_2$]= 1.5 mM.

Table 17.1 presents the Ni^{2+} ion occupancy number, before addition of aqueous NaOH, and the surface area provided by the colloidal nanoparticles. The mean ion occupancy number was calculated based on the given microemulsion properties and precursor salt concentration.[12,13,23–25] The ion occupancy number per reverse micelle increased as AOT concentration increased at constant R. As expected, the surface area/mass of colloidal nanoparticles decreased as the average particle size increased. However, the surface area/volume increased. Higher surface area/volume was obtained at higher nanoparticle uptake. A bout 60% increase in the surface area/volume of the microemulsions was achieved upon increasing the AOT concentration from 100 to 250 mM. Hence, in order to maintain a high surface area/volume of organic phase, in addition to small particle size, it is essential to maintain high nanoparticle uptake.

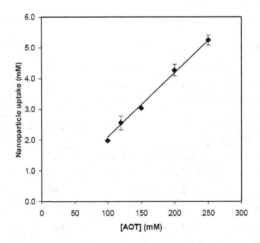

Fig. 17.1. Nanoparticle uptake as a function of AOT concentration obtained when mixing a stoichiometric amount of Ni(NO$_3$)$_2$-containing microemulsions {[Ni(NO$_3$)$_2$]= 11 mM, R= 3} for 1.5 h with NaOH.

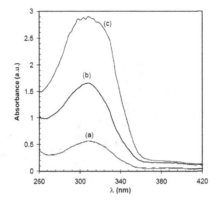

Fig. 17.2. UV-Vis absorption spectra of colloidal nickel oxide nanoparticles obtained when mixing a stoichiometric amount of Ni(NO$_3$)$_2$-containing microemulsions {[Ni(NO$_3$)$_2$]= 11 mM, R= 3; (a) [AOT]= 120 mM, (b) [AOT]= 150 mM, (c) [AOT]= 200 mM} for 1.5 h with NaOH.

17.3.3. Effect of Water to AOT Mole Ratio, R

R is considered one of the key parameters that affect the stability of the microemulsion,[26,27] the size of the nanoparticles,[25,28–31] and the nanoparticle uptake.[12–16] In this experiment, R was increased from 2.0 to 6.5 at constant AOT concentration of 200 mM, initial concentration of $N_i(NO_3)_2$ of 11 mM and stoichiometric amount of NaOH. The nanoparticle uptake at different R and the particle size distribution histograms for selected samples are depicted in Figs. 4 and 5, respectively.

Figure 17.4 shows a maximum uptake at $R = 3.0$. For $R > 3.0$, a sharp decrease in the nanoparticle uptake was observed. Measurements and calculations showed that as R increases, at fixed values of all other parameters, the number of reverse micelles decreases, while their waterpool size increases.[23,24,32,33] This trend has two opposing effects. On one hand, it increases the population of reverse micelles with high ion occupancy numbers, which in turn, improves nanoparticle uptake.

On the other hand, high water content reduces the rigidity of the surfactant layer, and hence contributes to particle aggregation upon collision of nanoparticle-populated reverse micelles.[24,25,33] In addition, high water content reduces the interaction between the surfactant head groups and the colloidal nanoparticles,[12,13] which also contributes to particle aggrega-

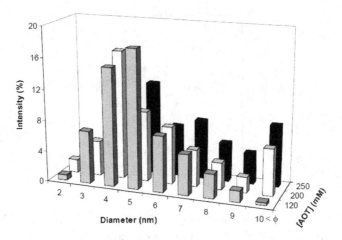

Fig. 17.3. Particle size distribution histograms obtained when mixing a stoichiometric amount of Ni(NO$_3$)$_2$-containing microemulsions {[Ni(NO$_3$)$_2$]= 11 mM, R= 3} at different AOT concentration for 1.5 h with NaOH.

tion. Fig. 17.5 shows a broader size distribution as R increased indicating particle aggregation. Similar trends were reported in our previous studies for iron and copper oxide nanoparticles.[12,13] It is worth noting that the maximum nanoparticle uptake occurred at different R values for each type of nanoparticles. For example the maximum nanoparticle uptake for iron oxide and copper oxide occurred at R of 6.5 and 5.0, respectively; while in this study it occurred at 3.0 for nickel oxide. In addition, when R exceeded the optimum value the nanoparticle uptake decreased gradually for iron and copper oxides and it decreased sharply for nickel oxide. These differences in response to the increase in R may be attributed to the degree of interaction between the surfactant head groups and the colloidal nanoparticles as well as the nature of the stabilized metal oxide particles.

Table 17.1 shows that the N_i^{2+} ion occupancy number increased as R increased. A comparison between $R = 3.0$ and the first entry of Table 17.1, where no bulk precipitation occurred, showed that about 170% increase in the N_i^{2+} ion occupancy number occurred upon saturating the microemulsion with the nanoparticles. This increase would promote the increase in nanoparticle uptake which favors the increase in surface area/volume. However, the less rigid surfactant layer led to particle aggregation and ultimately reduced the nanoparticle uptake and surface area.

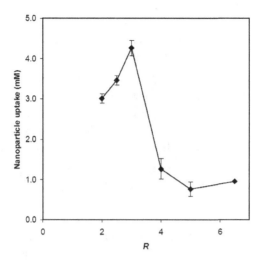

Fig. 17.4. Nanoparticle uptake as a function of water to AOT mole ratio, R, obtained when mixing a stoichiometric amount of $N_i(NO_3)_2$-containing microemulsions $\{[N_i(NO_3)_2]= 11$ mM, [AOT]= 200 mM$\}$ for 1.5 h with NaOH.

17.3.4. Effect of Concentration of $N_i(NO_3)_2$ Precursor

An increase in the concentration of $Ni(NO_3)_2$ favors the increase of Ni^{2+} ion occupancy number which is expected to increase the nanoparticle uptake.[12,13] In this experiment, the initial concentration of $Ni(NO_3)_2$ in the microemulsions was increased from 3.5 to 12.6 mM at constant mole ratio of NaOH to $N_i(NO_3)_2$ of 2.0, R of 3.0, and AOT concentration of 200 mM. The concentration of $N_i(NO_3)_2$ did not exceed 12.5, since further increase of this value resulted in a change of the microemulsion structure towards Winsor type II. It should be noted that the same microemulsions could only solubilize a maximum of 9.3 mM $FeCl_3$,[12] and 6.0 mM $CuCl_2$,[13] before a change of structure took place.

Figure 17.6 shows an increase in the nanoparticle uptake and Fig. 17.7 shows an increase in the particle size as the initial concentration of $N_i(NO_3)_2$ increased. The increase in the nanoparticle uptake and particle size is attributed to the increase of the nickel ion occupancy number and the improvement of intramicellar nucleation and growth.[34,35] A similar observation was reported in our previous studies.[12,13]

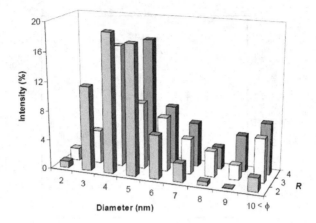

Fig. 17.5. Particle size distribution histograms obtained when mixing a stoichiometric amount of $N_i(NO_3)_2$-containing microemulsions $\{[N_i(NO_3)_2]=$ 11 mM, [AOT]= 200 mM$\}$ at different R values for 1.5h with NaOH.

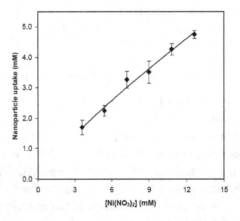

Fig. 17.6. Nanoparticle uptake as a function of $N_i(NO_3)_2$ concentration obtained when mixing a stoichiometric amount of $N_i(NO_3)_2$-containing microemuilsions $\{$R=3, [AOT] = 200 mM$\}$ for 1.5 h with NaOH.

Table 17.1 shows a decrease in the surface area/mass of nanoparticles as the concentration of $N_i(NO_3)_2$ in microemulsion increased due to the increase in particle size. Nonetheless, an increase in the surface area/volume accompanied the increase in $N_i(NO_3)_2$ concentration and nanoparticle uptake.

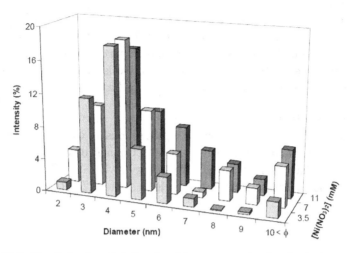

Fig. 17.7. Particle size distribution histograms obtained when mixing a stoichiometric amount of Ni(NO$_3$)$_2$-containing microemulsions {[AOT]= 200 mM, R= 3} at different [N$_i$(NO$_3$)$_2$] for 1.5 h with NaOH.

17.4. Conclusions

This work aimed at using the microemulsion approach for *in-situ* preparation of high concentration of stable colloidal nickel oxide nanoparticles. Parameters that promoted high concentration of stable colloidal nanoparticles, nanoparticle uptake, were identified. The nanoparticle uptake was found to be depended on the precursor salt concentration, water content and AOT concentration. At constant water to AOT mole ratio, R, the nanoparticle uptake increased lineally with the AOT concentration. This was attributed to the increase in the number of reverse micelles, which resulted in higher uptake and larger particle size due to the collision between nanoparticle-populated reverse micelles. A maximum value of nanoparticle uptake was obtained at R of 3.0. At relatively low water content, $R < 3.0$, increasing R increased the nanoparticle uptake due to the increase in the size of reverse micelle and the solubilization capacity of N$_i$(NO$_3$)$_2$. At $R > 3.0$, the nanoparticle uptake declined due to the decrease in the rigidity of the surfactant layer and the interaction between surfactant head groups and the stabilized nanoparticles. The nanoparticle uptake and size increased as the initibidyutmaster.al concentration of N$_i$(NO$_3$)$_2$ increased. This was due to the increase in ion occupancy number, which enhanced intramicellar nucleation and growth. Despite the decrease in surface area/mass as nanoparticle size increased, surface area/volume of the microemulsion increased in response to the increase in the nanoparticle uptake. Maintaining high surface area/volume of the organic phase is vital for ultradispersed catalysis and sorbents from within organic media.

Acknowledgments:
We would like to thank the following organizations for the financial support: Natural Science and Engineering Research Council of Canada (NSERC), Alberta Ingenuity Centre for *in Situ* Energy (AICISE), Centre for Environmental Engineering Research and Education (CEERE), and Alberta Ingenuity Fund.

References

1. N. Nassar, *A (w/o) Microemulsion Approach for In-Situ Preparation of High Concentration of Colloidal Metal Oxide Nanoparticles*, Ph.D. Thesis, University of Calgary, Calgary, AB, Canada (2007).
2. M.M. Husein, L. Patruyo, P. Pereira-Almao and N.N. Nassar, *J. Colloid Interface Sci.* 2009 (article in press).
3. A.D. Bianco, N. Panariti, S.D. Carlo, P.L. Beltrame and P. Carnit, *Energy Fuels* **8**, 593 (1994).
4. A.D. Bianco, A.D., N. Panarit, S.D. Carlo, J. Elmouchnino, B. Fixari and P.L. Perchec, *Appl. Catal.*, A **94**, 1 (1993).
5. S.B. Mirta, D. Wu and B.N. Holmes, *JADA*, **134**, 1382 (2003).
6. P. Tartaj, M.P. Morales, S. Veintemillas-Verdauguer, T. Gonzalez-Carreno and C.J. Serna, *J. Phys. D: Appl. Phys.* **36**, R182 (2003).
7. I. Brigger, C. Dubernet and P. Couvreur, *Adv. Drug Deliv. Rev.* **54**, 631(2002).
8. J. Yu, W. Ho, J. Lin, H. Yip and P. Wong, *Environ. Sci. Technol.* **37**, 2296 (2003).
9. M. Adachi, Y. Suzuki, N. Kashiwagi, T. Isobe and M. Senna, *Colloids Surf.*, A **153**, 617 (1998).
10. M.M. Husein, M. E. Weber and J.H. Vera, *Langmuir* **16**, 9159 (2000).
11. K. Holmberg, *Adv. Colloid Interface Sci.* **51**, 137 (1994).
12. N.N. Nassar and M. M. Husein, *Langmuir* **23**, 13093 (2007).
13. N.N. Nassar and M. M. Husein, *J. Colloid Interface Sci.* **316**, 442 (2007).
14. M.M. Husein, E. Rodil and J. H. Vera, *WJCE* **1**, 13 (2007).
15. M. Husein, E. Rodil and J. H. Vera, *Langmuir* **22**, 2264 (2006).
16. M. Husein, E. Rodil and J. H. Vera, *J. Colloid Interface Sci.* **288**, 457 (2005).
17. T.K. De and A. Maitra, *Adv. Colloid Interface Sci.* **59**, 95 (1995).
18. M. Husein, E. Rodil and J. H. Vera, *J. Colloid Interface Sci.* **273**, 426 (2004).
19. M. Husein, E. Rodil and J. H. Vera, *Langmuir* **19**, 8467 (2003).

20. T. Tanaka, H. Saijo and T. Matsubara, *J. Photogr. Sci.* **27**, 60 (1979).
21. C. Tojo, M. C. Blanco and M. A. Lopez-Quintela, *Langmuir* **13**, 4527 (1997).
22. C. Petit, M. P. Lixon and M. P. Pileni, *J. Phys. Chem.* **97**, 12974 (1993).
23. R.P. Bagwe and K. C. Khilar, *Langmuir* **16**, 905 (2000).
24. R.P. Bagwe and K. C. Khilar, *Langmuir* **13**, 6432 (1997).
25. M.P. Pileni, *J. Phys. Chem.* **97**, 6961 (1993).
26. M. Bourrel and R. S. Schechter, *Microemulsions and related rystems: formation, solvency, and physical properties,"* Surfactant Science Series, **30** (Dekker, New York, 1998).
27. P. D. Fletcher, *J. Chem. Soc., Faraday Trans.* **1**. 82, 2651 (1986).
28. F. Debuigne, L. Jeunieau, M. Wiame and J. B. Nagy, *Langmuir* **16**, 7605 (2000).
29. V. Pillai, P. Kumar, M. Hou, P. Ayyub and D. O. Shah, *Adv. Colloid Interface Sci.* **55**, 241 (1995).
30. M.A. Lopez-Quintela and J. Rivas, *J. Colloid Interface Sci.* **158**, 446 (1993).
31. S. Hingorani, V. Pillai, P. Kumar, M. S. Multani and D. O. Shah, *Mater. Res. Bull.* **28**, 1303 (1993).
32. M.L. Curri, A. Agostiano, L. Manna, M. D. Monica, M. Catalano, L. Chiavarone, V. Spagnolo and M. Lugara, *J. Phys. Chem. B.* **104**, 8391 (2000).
33. T. Hirai, H. Sato and I. Komasawa, *Ind. Eng. Chem. Res.* **33**, 3262 (1994).
34. A.R. Kumar, G. Hota, A. Mehra and K.C. Khilar, *AICHE* **50**, 1556 (2004).
35. R. Bandyopadhaya, R. Kumar and K.S. Gandhi, *Langmuir* **16**, 7139 (2000).

Chapter 18

A Brief Overview on Synthesis and Size Dependent Photocatalytic Behaviour of Luminescent Semiconductor Quantum Dots

A. Priyam, S. Ghosh, A. Datta, A. Chatterjee and A. Saha

UGC-DAE Consortium for Scientific Research,
Kolkata Centre, III/LB-8, Bidhannagar,
Kolkata -700 098, India

Supersaturation was found to play a pivotal role during nanoparticle-synthesis and its subtle variation helped to achieve two prime objectives: a) high photoluminescence quantum efficiency (PLQE) and b) narrow size distribution, thereby obviating the need for post-preparative treatments. The as-prepared CdS and CdTe QDs were found to have size distribution as low as 4% at higher supersaturation. Focussing of size distribution was also demonstrated in dendrimer-mediated synthesis of CdTe QDs by controlling the temperature. For a four-fold increase in supersaturation, PLQE of cysteine-capped CdTe QDs (4.3 nm) rose by 5 times to a remarkably high value of 54%. The focusing of size distribution with increasing supersaturation was found to work well even in the absence of any stabilizer. This study provides a simplified route for producing highly monodisperse, photoluminescent and surface-functionalized nanoparticles. Thiophenol capped CdS QD has been used as a model system to demonstrate the size dependence of photocatalysis of nitroaromatics. The catalytic efficiency of CdS quantum particles was quintupled with decrease in particle size from 5.8 nm to 3.8 nm. An empirical equation has been derived to correlate the catalytic efficiency of the nanoparticles with the twin factors operating in the quantum confinement regime: (i) change in surface to volume ratio and (ii) shift in conduction band edge. In addition, quenching of photoluminescence of thiophenol capped CdS QDs by nitromatics can be used for developing a protocol for a sensor. The Stern−Volmer constant of dinitrobenzene was about 15-fold higher than nitrobenzene, which indicates that introduction of nitro groups in the benzene ring increases the quenching efficiency.

18.1. Introduction

Plethora of synthetic methodologies has evolved over the last decade for preparation of II-VI and III-V colloidal quantum dots.[1,2] The widespread interest in these nanoparticles (NPs) has been primarily spurred by the excellent optical properties of II-VI semiconductor NPs that can be harnessed for opto-electronic device-making[3] and biomedical applications.[4,5] The luminescent quantum dots promise to be an attractive alternative for biolabelling and biosensing applications as they overcome the limitations of the organic dyes such as narrow excitation bands and broad red-tailing luminescence spectra, low resistance to photodegradation, photobleaching and random on/off light emission (blinking). QDs have high photobleaching threshold and emit bright and steady fluorescence.[6] A pre-requisite for the high-end applications is the obtainment of high quality nanoparticles. Amongst other things, the two attributes of a nanoparticle (NP) solution that define the 'high quality' are narrow size distribution and high photoluminescence quantum efficiency (PLQE). Controlling the surface, size distribution and PLQE had been the most nagging issues faced by the nanotechnologists around the globe. Surface of semiconductor nanocrystals plays a crucial role in determining PLQE. Passivation of electronic defects by modifying the surface with suitable capping agents can lead to higher quantum efficiency. In addition, these capping groups can impart required functionality for the intended applications.

In the framework of colloidal chemistry, there are basically two approaches for synthesis of high quality quantum dots (QDs), one involving high temperature thermolysis of organometallic precursors[7,8] and the other via aqueous route using polyphosphates[9,10] and thiols[11–13] as stabilizing agents. The aqueous route can use ionizing radiation like gamma radiation, electron beam, etc. as well.[14–17] Both of the synthetic strategies have some drawbacks as well as advantages over the other. In the organometallic method, a very effective separation of nucleation and growth achieved by 'hot injection technique'[7] leads to a narrower size distribution of QDs as compared to those prepared in aqueous solutions. However, the aqueous synthesis generally allows preparation of smaller particles and the post preparative treatments like size-selective precipitations work more efficiently and reproducibly in water.[11] Water-soluble NPs show an improvement in PLQE whereas TOPO-capped NPs show a decline after such treatment. The latter also exhibit unstable luminescence in air due to photo-oxidation, which limits its applicability.[8,11] Moreover, some of the key chemicals employed in organometallc synthesis are extremely toxic, pyrophoric, explosive and/or expensive. This problem has been largely addressed by Peng and co-workers,[18-20] who replaced the organometallic precursor, $(CH_3)_2Cd$, with $CdO/CdCO_3/Cd(Ac)_2$, and some additional ligands like fatty acids and phosphonic acids along with TOPO were also used. By controlling some of the principal parameters during growth, they could obtain NPs with very narrow size distribution (5%) and high PLQE (70% and 85% for CdTe[21] and CdSe[22], repectively) without any post-preparative procedure. By a little variation in this approach, Jang et al.[23] were able to obtain PLQE of 75% after treating the as-prepared CdS QDs with sodium borohydride. Thus synthesized NPs are only soluble in organic solvents and for biomedical applications, the surface-ligands need to be replaced to make them water-soluble. However, some of these NPs become non-emitting on carrying out ligand-exchange with hydrophilic thiols.[21] Even CdSe/CdS and CdSe/ZnS core/shell NPs were found to lose PLQE substantially on ligand-replacement.[24,25] So, the NPs obtained via organometallic/alternative routes are not fully compliant, or rather fail the criteria for biomedical applications. In this regard, the aqueous synthesis using hydrophilic thiols is of prime importance as NPs with high stability and biocompatibility could be obtained. Additionally, surface charge and other surface properties of the NPs can be easily tuned by the choice of thiol-compounds with free functional groups, which is vital for luminescent biolabeling and biosensor application. In this approach, TGA (thioglycollic acid)-capped CdTe QDs with PLQE of 40% were obtained.[11] But, the as-prepared thiol-capped QDs were made to undergo rigorous post-preparative treatments like size selective precipitation and size-selective photoetching to enhance the particle characteristics. By changing the Cd^{2+}:TGA ratio form 1:2.43 to 1:1.3, PLQE as high as 65% was attained for CdTe QDs without any further treatment.[26] However, a reduced ratio of surface-ligands severely compromises the particle stability.

In this report, we present a facile technique to improve the quality of colloidal nanocrystals during the aqueous synthesis itself. It underscores the enormity of supersaturation, a key parameter, which if properly exploited, can better the PLQE and size distribution obviating the need for post-preparative treatment. Evolution of the colloidal nanoparticles essentially involves three steps: supersaturation, nucleation and growth. If a proper control is developed at the very first step, it could have manifold ramifications for the subsequent stages leading to attainment of particles with improved properties. Here, we show that a remarkable focusing of size distribution can be achieved simply by increasing the degree of supersaturation of the initial synthetic mixture. Size distribution as low as ~4% was obtained for the as-prepared semiconductor NPs prepared at higher supersaturation. Additionally, it also augments the PLQE of the nanoparticles to a great extent. Our synthetic approach was quite simple: keep the molar ratio of reagents, Cd^{2+}:Cysteine:Chalcogenide, constant at 1:2.5 :0.5; and increase the absolute concentration of each of the reagents. It raises the supersaturation of initial reaction mixture and, at the same time, also ensures that number of ligands per monomer remains same. By deftly manipulating the supersaturation in this manner, PLQE as high as 54% was attained for cysteine-capped CdTe nanoparticles, which is amongst the highest PL efficiencies reported for the colloidal semiconductor nanoparticles prepared by aqueous route.[11,26] In contrast, earlier workers could obtain PLQE of 15-20% with nearly identical molar ratio of reagents, 1:2.43:0.47 for CD^{2+}:Cysteine:Te.[2-27] It has also been shown that a good control over size distribution can be achieved in dendrimer-mediated synthesis through variation of temperature. Polyamidoamines (PAMAM) dendrimer are highly branched macromolecules possessing both the solvent filled interior core (nanoscale container) as well as a homogeneous, exterior surface functionality (nano-scaffold)[28,29] making them potentially useful for gene transfection and drug delivery.[30] Hence, from functional point of view, the amalgamation of biomimetic properties of dendrimer with excellent luminescence properties of semiconductor NPs like CdS, CdTe *etc.*, can yield novel hybrid materials ideally suited for various biomedical applications. These soft chemical approaches circumvent the problem of using high temperature and toxic chemicals in organometallic/alternative approach, and the entire process gets speeded up as umpteen steps like ligand-exchange, silica coating, size sorting, photoetching *etc.* become redundant. The effect of initial supersaturation on growth kinetics of the QDs has also been examined.

We have endeavoured to use our surface-functionalized semiconductor quantum dots for a variety of applications, like in disease diagnosis, biosensing etc.[5,31] In addition, catalytic properties of these NPs can be utilized in environmental remediation of pollutants.[32] Among semiconductors, TiO_2 is considered to be the most suitable for photocatalytic degradation of pollutants due to its chemical inertness and greater resistance to photocorrosion.[33] However, TiO_2 has several limitations.[34] CdS NPs have been chosen as model system for studying the size dependent photocatalytic activity. Further, the effect of change in surface to volume ratio coupled with shift in conduction/valence band edge on photocatalysis can be easily investigated for CdS nanoparticles due to availability of quantitative correlation of the energy shift with particle size.[35–37] In addition, its fluorescence properties can be utilized for possible detection of toxic contaminants in the environment.

This article, in general, focuses on understanding the role of supersaturation in determining the quality of nanoparticles prepared through colloidal route. It has also demonstrated the size dependence of photocatalysis of nitroaromatics. In addition, the present paper illustrates the utilization of photoluminescence of the synthesized CdS NPs for possible NP-based sensor for nitroaromatics, which are often detected in industrial effluents, freshwater, ambient environments and atmosphere.[38,39]

18.2. Experimental

18.2.1. *Synthesis of Thiol-capped CdS and CdTe Quantum Dots*

CdS and CdTe QDs were synthesized following the method as reported earlier for cysteine capped nanoparticles[11,12] with some modification. The effect of supersaturation was studied by preparing different sets (5 ml each) of Cd^{2+}-cysteine solution (molar ratio, 1:2.5) with following concentrations of Cd^{2+} ions: 11.7, 23.4, 35.1, 46.8, 70.2 and 93.6 mM. *pH* of the solutions was adjusted to 11.2-11.8 by adding NaOH solution (0.1M) and argon was bubbled through the solution to remove dissolved oxygen. For each set, a suitable amount of Na_2S (or NaHTe) solution was added at room temperature to make the final molar ratio of Cd^{2+}:cysteine:S^{2-} (or Te^{2-}) as 1: 2.5: 0.5. Test tubes containing the resultant solution were put together in a water bath (100°C) for 10 minutes and their UV-Vis absorption and photoluminescence measurements carried out. Here, NaHTe was synthesized following the method as reported earlier.[12]

18.2.2. *Dendrimer-mediated Synthesis of CdTe Quantum Dots*

A typical preparation of CdTe NPs in Dendrimer matrix with an initial Cd^{2+}/Te^{2-} molar ration of 1:1 was as follows: 5 ml aliquot of Cd^{2+} stock solution was added to 10 ml of dendrimer stock solution at 100C, then NaHTe solution was taken in a syringe and injected to the argon purged solution mixture of Cd^{2+} and dendrimer. The resulting solution was bright orange and did not show any evidence of precipitates. The binding of dendrimer on the particle surface has been followed by FTIR.

18.2.3. *Determination of the Average Particle Size and Size Distribution*

Band gap was calculated from the point of inflection in the first derivative plot of the UV-Vis absorption spectra of nanoparticle solutions. The average particle size of quantum dots was obtained from the correlation of bandgap and particle size deduced by tight-binding approximation.[36] The relative percentage distribution of particle sizes within a QD-solution was determined from the absorption spectra by following the method developed by Sarma and co-workers.[40] The particle size and distribution determined from UV-Vis absorption spectra has been found to be reliable and in good agreement with TEM measurements.[12,41]

18.3. Determination of PLQE

The room temperature photoluminescence quantum efficiency (PLQE) of CdS and CdTe nanoparticles was estimated following the procedure[42] by comparison with quinine sulphate (analytical grade, BDH) in 0.1 M aqeous H_2SO_4 (Merck, India) taking its PLQE as 57%.

18.4. Results and Discussions

18.4.1. *Supersaturation Driven Tailoring of Size Distribution and PLQE*

CdS and CdTe QDs, with surface capping of cysteine, were synthesized by controlling supersaturation of initial reaction mixture. For doing so, the molar ratio of Cd^{2+}: Cysteine: S^{2-} (or Te^{2-}), was kept constant at 1:2.5:0.5, but the absolute concentration of each reagent was increased for different sets. It causes the initial supersaturation to increase, however, the number of ligands available for each monomer remains same. The degree of supersaturation corresponding to the 23.4 mM Cd^{2+} ions has been denoted as SCdS and SCdTe for CdS and CdTe system, respectively and has been used for relative scaling of supersaturation.

18.4.2. Size Distribution

The effect of supersaturation on the size distribution of nanoparticles can be easily discerned from Fig. 18.1a-b. For the same time of heating, only a broad absorption shoulder was observed at $0.5S_{CdS}$, and $0.5S_{CdTe}$, which transformed into a peak at S_{CdS} and S_{CdTe} respectively. With further increase in the supersaturation value the peak gradually became sharper indicating narrowing of the size distribution. In order to gauge the extent of focusing, a simple, but accurate, approach has been used to estimate size distribution of nanocrystals from the UV-Vis absorption spectrum.[40] This approach has been found to work well for a variety of semiconductor nanocrystals in the quantum confinement regime.[41] As shown in Fig.1b, the relative percentage distribution decreased from 11% to 4% on going from $0.5S_{CdS}$ to $4S_{CdS}$. This corresponds to a sharp focusing of size distribution by 64% for an eight-fold change in supersaturation. At the same time, average particle size changed marginally from 3.2 nm to 2.8 nm. For CdS QDs, the particle size and distribution appear to vary linearly with supersaturation. However, the reason for linearity is not clear at his stage.

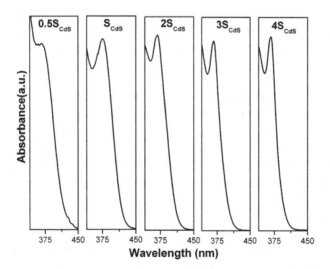

Fig. 18.1a. Absorption spectra of cysteine capped CdS quantum dots synthesized at different degrees of supersaturation. (heating time:10 min.) S_{CdS} is the supersaturation of CdS corresponding to $[Cd^{2+}] = 23.4$ mM.

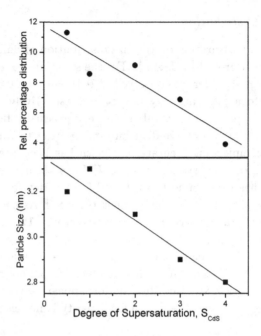

Fig. 18.1b. Effect of supersaturation on average particle size and the size distribution of CdS QDs. (Average size and distribution was measured by UV-Vis absorption as described in sec. 2.3.)

The aforementioned trend can be understood in terms of phenomenological cluster nucleation model as described by Belloni et al.[43] The kinetics of particle formation is controlled by the formation of primary critical nuclei, which are essentially neutral CdS monomers. A high concentration of reagents favors the formation of these primary monomers and the final particle size is smaller due to the generation of large number of nucleaton centers. Under these conditions, the size distribution is narrow as the conditions for homogeneous nucleation are also achieved.

The focusing of size distribution with increasing degree of supersaturation is also in consonance with the theoretical model developed by Talapin et al.[44] At any given time of growth, there exists a critical radius, which has zero growth rate and is in equilibrium with respect to particle growth,

$$r_{cr} = \frac{2\gamma V_m}{RT \ln S} \quad (1)$$

Here, supersaturation is expressed as $S = [M]_{bulk}/C_0^{flat}$, r_{cr} is the critical radius, $[M]_{bulk}$ is the concentration CdS or CdTe monomers in the bulk of solution, C_0^{flat} is the solubility of the bulk material, V_m is the molar volume of solid, γ is the specific surface energy of the particles, T is the temperature of the reaction mixture. At any intervening time during the particle growth, the critical radius (r_{cr}) would be smallest for the NP-solution with highest degree of supersaturation. If this critical size happens to be smaller than the smallest particle present in the ensemble, all particles will show positive growth rate and smaller crystals within the ensemble grow faster than the larger ones leading to focusing of size distribution.[12,45]

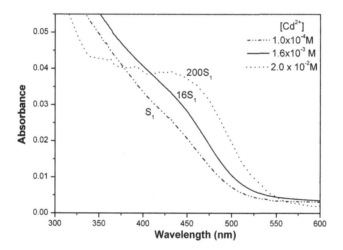

Fig. 18.2. Absorption spectra of uncapped CdS NPs at varying supersaturation. $[Cd^{2+}] : [S^{2-}]$, 1:0.5, S_1 is the supersaturation of CdS corresponding to $[Cd^{2+}] = 10^4 M$.

The fundamentality of the phenomena is established from our results which show that polydispersity of nanoparticles gets lowered even in the absence of stabilizer/capping agent. As displayed in the inset of Fig. 18.2, the absorption profile of bare CdS NPs at a lower value of supersaturation (S_1) was rather diffuse, but was modified to a more structured one at a higher supersaturation ($200S_1$). However, the effect of supersaturation was apparently less pronounced in the absence of capping agent as there was hardly any change in the absorption spectra on moving from S_1 to $16S_1$. For a similar change in supersaturation, a sharp focusing of size distribution was observed with cysteine as capping agent.

18.4.3. *Photoluminescence Quantum Efficiency*

The most interesting outcome of the present study is the supersaturation dependent PL efficiency of the semiconductor QDs; higher the degree of supersaturation, greater is the PLQE. For a given time of heating (10 min.), both CdS and CdTe QDs exhibited a consistent rise in the PLQE with increasing supersaturation, the extent of enhancement being much higher for CdTe QDs (Fig. 18.3). For an eight-fold change in the degree of supersaturation (0.5S → 4S), the PLQE of CdTe QDs gets quadrupled (5% → 20%) while that of CdS QDs goes up by 45% only (11% → 16%).

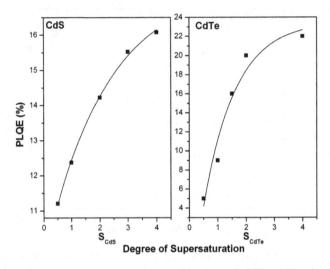

Fig. 18.3. Variation of PLQE of CdS and CdTe NPs with change in degree of supersaturation (heating time = 10 min.) S_{CdS} and S_{CdTe} are the supersaturation of CdS and CdTe respectively, correspoinding to $[Cd^{2+}] = 23.4$ mM.

In order to comprehend the role of supersaturation in determining the PLQE of semiconductor QDs, the entire growth process of CdTe QDs was followed at two different degrees of supersaturation, S_{CdTe} and $4S_{CdTe}$, and particles of different sizes collected at different time of reflux. As shown in Fig. 18.4, initially, there is a sharp rise in PLQE with increasing particle size and starts declining after reaching its zenith. This position of maximum PLQE during growth would be referred to as 'PL bright point' following the mnemonics suggested by Qu et al.[22] As the initial supersat-

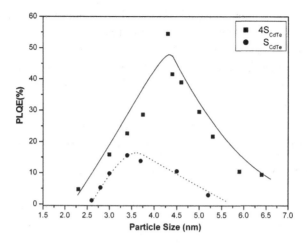

Fig. 18.4. A plot of PLQE of CdTe QDs against particle size at two different degrees of supersaturation, S_{CdTe} and $4S_{CdTe}$. Inset : UV-Vis absorption and PL spectra of CdTe QDs prepared at $4S_{CdTe}(d_{av} = 2.8nm)$.

uration is increased, the PL bright point moves to a larger value, both in terms of particle size as well as PLQE. At S_{CdTe}, the 'PL bright point' was characterized by 3.4 nm sized particles having PLQE of 16%. With a four-fold rise in supersaturation ($S_{CdTe} \to 4S_{CdTe}$), the 'PL bright point' shifts to a higher particle size of 4.3 nm and PLQE reached a remarkably high value of 54%. However, similarly sized particles at lower supersaturation (S_{CdTe}) exhibited merely 11% PLQE. It is quite evident from these results that for any given particle size, the one synthesized at higher supersaturation ($4S_{CdTe}$) possess substantially high PLQE as compared to the particles at lower supersaturation (S_{CdTe}). As shown in the inset of Fig. 18.4, the as-prepared QDs at $4S_{CdTe}$ exhibit sharp band-edge emission and FWHM as narrow as 28 nm was obtained. Without any size sorting, the UV-Vis and PL spectra are at least comparable to the best optical spectra of CdTe QDs prepared by any method.[11,21]

18.5. Synthesis of CdTe QDs in Dendrimer Matrix

CdTe/Dendrimer nanocomposite essentially comprises of two components, dendrimer and CdTe nanoparticles. Initially, the reaction vessel contains a solution of dendrimer and Cd^{2+} ions and subsequent addition of NaHTe leads to formation of CdTe nanoparticles. The dendrimer mediated formation and growth of CdTe NPs has been followed by UV-Visible spectroscopy. The effects of various experimental parameters like reagent ratio, pH and temperature on the particle size, distribution and luminescence have been extensively investigated.

18.5.1. *Effect of variation in $Cd^{2+}:Te^{2-}$ Molar Ratio*

The molar ratio of $Cd^{2+}:Te^{2-}$ in the synthetic mixture was varied from 1:1 to 1:0.25 for various sets at different temperatures. On going from $Cd^{2+}:Te^{2-}$ ratio of 1:1 to 1: 0.5, gradual decrease in the average particle size from 5.0 nm to 3.4 nm is observed and the size distribution also shows an improvement. The size distribution also shows an improvement as indicated by the corresponding decrease in $\Delta\lambda_{abs}$ values (Fig. 18.5a-b). $\Delta\lambda_{abs}$, here, denotes the difference between absorption onset (λ_{onset}) and peak (λ_{peak}) which can give some idea about polydispersity of nanoparticle samples. In an ensemble, larger particles are expected to absorb in longer wavelength region due to their smaller bandgap and smaller ones in short wavelength region. So, large differences in absorption onset and peak indicates larger size distribution and vice versa.[12,46] As the concen-

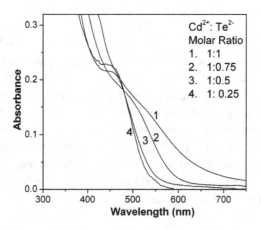

Fig. 18.5a. UV-Visible absorption spectra of CdTe/Dendrimer nanocomposites dendrimer with varying molar ratio of $Cd^{2+}:Te^{2-}$ in water. (Synthesis Temp : 10°C.)

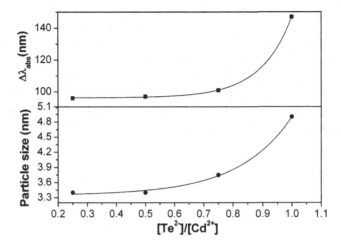

Fig. 18.5b. A plot of average particle size and $\Delta\lambda_{abs}$, against molar ratio of $[Te^{2-}]$ $[cd^{+2}]$. Here, $\Delta\lambda_{abs} = \lambda_{onset} - \lambda_{peak}$. (Synthesis Temp. : 10°C.)

tration of telluride is reduced, the relative amount of Cd^{2+} ions available for complexations with surface amine groups of dendrimer increases. The enhanced complexation on the particle surface hinders that process i.e., growth process is moderated better leading to attainment of smaller particles with narrower size distribution. With further decrease in Cd^{2+}:Te^{2-} ratio to 1:0.25, no change in absorption onset or $\Delta\lambda_{abs}$ was observed i.e., particle size and distribution remain unchanged.

Here the formation of cadmium amine complex is supported by the results of FTIR measurement. Fig. 18.6. Displays the FTIR spectra of pure PAMAM Dendrimer Generation 5[47,48] and CdTe/dendrimer nanocomposites. On comparison of the two spectra the most significant point to be noted is that the band at 3278 cm^{-1} in dendrimer get shifted to 3376 cm^{-1} for nanocomposites and the band becomes wider. This particular band can be assigned for *stretching* mode of surface amine group (NH_2) of the dendrimer and the observed shift due to the coordination of NH_2 to the Cd atom present on the particle surface. It can also be seen that the $-NHCO-$ bands at 1634 and 1556 cm^{-1} correspond to $C = O$ vibration and $N - H$ bending vibration shifted to 1630 and 1568 cm^{-1}. This suggests NPs are attached to the dendrimer through the amino (NH_2) surface group as well as the $-NHCO-$ moieties lying in the interior of dendrimer structure. A typical TEM image of CdTe/Dendrimer nanocomposite in methanol at 5°C with molar ratio of $Cd^{2+} : Te^{2-}$, 1:0.5 is shown in Fig. 18.7, which gives average particle size of 4nm for CdTe NPs. Selected area electron diffraction (SAED) pattern of CdTe nanoparticles is shown in the inset of Fig 18.7. The SAED pattern displays bright rings at a distance of 3.3, 2.1 and 1.5 A^o corresponding to 200, 220 and 420 lattice planes of the cubic crystal phase of CdTe (JCPDS no. 75-2083). The overall size of the composite was obtained from DLS measurements. The average size of nanocomposites was found to be 182 nm and 23 nm in water and methanol respectively.

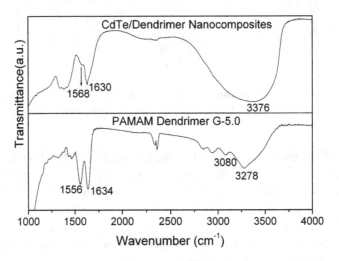

Fig. 18.6. FTIR Spectra of PAMAM - DENDRIMER (G-5) and CdTe/dendrimer.

18.5.2. pH Effect

The absorption spectra of CdTe/dendrimer nanocomposites synthesized at various pH is displayed in Fig. 18.8. The sharp change in absorption profile of CdTe NPs with corresponding change in synthesis pH could easily be noted. At pH 8, absorption spectrum is quite diffuse and only an absorption shoulder is obtained instead of excitonic peak. On moving towards higher pH, there is gradual blue shift in the absorption onset and a sharp peak at 454 nm is observed at pH 10. The particle size appears to be quite sensitive to change in pH as discerned from Table 18.1. For 2 units change in pH (8-10), the average particle size decreases by "25%. At the same time, the

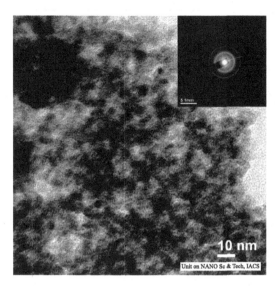

Fig. 18.7. A typical TEM image.

particle size distribution gets focused by nearly 30%. It seems that pH 10 is most appropriate for the synthesis of CdTe NPs having smaller size and narrow size distribution.

The strong influence of pH on the particle size and distribution could be attributed to the differential behavior of amine groups on the surface and interior of the dendrimer. The peripheral amines are more basic than the interior amines[49] (pKa = 9.23, pKa = 6.30, respectively) and therefore act as a better coordinating ligand for the Cd^{2+} ions present in the solution. Stronger the Cd^{2+}- amine complexation, greater will be the moderation of the particle growth. At pH lower than 9, most of the peripheral amines remain protonated and coordination to Cd^{2+} ions gets severely affected. However, on increasing the pH of the synthetic mixture to 10, these surface sites on dendrimer become available for coordination and an effective moderation of growth process is achieved which results in formation of smaller CdTe NPs with narrow size distribution.

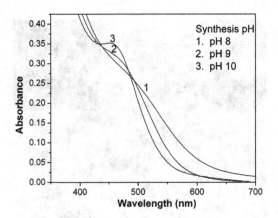

Fig. 18.8. Optical absorption spectra of CdTe/Dendrimer nanocomposites synthesized at different pH in water. (Synthesis Temp.: 10°C, molar ratio of $Cd^{2+} : Te^{2-}$ 1:0.5.)

Table 18.1. Effect of PH on particle size and $\Delta\lambda_{abs}$ of CdTe NPs in water

pH	Particle Size* (nm)	$\Delta\lambda_{abs}$ (nm)
8.0	4.3	134
9.0	3.6	111
10.0	3.2	94

*determined from optical absorption spectra using the correlation between band gap and particle size (sec. 18.2.3).

18.5.3. Temperature Effect

Temperature had a profound influence on the synthesis of the CdTe nanocrystals in both the solvents, methanol and water. Lowering of the synthesis temperature leads to a gradual improvement in the absorption profile and the broad absorption shoulder obtained at 25°C gets transformed into a sharp excitonic peak at 5°C (Fig. 18.9). In both the solvents, a sharp focusing of size distribution could be easily discerned from the $\Delta\lambda_{abs}$ values given in Table 18.2. However, the extent of focusing was a little higher in methanol (40%) as compared to that in aqueous system (30%). This 'temperature dependent focusing' of size distribution bears great significance as it offers a tool to control the distribution of nanocrystals during the synthesis itself obviating the need for any post-preparative treatment like 'size selective precipitation' etc.[7,11] This is due to the fact that the solubility of bulk CdTe would be different at these temperatures leading to different degree of supersaturation at each of these temperatures, which in turn affects the critical radius for these growing NP-solutions, as explained in earlier sections. On considering any intervening stage during the given time span of particle growth, the NP solution at 5°C will have smallest critical radius, which could be smaller than the size of smallest particle present in the ensemble leading to focusing of size distribution.

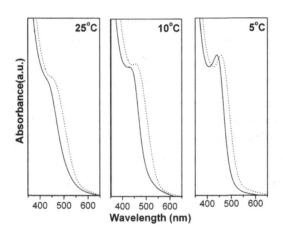

Fig. 18.9. Effect of synthesis temperature on optical absorption of CdTe/Dendrimer nanocomposites in methanol (−) and water (···). (Synthesis Temp.: 10°C, molar ratio of Cd^{2+}: Te^{2-} 1:0.5.)

Table 18.2. Effect of synthesis temperature on particle size and $\Delta\lambda_{abs}$ of CdTe NPs in water and methanol

Temp (°C)	Particle Size* (nm)		$\Delta\lambda_{abs}$		Avg. size of Nanocomposites# (nm)	
	Water	Methanol	Water	Methanol	Water	Methanol
25	3.4	3.1	97	103	-	-
10	3.4	3.0	95	84	-	-
5	3.1	2.8	71	60	182±17	23 ± 2

* determined from optical absorption spectra using the correlation between band gap and particle size [sec. 18.2.3], # determined from dynamic light scattering.

18.6. Size Dependent Luminescence Quenching of CdS QDs and Photocatalytic Degradation of Nitroaromatics

We studied the effect of nitroaromatics namely, dinitrobenzene, nitrotoluene and nitrobenzene on the photoluminescence of thiophenol capped CdS nanoparticles. PL intensity of CdS nanoparticles was quenched on gradual addition of methanolic solution of nitroaromatics. Nitroaromatics in gaseous phase also influenced the luminescence of nanocrystalline porous silicon film in a similar way as observed by earlier workers.[50] Fig. 18.10 illustrates the effect on the luminescence of CdS NPs on gradual addition of nitrobenzene.

Table 18.3. Stern–Volmer fluorescence quenching constants of CdS NPs ($d_{av} = 3nm$) for different nitroaromatics and linear detection range

Nitroaromatics	$K_{S-V}(10^4 M^{-1})$	Linear Range
Nitrobenzene	1.4	0.55-12 mM
Nitrotoluene	1.5	0.55-10 mM
Dinitrobenzene	21.2	0.40-7 μM

Other nitroaromatics selected in our studies also showed similar pattern. It is evident from the inset of Fig. 18.8 that the quenching of CdS luminescence follows the Stern–Volmer equation:

$$I_0/I = 1 + K_{S-V}[\text{Nitroaromatics}] \qquad (2)$$

where I_0 and I are the fluorescence intensities in the absence and the presence of nitroaromatics and K_{S-V} is the Stern–Volmer quenching constant. The fluorescence lifetime of CdS nanoparticles both in the absence as well as in the presence of nitrotoluene were measured by the TCSPC method. The lifetime (τ) of CdS nanoparticles in our experimental conditions was determined to be 8.3 ns and this was not affected by the addition of nitrotoluene, suggesting static nature of quenching. To establish the proposed mechanism, quenching of CdS luminescence with nitrotoluene was studied by varying the dielectric constant of the medium. The Stern–Volmer constant also increased with increase in polarity of the solvent (Fig. 18.11).

The observed trend suggests that charge transfer takes place between the photo-excited CdS NPs and the nitroaromatic compounds. The electron-transfer process and subsequent stabilization of the charge-transfer com-

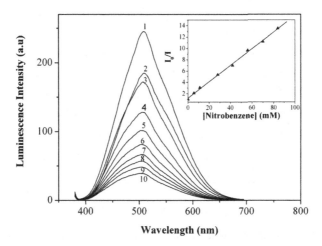

Fig. 18.10. Effect of varrying concentration of nitrobenzene on the PL spectra of CdS nonoparticles. [Nitrobenzene] : (1) 0.00 M, (2) 0.37 mM, (3) 1.4 mM, (4) 2.8 mM, (5) 4.2 mM, (6) 5.6 mM, (7) 7.0 mM, (8) 8.4 mM (9) 9.8 mM and (10) 11.2 mM. Inset : The corresponding Stern−Volmer plot (λ_{em} = 505 nm and λ_{ex} = 370 nm).

plex is more favored in polar solvents. A consistent rise, therefore, in the quenching efficiency is observed with increasing solvent polarity. This is also reflected in the quenching pattern as the nature of substituents on the aromatic ring is varied. On substitution of methyl group in nitrobenzene, K_{S-V} value remains virtually unchanged as given in Table 18.3. However, introduction of nitro group in the ring leads to an increase in K_{S-V} value by 14 times. The remarkably higher value of Stern−Volmer constant for DNB can be attributed to the electron withdrawing effect of the additional nitro group, which facilitates the electron transfer from excited CdS NPs to the nitroaromatic molecules.

Fluorescence quenching of CdS nanoparticles by nitrotoluene has been used to study the size-dependent behavior. Fig. 18.12 shows the Stern − Volmer plot for the CdS NPs of different sizes. K_{S-V} values increased considerably from $1.0 \times 10^4 M^{-1}$ to $4.0 \times 10^4 M^{-1}$ on moving from 5 nm sized particles to 4.2 nm. However, only a small change in K_{S-V} value ($5.33 \times 10^4 M^{-1}$) was observed when particle size was further lowered to 3.8 nm. The increase in Stern−Volmer quenching constants with decrease in particle size could be explained by considering the change in electronic

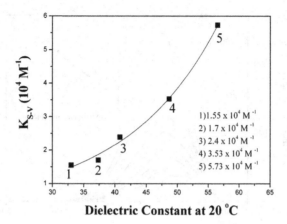

Fig. 18.11. A plot of Stern–Volmer constant (K_{S-V}) as a function of dielectric constant of the medium obtained by varying solvent composition : (1) 100% methanol, (2) 83% DMF + 17% methanol, (3) 33% water + 67% methanol mixture, (4) 50% water +50% methanol.

energy levels of the semiconductor NPs. As the size of nanoparticles decreases, the energy of conduction band shifts to higher energy due to quantum confinement effect. Redox potentials of the conduction band become more negative,[51] thereby enhancing the reducing power with decrease in particle size. Due to higher surface to volume ratio in smaller nanoparticles, most of the constituent atoms reside on the surface of particles[52] and can have more efficient transfer of electrons to the suitable species adsorbed on the surface reducing the chances of radiative recombination of $e^- - h^+$ pairs. So, enhanced reducing power coupled with higher ratio of surface to core atoms could account for the observed increase in the value of Stern–Volmer constants with decreasing particle size.

The trend emerging from the quenching studies suggests that photoexcited CdS NPs can cause reduction of toxic nitroaromatics. Attempts were made here to look into possible remediation of nitroaromatics by UV irradiation using CdS nanoparticles as an active photo-catalyst. Earlier TiO_2 was used as photo-catalyst for degradation of trinitrotoluene (TNT) in methanol on UV light exposure.[53] However, the effect of particle size towards degradation of nitroaromatics was not investigated. In the present study, CdS NPs were found to act as an effective catalyst and also showed size dependent photocatalysis when nitroaromatics were irradiated with photons

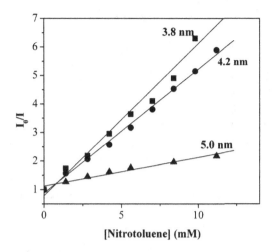

Fig. 18.12. Size dependent Stern–Volmer plot of PL quenching of CdS NPs in presence of nitrotoluene. ($\lambda_{em} = 505$ nm and $\lambda_{ex} = 370$ nm)

of 370 nm. Stability of CdS nanoparticles was checked after each set of UV irradiation. Neither any change in absorbance values at excitonic peak nor any shift in absorption onset was observed under our experimental conditions. The degradation of nitrotoluene was followed by measuring the decrease in absorbance at 265 nm as shown in Fig. 18.13. Fig. 18.14a shows that with decreasing particle size, the degradation efficiency increased substantially. On moving down from 5.8 nm to 3.8 nm sized particles, the catalytic efficiency was quintupled *i.e.*, for a 34% decrease in particle diameter, the reaction rate increases by about 400% [Table 18.4]. In contrast, simple photolysis of nitrotoluene carried out in the absence of CdS NPs by irradiating samples at 370 nm (where, only CdS absorbs) as well as at 265 nm (where, nitrotoluene absorbs) resulted in less than 5% degradation in each case.

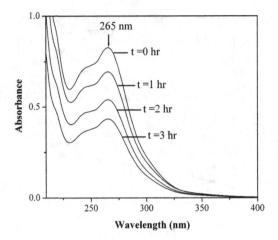

Fig. 18.13. Absorption of nitrotoluene in presence of CdS nanoparticles at different time intervals of irradiation.

Table 18.4. Correlation of rate acceleration with relative shift in conduction band edge (E) and change in surface to volume ratio (S) of CdS NPs

Particle size, d (nm)	Rate Acc., k ($Kcat/kuncat$)	$R = \frac{k(d)}{k(d_1)}$	$S = \frac{A(d)/V(d)}{A(d_1)/V(d_1)}$	$E = \frac{\Delta E_c(d)}{\Delta E_c(d_1)}$ #	$R' = Ex$
5.8	8	1	1	1	1
5.0	18	2.25	1.2	1.8	2.16
4.2	24	3.0	1.4	2.4	3.36
3.8	40	5.0	1.5	3.0	4.80

change in the energy of conduction band edge of CdS NPs, ΔE_c (taken from ref. 35).

The concentration of nitrotoluene decreased exponentially with increasing irradiation time as shown in Fig. 18.14a. The logarithmic plot of nitrotoluene concentration against time follows a linear equation suggesting first order kinetics (inset of Fig 18.14a). Rate constant for degradation was estimated from the respective slopes of such logarithmic plots in each case. For the uncatalysed reaction, rate constant, k_{uncat} is 2.2×10^{-3} sec^{-1}. The catalytic efficiency of the CdS NPs can be gauged by comparing the rate acceleration, i.e. k_{cat}/k_{uncat}. It was found to decrease with increasing particle size. In the size range studied, 3.8 nm sized CdS NPs were found to be most efficient with a rate acceleration factor of 40. On extrapolation

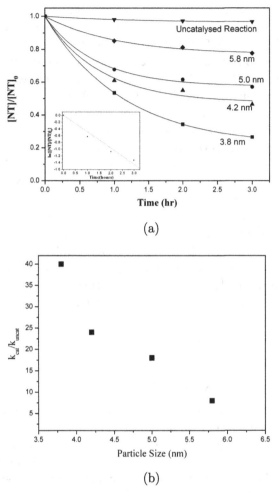

Fig. 18.14. Size dependent photocatalytic behaviour of CdS NPs towards reductive degradation of nitrotoluene: (a) A plot of $[NT]/[NT]_0$ as a function of irradiation time in presence and in absence of CdS NPs. ($[NT]_0$ and [NT] denotes the concentration of nitrotoluene before and after irradiation, respectively; $[NT]_0$ =33.6 mM). [NT] calculated from the respective absorbance values at 265 nm, the absorption of the possible product, aminotoluene, has also been taken into account. Inset: A typical plot of $ln\{[NT]/[NT]_o\}$ as a function of irradiation time for CdS of 3.8 nm. (b) Plot of rate acceleration (k_{cat}/k_{uncat}) of reaction as a function of particle size.

of our data as shown in Fig. 18.14b, it is apparent that catalytic efficiency for CdS particles > 6 nm in size (Bohr exciton diameter) will be negligible. Hence, it can be inferred that only size-quantized CdS particles can catalyze the photodegradation of nitroaromatics. The rise in the energy of the conduction band edge due to quantum confinement effects makes the energy gradient much steeper for transfer of electrons from photo-excited CdS NPs to nitroaromatic molecules. On going from particle size of 5.8 nm to 3.8 nm, there is almost 3 times increase in the energy of conduction band edge (Table 18.4). However, this alone could not explain the relative rise in rate acceleration, which goes up by a factor of 5. If the relative increase in surface to volume ratio is also taken into account, a good quantitative correlation is achieved. So, it is the combined effect of the two factors, $viz.$, rise in the energy of conduction band edge and increase in surface to volume ratio, makes the smaller particles a better photocatalyst. It is also evident from these data that the shift in the conduction band edge plays a dominant role for the increase in catalytic efficiency of the size-quantized nanoparticles. The empirically derived expression relating the rate acceleration with the aforementioned two factors can be given as follows:

$$k(d_2) = k(d_1) \times E \times S, \quad (3)$$

where,

$$E = \frac{\Delta E_C(d_2)}{\Delta E_c(d_1)} \text{ and } S = \frac{A(d_2)V(d_2)}{A(d_1)V(d_1)}.$$

k is the rate of acceleration, ΔE_c is increase in the energy of conduction band edge, A is surface area, V is volume, d_2 and d_1 in the parenthesis denote the average particle diameter of the respective sample. If the photocatalytic activity of CdS NPs of a given size, $k(d_1)$, is known, the catalytic efficiency of CdS NPs of other sizes in the quantum confinement regime can be predicted using the above mentioned equation.

Further, we have also observed that substitution with nitro-group in the aromatic ring enhances the degradation rate. The additional nitro-group exerts electron-withdrawing effect causing a reduction of electron density on the ring. Consequently, the electron transfer from photoexcited CdS NPs to the aromatic ring gets facilitated leading to higher rate of degradation.

18.7. Concluding Remarks

It is established that by a deft manipulation of supersaturation, one can achieve two prime objectives in nanoparticle synthesis: a) narrow size distribution and *b*) high PL efficiency. This indicates a paradigm shift in the synthetic approach making the process rather hassle free, as it overrides the complications of post-preparative treatments. At a suitably chosen reagent ratio, we simply need to raise the respective concentrations to increase the degree of supersaturation to get high quality NPs. It also makes possible to obtain a wider range of particle-sizes. Similiar conditions can also be achived in radiation-induced synthesis of semiconductor or metal nanoclusters by using the high dose rates, which play the role of enhancing the instantaneous concentrations of reactants and probability of forming primary nuclei, e.g., CdS monomers or metal dimers.[17,54] As supersaturation is the most basic phenomena occurring in all the colloidal syntheses, this technique should be universally applicable to any NP-system, be it semiconductor, magnetic, ceramic or any other NPs in colloidal phase. It can control the size distribution even in the absence of capping agent. On the contrary, size-selective precipitation can only be applied to surface-capped NPs. The augmentation of PL efficiency of the quantum dots with increasing supersaturation is quite intriguing and the mechanistic details of the process need to be worked out. Due to the enhanced luminescence high monodispersity and bioactive surface-functionality, the cysteine-capped CdS and CdTe QDs prepared at higher supersaturation are ideally suited for biosensing and bioimaging works. In this context, it would be pertinent to see the effect of supersaturation in other systems of prime importance for biomedical applications, specifically, super-paramagnetic iron oxide nanoparticles (SPION).[55,56] Does a change in supersaturation and concomitant changes in nucleation and growth kinetics manifest itself into betterment of magnetic properties as well?

The present article also demonstrates the significance of the role of temperature, pH and reagent ratio in determining the quality of semiconductor/dendrimer nanocomposites. For the dendrimer of a particular generation, the average particle size of the semiconductor NPs could be tuned by varying the reagent ratio, which in turn can control the luminescence properties of the NPs in the quantum confinement regime. The 'temperature dependent focusing' shows a unique way to narrow down the size distribution with negligible change in average size of the NPs.

Possibility of developing a fluorescence-based NP-sensor can be explored based on the observed quenching of CdS photoluminescence by nitroaromatics. On introduction of another nitro group in the aromatic ring of nitrobenzene, the detection limit was improved by 1000 times. Extrapolation of the observed data suggests that a higher sensitivity and detection limit upto nanomolar range is likely to be achieved for trinitrotoluene (TNT), the most nefarious of all nitroaromatics. Size dependent quenching of CdS luminescence indicates that employing smaller particles can enhance the sensitivity of detection. It is also demonstrated that only those CdS particles lying in size-quantization regime were found to catalyze the degradation of these toxic nitro-compounds. This study suggests that the conduction band edge shift plays a major role in determining the photocatalytic behaviour of these size-quantized particles. The degradation of nitroaromatics was enhanced with increasing number of nitro-substituents in the aromatic ring because of electron withdrawing effects of nitro-groups. This indicates that TNT can be degraded more efficiently.

Acknowledgements

The authors thank Dr. S. Pal, S. N. Bose National Centre for Basic Sciences for help in DLS measurements and Saha Institute of Nuclear Physics for support in fluorescence lifetime and FTIR. TEM was carried out in Indian Association for the Cultivation of Science supported by Nanoscience and Technology Initiation Programme of DST, New Delhi. One of the authors (A. Priyam) is thankful to UGC (Govt. of India) for the award of Senior Research Fellowship.

References

1. D. Crouch, S. Norager, P. O'Brien, J. H. Park and N. Pickett, *Philos. Trans. R. Soc. A*, **361**, 297 (2003).
2. A. L. Rogach, A. Eychmller, S. G. Hickey and S. V. Kershaw, *Small*, **3**, 536 (2007).
3. S. V. Kershaw, M. T. Harrison and M. G. Burt, *Phil. Trans. R. Soc. Lond. A*, **361**, 331 (2003).
4. X. Michalet, F. F. Pinaud, L. A. Bentolila, J. M. Tsay, S. Doose, J. J. Li, G. Sundaresan, A. M. Wu, S. S. Gambhir and S. Weiss. *Science*, **307**, 538 (2005).
5. A. Priyam, A. Chatterjee, S. K. Das and A. Saha, *Chem. Commun.*, **32**, 4122 (2005).
6. X. Michalet, F. Pinaud, T. D. Lacoste, M. Dahan, M. P. Bruchez, A. P. Alivisatos and S.Weiss, *Single Mol.*, **2**, 261 (2001).
7. C. B. Murray, D. J. Norris and M. G. Bawendi, *J. Am. Chem. Soc.*, **115**, 8706 (1993).
8. D. V. Talapin S. Haubold, A. L. Rogach, A. Kornowski, M. Haase and H. Weller, *J. Phys. Chem. B*, **105**, 2260 (2001).
9. R. Mahtab, J. P. Rogers and C. J. Murphy, *J. Am. Chem. Soc.*, **117**, 9099 (1995).

10. J. R. Lackowicz, I. Gryczynski, Z. Gryczynski and C. J. Murphy, *J. Phys. Chem. B*, **103**, 7613 (1999).
11. N. Gaponik, D.V. Talapin, A. L. Rogach, K. Hoppe, E. V. Shevchenko, A. Kornowski, A. Eychmuller and H. Weller, *J. Phys. Chem. B*, **106**, 7177 (2002).
12. A.Priyam, A.Chatterjee, S.C. Bhattacharya and A. Saha, *J. Cryst. Growth*, **304**, 416 (2007).
13. D. V. Talapin, A. L. Rogach, E. V. Shevchenko, A. Kornowski, M. Haase and H. Weller, *J. Am. Chem. Soc.*, **124**, 5782 (2002).
14. A. Chatterjee, A. Priyam, S. K. Das and A. Saha, *J. Colloid Interface Sci.*, **294**, 334. (2006).
15. D. Hayes, O. I. Micic, M. T. Nenadovic, V. Swayambunathan and D. Meisel, *J. Phys. Chem.*, **93**, 4603 (1989).
16. V. Swayambunathan, D. Hayes, K. H. Schmidt, Y. X. Liao and D. Meisel, *J. Am. Chem. Soc.*, **112**, 3831 (1990).
17. M. Mostafavi, Y. P. Liu, P. Pernot and J. Belloni, *Rad. Phys. Chem.*, **59**, 49 (2000).
18. L. Qu, Z.A. Peng and X. Peng, *Nano Lett.*, **1**, 333 (2001).
19. W. W. Yu and X. Peng, Angew. *Chem. Int. Ed. Engl.*, **41**, 2368 (2002)
20. Z. A. Peng and X. Peng, *J. Am. Chem. Soc.*, **123**, 183 (2001).
21. W. W. Yu, Y.A. Wang and X. Peng, *Chem. Mater.*, **15**, 4300 (2003).
22. L. Qu and X. Peng, *J. Am. Chem. Soc.*, **124**, 2049 (2002).
23. E. Jang, S. Jun, Y. Chung and L. Pu, *J. Phys. Chem. B*, **108**, 4597 (2004).
24. M. Bruchez, M. Moronne, P. Gin, S. Weiss and A.P. Alivisatos, *Science*, **281**, 2013 (1998).
25. W. Guo, J.J. Li, Y.A. Wang and X. Peng, *J. Am. Chem. Soc.*, **125**, 3901 (2003).
26. C. Li and N. Murase, *Chem. Lett.*, **34**, 92 (2005).
27. N..N. Mamedova, N.A. Kotov and J. Studer, *Nano Lett.*, **1**, 281 (2001).
28. D. A. Tomalia, A. M. Naylor and W. A. Goddard III, *Angew. Chem. Int.Ed. Engl.*, **29**, 138 (1990).
29. A.W. Bosman, H.M.Janssen and E.W.Meijer, *Chem. Rev.*, **99**, 1665 (1999).
30. S. Svenson and D. A. Tomalia, *Adv. Drug Delivery Rev*, **57**, 2106 (2005).
31. A. Chatterjee A. Priyam, S. C. Bhattacharya and A. Saha, *J. Luminescence*, **126**, 764 (2007).
32. P.V. Kamat and D. Meisel, Curr. Opin. *Colloid Interface Sci*, **7**, 282 (2002).
33. T. Tachikawa, M. Fujitsuka and T. Majima, *J. Phys. Chem. C.*, **111**, 5259 (2007).
34. P. J. G. Coutinho, M. Teresa and C. M. Barbosa, *J. Fluoresc.*, **16**, 387 (2006).
35. P.E. Lippen and M. Lannoo, *Phys. Rev. B*, **39**, 10935 (1989).
36. S. Sapra and D. D. Sarma, *Phys. Rev. B*, **69**, 125304 (2004).
37. M.V. Rama Krishna and R.A. Friesner, *J. Chem. Phys.*, **95**, 8309 (1991).
38. Z.M. Li, P.J. Shea and S.D. Comfort, *Chemosphere*, **38**, 1849 (1998).
39. C. Galino, P. Jacques and A. Kalt, *J. Photochem. Photobiol. A : Chem.*, **130**, 35 (2000).
40. R. Viswanatha and D. D. Sarma, *Chem. Eur. J.*, **12**, 180 (2006).

41. N. S. Pesika, N. J. Stebe and P. C. Searson, *J. Phys. Chem. B*, **107**, 10412 (2003).
42. J. R. Lackowicz in *Principles of Fluorescence Spectroscopy*, Kluwer Academic/Plenum Publishers, New York, 1999), p. 52.
43. J. Belloni and P. Pernot, *J. Phys. Chem. B*, **107**, 7299 (2003).
44. D.V. Talapin, A.L. Rogach, M. Haase and H. Weller, *J. Phys. Chem. B*, **105**, 12278 (2001).
45. X. Peng, J. Wickham and A.P. Alivisatos, *J. Am. Chem. Soc.*, **120**, 5343 (1998).
46. S. Modes and P. Lianos, *J. Phys. Chem*, **93**, 5854 (1989).
47. N. Kangming, H. Jinlian, P.Wenmin and Z.Qingren, *Mater. Lett.*, **11**, 115 (2006).
48. P.Bifeng, G.Feng, H.Rong, C.Daxiang and Y.Zhang, *J. Colloid Interface Sci.* **151**, 297 (2006).
49. W. J. Scoot, O. M. Willson and R. M. Crooks, *J. Phys. Chem. B*, **109**, 692 (2005).
50. S. Content, W.C. Trogler and M.J. Sailor, *Chem. Eur. J.*, **6**, 2205 (2000).
51. S. K. Haram, B. M. Quinn and A. J. Bard, *J. Am. Chem. Soc.*, **123**, 8860 (2001).
52. Y. Chen and Z. Rosenzweig, *Anal. Chem.*, **74**, 5132 (2002).
53. M. Nahen, D. Bahnemann, R. Dillert and G. Fels, *J. Photochem. Photobiol A: Chem.* **110**, 191 (1997).
54. H. Remita, J. Khatouri, M. Mostafavi, J. Amblard and J. Belloni, *J. Phys. Chem. B*, **102**, 4310 (1998).
55. J. Dobson, *Drug. Dev. Res.*, **67**, 55 (2006).
56. F. Cengelli, D. Maysinger, F. Tschudi-Monnet, X. Montet, C. Corot, A. Petri-Fink, H. Hofmann and L. Juillerat-Jeanneret, *J. Pharmacol. Exp. Ther.*, **318**, 108 (2006).

Chapter 19

Dramatic Enhancement in the Cation Sensing Efficiency in Anionic Micelles: A Simple and Efficient Approach Towards Improving the Sensor Efficiency

Paramita Das, Deboleena Sarkar and Nitin Chattopadhyay

Department of Chemistry,
Jadavpur University,
Kolkata - 700 032, India

A simple, convenient and efficient strategy has been introduced, using commercially available anionic surfactants, to enhance the efficiency of a quenching-based cationic fluorosensor for Cu^{2+}, an essential metal ion, by several order of magnitude. Length of the hydrophobic chain and optimum concentration of the surfactant are the key factors playing together to boost the sensing efficiency. This simple strategy is, in general, applicable to quenching-based fluorosensors, new or established, in aqueous solution. The technique transforms a virtually non-sensor fluorophore to a sensor with a commendable efficiency.

19.1. Introduction

In the field of chemistry and biology, development of fluorescent chemosensors for sensing of essential metal ions as well as inorganic anions is an important goal to achieve.[1-9] This area receives attention to combat ever-increasing air and water pollution. Sensing of the essential and/or fatal metal ions in foodstuff, drinking water and more importantly in the blood stream of living beings is essential to ensure normal functioning of the body. Various methods, like chromatography, potentiometry, capillary electrophoresis etc., are used for the detection of the analytes. Over these existing methods, fluorosensing proves itself to be the most sensitive and powerful technique towards the detection of specific analytes in micromolar or submicromolar concentration range.[10] Micellar solutions prepared from synthetic surfactants are a promising environment for tuning fluorosensors towards various metal ions found in physiological fluids.[11-14] Improvement in the fluorosensing ability follows several approaches. A major effort consists of realistic design, synthesis and modification of new fluorescent probes. For example, chemical redesigning of a particular sensor molecule can be exploited to modify the emission wavelength and fluorescence quantum yield. However, this approach is time-consuming, costly, and in many cases involves complicated synthetic efforts. Hence, it is appreciated always to develop a suitable and simple technique, avoiding complicated synthetic routes, to tune the sensing ability of fluorosensor particularly in aqueous medium and physiological fluids. Choice of quenching-based fluorosensors with acceptable efficiency is limited for the detection and/or estimation of the metal ions. It is pertinent to mention here that use of cationic fluorosensor in this respect is yet to be successful, although sometimes they can serve as complexation-based sensors.[15] Mutual repulsive electrostatic interaction between the metal ions and the sensor molecules thus leads to exclusion of a large number of available cationic fluorophores as fluorosensors. We, therefore, sought to develop a simple and economic technique to encompass maximum number of fluorophore molecules and provide easy detection of the analyte at a largely enhanced sensitivity level using readily available surfactants.[13,16] This work proposes that coupling of the two important parameters, namely, variation of the concentration of anionic surfactants and variation in the surfactant chain length provides a simple and convenient strategy to tune, and hence maximize, the sensing ability of quenching-based fluorescence sensors in aqueous environments.

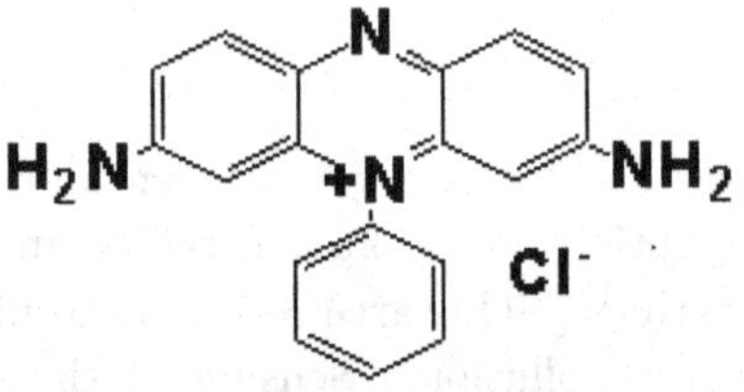

Scheme 19.1. Structure of PSF.

In the present study, we have used a cationic dye, phenosafranin (PSF) (scheme 1) for sensing Cu^{2+} ion. Sodium alkyl sulfates (S_nS, n being an even number ranging from 8 to 14) plays an important role towards increasing the sensing ability of the chosen fluorosensor. In a recent report, we tuned and maximized the fluorosensing activity of a neutral fluorophore, through the introduction of sodium dodecyl sulfate micellar environment.[14] The present work extends this idea through use of a cationic fluorophore as an efficient cation sensor. In the field of fluorosensing, both the sensitivity and selectivity are important parameters. However, in this study, we put emphasis on using suitable surfactants towards improving the sensitivity of a quenching-based fluorosensor.

19.2. Materials and Methods

PSF was purchased from Sigma (USA) and used as received. Its purity was confirmed from absorption and emission spectra in standard solvents. Sodium octyl sulfate (S_8 S), sodium decyl sulfate (S_{10} S), sodium dodecyl sulfate (S_{12} S) and sodium tetradecyl sulfate ($S_{14}S$) were procured from Aldrich (USA) and were used as received. The surfactant solutions were freshly prepared prior to the experiments to avoid hydrolysis of the alkyl sulfates that might lead to a change in the micellar behavior. Before using them, it was checked that the surfactants did not contribute to either absorption or fluorescence in the region of interest. AR grade copper sulfate was purchased from SRL (India). Triply distilled water was used to make the experimental solutions. The concentration of PSF was 4.0×10^{-6} mol dm^{-3} throughout.

Absorption and steady state fluorescence measurements were carried out using a Shimadzu MPS 2000 spectrophotometer and a Spex fluorolog-2 spectrofluorimeter equipped with DM3000F software respectively. Fluorescence lifetime measurements were performed in a time resolved intensity decay analyses by the method of time correlated single-photon counting (TCSPC) using a nanoLED at 403 nm (IBH, UK nanoLED-07) as the light source. The decay curves were analyzed using IBH DAS-6 decay analysis software. Quality of fits was evaluated by χ^2 criterion and visual inspection of the residuals of the fitted function to the data. All the experiments were performed at ambient temperature (300K).

19.3. Results and Discussion

In aqueous medium, the absorption spectrum of PSF consists of a broad, unstructured band with a maximum at 520 nm. The fluorophore exhibits a broad emission with a maximum at 585 nm and binds to the anionic micelles.[17] Study of quenching of the fluorophore as a function of concentration of Cu^{2+} has been performed in anionic micelles of varying hydrophobic chain length. In pure aqueous solution, the fluorescence of PSF is almost unquenched by Cu^{2+}. This inefficient quenching makes this fluorophore a poor sensor in aqueous solution. Introduction of anionic micelles enhances the fluorescence quenching of PSF by Cu^{2+}. Fig. 19.1 shows such plots showing the effect of the addition of Cu^{2+} on the fluorescence intensity of PSF in the presence of sodium alkyl sulfate micelles. The quenching experiments using the cationic fluorophore have been performed at various concentrations of each of the surfactants within the series S_8 S, S_{10} S, S_{12} S and S_{14} S.

The quenching of the fluorescence of PSF upon addition of Cu^{2+} in aqueous and aqueous S_nS media follows Stern-Volmer relation

$$F_0/F = 1 + K_{SV}[Cu^{2+}] = 1 + k_q\tau[Cu^{2+}] \qquad (1)$$

where F_0 and F are the fluorescence intensities in the absence and presence of the quencher respectively, $[Cu^{2+}]$ is the molar concentration of the quencher and K_{SV} is the Stern−Volmer quenching constant,[3,18] and k_q, τ are the quenching constant and fluorescence lifetime in the absence of the quencher respectively. The slope of the plot of F_0/F against $[Cu^{2+}]$ gives K_{SV}, which acts as an indicator of the sensitivity of the fluorophore (PSF) towards the quencher (Cu^{2+}). In each micellar system, plot of K_{SV} against the concentration of the surfactant passes through a maximum reflecting a tuning in the fluorosensing activity. Figure 19.2 depicts such plots for PSF in the S_n S environments. Table 19.1 presents the maximum K_{SV} values in the micellar environments. It is important to mention here that for all the surfactant systems K_{SV} attains its maximum near the corresponding literature values of critical micellar concentrations (CMC) (Table 19.1). Slightly lower experimental values possibly indicate the lowering of the CMCs of S_nS micelles in the presence of the added Cu^{2+} ion. It is already known that depending on the concentrations of the free metal ions, CMC value of a surfactant decreases. Du et al. observed that 9.3 mM Cu^{2+} decreases the CMC of S_{12} S from 8.2 to 6.5 mM.[19] Since our systems contain copper ion at a very low concentration the decreases are expected to be small.

It is important to note that the increments in the K_{SV} values for the dye in the presence of optimum concentrations of the surfactants relative to that in pure aqueous medium are not the same for all the micellar systems. As we follow the series from S_8S to S_{14} S the enhancement factor in K_{SV} and hence the sensing ability of the dye increases dramatically. Figure 19.3 depicts the maximized K_{SV} values as a function of carbon number in the surfactant chain for different micelles.

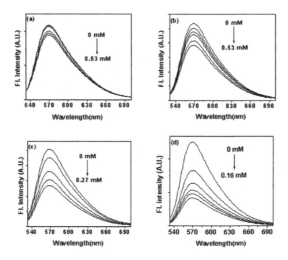

Fig. 19.1. Emission spectra of PSF as a function of [Cu^{2+}] in (a) S$_8$S (128 mM) (b) S$_{10}$S (34 mM) (c) S$_{12}$S (7.5 mM) and (d) S$_{14}$S (1.8 mM) micellar solutions.

Table 19.1. Maximized Stern-Volmer quenching constants (K$_{SV}$) of PSF in various S$_n$ S micellar environments at optimum concentrations and literature CMC values and aggregation numbers of the corresponding micelles (CMC and aggregation numbers are from ref.[20])

Medium	Optimum Concentration (mM)	K$_{sv}(M^{-1})$	Literature CMC values (mM)	Aggregation number
Water	....	4.5 ± 0.5	....	....
S$_8$ S	128	240±20	134	46
S$_{10}$ S	34	800±20	30	64
S$_{12}$ S	7.5	3260±80	8.0	92
S$_{14}$ S	1.8	14100±200	2.0	120

Figure 19.3 and Table 19.1 reveal that choice of a surfactant of proper hydrophobic chain length at a proper concentration dramatically improves the fluorosensing ability of PSF towards Cu^{2+}.

In aqueous medium K$_{sv}$ for the quenching of fluorescence of PSF by Cu^{2+} comes out to be 4.5 ± 0.5 M^{-1} while the same attains a value of

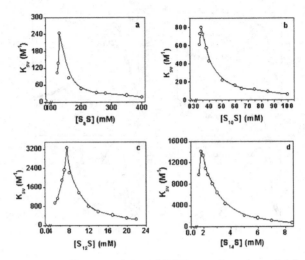

Fig. 19.2. Variation of K_{sv} in case of PSF as a function of (a) S_8S (b) S_{10} S (c) S_{12} S and (d) S_{14}S concs. Each point in the graphs corresponds to a full set of quenching experiments with various concentrations of Cu^{2+} (in millimolar range).

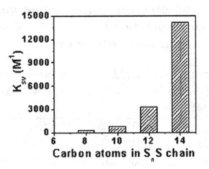

Fig. 19.3. Variation of maximized K_{sv} in the micellar systems as a function of carbon number in the surfactant chain.

14100 ± 200 M^{-1} in 1.8 mM S_{14} S solution. The results indicate an enhancement of ~ 3000 in the efficiency of quenching (hence sensing). The technique is thus capable of increasing the sensing efficiency of PSF towards Cu^{2+} ion by more than three orders of magnitude within the S_n S series studied. Solubility problem prevents us from using surfactants with chain lengths longer than that of S_{14} S. To assess the sensitivity of this quenching-based fluorosensor it is pertinent to mention here that at 1.8 mM S_{14} S environment 70 μM Cu^{2+} quenches the fluorescence of PSF to 50% of the original. Comparing with the poor or negligible sensitivity of PSF towards Cu^{2+} in aqueous environment, the quenching in optimum concentration of S_{14} S is large enough to establish that the simple strategy is extremely powerful in enhancing the fluorosensing activity of a quenching-based fluorosensor, new or existing. It further substantiates that the technique can turn an essentially non-sensor fluorophore to an efficient sensor. Similar quenching experiment performed with interfering ions like Ca^{2+} reveals that the fluorescence of PSF is not affected by the presence of such ions even at moderate concentrations (e.g., 20 mM Ca^{2+}) and reflects that the chosen sensor molecule is blind to these interfering ions commonly present in the physiological fluids.

To address the mechanism of the phenomenon we have performed the time-resolved fluorescence measurements of PSF in the absence and the presence of the quencher. In aqueous solution the fluorescence shows a single exponential decay with a decay time of 0.93 ns.[17] In micellar environments the fluorescence lifetime increases to (∼ 2.0 ns. In aqueous as well as micellar systems the fluorescence lifetime decreases gradually with the addition of Cu^{2+}. This observation confirms that the quenching process is a dynamic one.[3,14] Figure 19.4 depicts such representative plots of fluorescence decay of PSF in 7.5 mM S_{12} S as a function of concentration of the quencher. Because of the low concentrations (in the micromolar range) of both the fluorophore and the quencher in the aqueous solution, the interacting partners reside reasonably far from each other. Therefore the collisional interaction between them within the short lifetime of the excited fluorophores becomes insignificant. In the present case, the mutual electrostatic repulsion between the cationic fluorophore (PSF) and the cationic quencher (Cu^{2+}) retards the process even more.

Introduction of anionic micelles into the environment assists in bringing the quenching partners in proximity, making the quenching process effective. Because of the cationic charge characteristics of both the dye molecule and the quencher they will be restricted to reside in the interfacial region of anionic micelles (since they will not dissolve in the hydrocarbon core).[17,21–23] The local concentrations of these two interacting partners can, therefore, be much greater in the interfacial region than that in the bulk solution. This renders the quenching partners in proximity.[24] Both the enhanced lifetime of the photoexcited state in the micellar environment and the proximity between the fluorophore and the quencher ensures effective communication between the active components responsible for quenching and facilitates the process. This enhances the efficiency in sensing the metal ion.[3]

The nature of variation of the plot of K_{sv} as a function of S_n S concentrations (Figure 19.2) is consistent with the pseudophase ion exchange model assuming that the micelles act as a separate phase from water.[25-27] According to the model, when the surfactant concentration is close to or exceeds its CMC value, the interacting components (PSF and Cu^{2+}) distribute themselves between micelle and water. The distribution remains at dynamic equilibrium so long as the solutions are fluid. In the present case, the micellar pseudophase binds both the cationic species (PSF and Cu^{2+}) in the micellar surface and their local concentrations are likely to be much higher than their stoichiometric concentrations in solution. This leads to an increase in the K_{sv} value with an increase in the surfactant concentration.

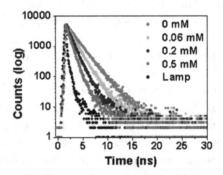

Fig. 19.4. Fluorescence decays of PSF in S_{12} S (7.5 mM) micellar solution. Inset shows respective Cu^{2+} concentrations.

At high surfactant concentrations, e.g., well above the CMC, both PSF and Cu^{2+} will be completely associated with the micellar pseudophase. A growing volume of the micellar pseudophase with addition of more and more surfactant leads to a dilution of both the cations (interacting partners) resulting in a decrease in the KSV values. Attainment of maximum K_{sv} near the CMCs of the respective surfactants results from the transition between the above mentioned two opposing phenomena.

A further enhancement in the quenching efficiency with an increase in the chain length of the surfactant as we proceed through the S_n S series can be rationalized considering a gradual decrease in the volume of the micellar pseudophase.[28] Since binding of both the reactants with the micelles is strong (because of the electrostatic factor) the aspect leads to an enhancement in the interfacial concentrations of the interacting species (Cu^{2+} and PSF). This results in an enhancement in the K_{sv} values as we move from S_8 S to S_{14} S through the series.

The strategy introduced through this work provides a way (i) to modulate the sensitivity of a quenching-based fluorosensor as a function of concentration of the surfactant and hence finding the condition for the maximum sensing ability in a particular micellar environment and (ii) to further enhance the maximized efficiency in a micellar system with an increase in the chain length of the surfactant in the S_n S series.

19.4. Conclusion

The objective of the present work is to offer a simple and efficient strategy for improving the sensing efficiency of a quenching-based fluorosensor by several orders of magnitude using commercially available surfactants. Here the cationic dye (PSF) has been used for a demonstration only. The study with PSF substantiates that the strategy is powerful enough to transform a cationic fluorophore to an efficient sensor for cationic species; selection of an anionic surfactant of proper chain length and a proper concentration being the key controlling factors. The strategy is applicable in aqueous medium and should be applied to physiological fluids as well.

Acknowledgment
D.B.T. and C.S.I.R., Government of India supported this research. P.D. and D.S. thank C.S.I.R. for the research fellowships.

References

1. A. W. Czarnik, *Fluorescent chemosensors for ion and molecule recognition* (ACS, Washington, 1992).
2. A. P. De Silva, H. Q. N. Gunaratne, T. Gunnlaugsson, A. J. M. Huxley, C. P. McCoy, J. T. Rademacher and T. E. Rice, *Chem. Rev.*, **97**, 1515 (1997).
3. J. R. Lakowicz, *Principles of fluorescence spectroscopy* (Plenum, New York, 1999).
4. V. O. Nikolaev, S. Gambaryan and M. J. Lohse, *Nature Methods*, **3**, 23 (2006).

5. U. E. Spichiger-Keller, *Chemical Sensors and Biosensors for Medical and Biological Applications* (Wiley-VCH, Berlin, 1998).
6. N. Chattopadhyay, A. Mallick and S. Sengupta, *J. Photochem. Photobiol.* A **177**, 55 (2006).
7. K. Ray, R. Badugu and J. R. Lakowicz, *J. Am. Chem. Soc.*, **128**, 8998 (2006).
8. R. Badugu, J. R. Lakowicz and C. D. Geddes, *J. Am. Chem. Soc.*, **127**, 3635 (2005).
9. S. Banthia and A. Samanta, *J. Phys. Chem.*, B **110**, 6437 (2006).
10. O. S. Wolfbeis, *Fluorescence spectroscopy : New methods and applications* (Springer-Verlag, New York, 1993).
11. G. B. Behera, B. K. Mishra, P. K. Behera and M. Panda, *Adv. Colloid Interface Sci.*, **82**, 1 (1999).
12. M. Panda, P. K. Behera, B. K. Mishra and G. B. Behera, *J. Photochem. Photobiol.* A **90**, 69 (1995).
13. Y. D. Fernandez, A. P. Gramatges, V. Amendola, F. Foti, C. Mangano, P. Pallavicini and S. Patroni, *Chem. Comm.*, **14**, 1650 (2004).
14. (a) A. Mallick, M. C. Mandal, B. Haldar, A. Chakrabarty, P. Das and N. Chattopadhyay *J. Am. Chem. Soc.*, **128**, 3126 (2006). (b) *Ibid J. Am. Chem. Soc.*, **128**, 10629 (2006).
15. S. P. Gromov, E. N. I. Ushakov, A. Vedernikov, N. A. Lobova, M. V. Alfimov, Y. A. Strelenko, J. K. Whitesell and M. A. Fox, *Org. Lett.*, **1**, 1697 (1999).
16. P. Grandini, F. Mancin, P. Tecilla, P. Scrimin and U. Tonellato, *Angew. Chem. Int.*, **38**, 3061 (1999).
17. P. Das, A. Chakrabarty, A. Mallick and N. Chattopadhyay, *J. Phys. Chem.* B **111**, 11169 (2007).
18. A. Mallick, B. Haldar, S. Maiti and N. Chattopadhyay, *J. Colloid Interface Sci.*, **278**, 215 (2004).
19. J. Du, B. Jiang, J. Xie and X. Zeng, *J. Dispersion Sci. Tech.*, **22**, 529 (2001).
20. R. Ranganathan, M. Peric and B. L. Bales, *J. Phys. Chem.*, B **102**, 8436 (1998).
21. A. Chakrabarty, A. Mallick, B. Haldar, P. Purkayastha, P. Das and N. Chattopadhyay *Langmuir*, **23**, 4842 (2007).
22. B. Gohain, P. M. Saikia, S. Sarma, S. N. Bhat and R. K. Dutta, *Phys. Chem. Chem. Phys.*, **4**, 2617 (2002).
23. P. Pal, H. Zeng, G. Durocher, D. Girard, R. Giasson, L. Blanchard, L. Gaboury and L. Villeneuve, *J. Photochem. Photobiol.* A **98**, 65 (1996).
24. L.S. Romsted, *Langmuir*, **23**, 414 (2007).
25. C.A. Bunton, F. Nome, F. H. Quina and L. S. Romsted, *Acc. Chem. Res.*, **24**, 357 (1991).
26. C.A. Bunton and L. S. Romsted, *Organic Reactivity in Microemulsions*, in *Handbook of Microemulsion Science and Technology* (Marcel Dekker, New York), p. 457.
27. C.A. Bunton, *Adv. Colloid Interface Sci.*, **123**, 333 (2006).
28. R. Bacaloglu, A. Blasko, C. A. Bunton, G. Cerichelli and F. Ortega, *J. Phys. Chem.*, **94**, 5062 (1990).

Chapter 20

Organic Reactivity in AOT-Based Microemulsions: Pseudophase Approach to Transnitrosation Reactions

G. Astray, A. Cid, J.C. Mejuto

Departamento de Quimica Fisica,
Facultad de Ciencias,
Universidad de Vigo,
Ourense (SPAIN)

and

L. García-Río

Departamento de Quimica Fisica,
Facultad de Qumica, Universidad de Santiago de Compostela,
Santiago de Compostela (SPAIN)

Microemulsions are very versatile reaction media, which now-a-days find many applications in the field of chemical kinetics and catalysis, ranging from nanoparticle templating to preparative organic chemistry. The thermodynamically stable and microheterogeneous nature of microemulsions, used as reaction media, induces drastic changes in the reagent concentrations, and this can be specifically used for tuning the reaction rates. In particular, amphiphilic organic molecules can accumulate and orient at the oil-water interface, inducing regiospecificity in organic reactions. As well as, microemulsions can constitute a scalable microreactor, which can be used as template in numerous synthetic processes, as i.e. the synthesis on monodisperse nanoparticles. In this review, we will show the modelization of the chemical kinetics in this media, using as example the trasnitrosation reactions.

20.1. Introduction

Microemulsions are stable, transparent solutions of water, oil, and surfactant, with or without a cosurfactant. They have been described as consisting of spherical droplets of a disperse phase separated from a continuous phase by a film of surfactant.[1-3] Because they provide both organic and aqueous environments, microemulsions can simultaneously dissolve both hydrophobic and hydrophilic compounds, each compound being distributed among water, organic solvent, and surfactant film in accordance with its physicochemical nature. One of the microemulsion forming surfactants receiving much attention in recent years has been sodium bis (2-ethylhexyl) sulfosuccinate (AOT).

Microemulsions have found a growing number of scientific and technological applications: they afford control over the size of synthesised microparticles;[4] they have numerous applications in the fields of solubilization and extraction;[5-8] and when the surfactant interface is stereoselective for certain reagents, they can be used for stereoselective synthesis.[9] They have been used to simulate complex biological structures (in particular as regards the behaviour of trapped water).[10-15] In keeping with this proliferating range of applications, there is an increasing interest in studying the details of chemical[1], photochemical,[16] and enzymocatalytic,[17,18] processes in microemulsions. In particular, since microemulsions are able, like phase transfer catalysis systems, to enhance reactions between non-hydrosoluble organic substrates and hydrosoluble reagents, the kinetics of numerous reactions in microemulsions have been studied.[2,19-30]

We now present a review of results supporting the validity of the pseudophase model for reactions taking place simultaneously in two of the pseudophases of an AOT/isooctane/water microemulsion. The type of reaction showed in the present revision was different transnitrosation reactions, a process that has been extensively studied in aqueous media. In particular we revised the nitrosation of amines by two different nitrosating agents: alkyl nitrites and nitroso sulfonamides.

20.2. Microemulsions as Reaction Media

Microemulsions, just like normal micelles, have aroused great interest as reaction media. This interest is justified by the great variety of compounds that they can solubilize simultaneously, offering at the same time a possibility of compartmentalization of the reagents, which represents an alternative use to the phase-transfer catalysts and other techniques used in order to obtain the contact between water-insoluble organic substrates and water-soluble species. Despite this interest, there are few studies, which realise a rigorous quantitative interpretation of the influence of microemulsions on the chemical reaction rates, in contrast to what happens in micellar media.

The studies on the influence that the structure of microemulsions exercises on the reaction rate fundamentally focus on w/o microemulsions, although there are numerous examples of reactions in microemulsions[24] that can be treated, to a certain extent, as direct micelles, using the traditional pseudophase model and the pseudophase ion-exchange model. These models have allowed the realisation of quantitative interpretations of the effect that the microemulsion structure has on the reaction rate.

The first studies were carried out by Fendler et al.[23] and Menger et al.[19] The formers studied the hydration of $Cr(C_2O_4)_3^{3-}$, finding that at a constant water concentration, the pseudo-first order rate constant decreases as the concentration of the tensioactive increases, whereas at a constant concentration of the tensioactive the rate constant increases on increasing the water concentration. These researchers interpreted the observed inhibition supposing that the microemulsion was saturated with water in such a way that the addition of a major quantity of tensioactive agent would have simply decreased the effective water concentration in the aggregate. The most surprising effect that they found was the sudden enhancement of reactivity. Thus, the rate was 5×10^6 times faster in the microemulsion than in "normal" water, supposing that this sudden increase was caused by the distribution of $Cr(C_2O_4)_3^{3-}$ between the benzene and the core of the microemulsion, and also by the possible existence of hydrogen bound between the substrate oxygen atom and the ammonium ion of the polar head of the tensioactives, which facilitates the proton transfer. In the same way, they also proposed that the increase of the water activity (highly structured inside the microemulsion) could be responsible for the activity increase. The effect of the water structure was also proposed by El Seoud et al.[24] in order to explain the increase of the hydration rate of 1,3-dichoroacetone in $AOT/CCl_4/H_2O$ microemulsions.

The works of Menger[19] on the alkaline hydrolysis of esters led to the conclusion that the transfer of ester from the organic phase to the aqueous phase was not rate limiting, although they recognised the importance as a pre-equilibrium of the substrate distribution among water droplets. In fact, Menger et al.[19] also observed that it was possible to establish a correlation between the hydrolysis rates of various esters in microemulsions and their coefficient of distribution between water and isooctane. Jaeger et al.[25] studied the nucleophilic substitution of benzyl and p-alkylbenzyl chlorides in $CTABr/nC_6/1\text{-}butanol/H_2O$ microemulsions, proposing that the reaction was located in the interface (the only possible contact zone among reagents) and interpreted the results assuming the distribution, that is compartmentalisation, of the reagents in the three pseudophases (alkane, water and interface).

Studies on reactivity in microemulsions may be classified according to the nature of the reagents involved. Thus, we can consider reactions between two ionic species, that will take place inside the core of the microemulsion droplet due to solubility reasons, and reactions between two non-ionic species, where the solubility factors will have a great importance at the moment of determining the reactivity. Among the examples of ionic reactions we can mention those of the oxidation of iodide by peroxodisulfate in microemulsions,[29-31] or the reactions of oxidation of hexacyanoferrate (II) by peroxodisulfate in $AOT/nC_{10}/H_2O$,[32] where both reagents are anionic. In these examples, the influence of the structural changes in the microemulsion on the rate reaction may be interpreted taking into account the dilution factors, since reagents are concentrated in the aqueous core due to the electrostatic repulsions between the AOT anionic heads. The most important effect found in these studies is that in all cases the "intrinsic" rate constant of the reaction in the microemulsion is superior to that corresponding to the reaction in "normal" water. This efficiency of the microemulsion water as reaction medium may be due to the high ionic concentration existing inside the droplet, and especially to the high structuring of water for low values of W, which implies a decrease of its activity. This result is clearly contradictory to the one obtained by Fendler (*vide supra*). In fact, various authors justify the changes in the reaction rate as a consequence of the changes in water activity (increase or decrease). Nevertheless, these changes in activity have been subsequently questioned.

The basic hydrolysis reaction of crystal violet (CV) represents a special case where both reagents are ionic species of opposite charge. In 1984, Blandamer et al.[33] realised a study of basic hydrolysis of CV and malachite green (MG) in microheterogeneous media, finding a significant catalysis with regard to the reaction in water. Subsequently, Rodenas et al.[26] studied the same reaction in microemulsions of $CTABr/1$-*hexanol*$/H_2O$, also finding a rate increase that they attributed to the lower dielectric constant in the microemulsion water in comparison with that of "normal" water. These researchers used kinetic data in order to estimate the dielectric constant in the aqueous microdroplet, obtaining values between 56 and 63.[34-36] Zanette et al.[37] studied the same reaction at various pressures, using different buffers as OH^- source and they found a strong inhibition for low W values, interpreting it on the basis of the absence of CV in the aqueous pseudophase, which resides exclusively at the interface without coming into contact with the alkaline medium. They also found a decrease in the reaction rate on having increased the pressure. Casado et al.[27] studied the reaction in the $AOT/nC_{10}/H_2O$ system, finding a decrease of the bimolecular rate constant in relation to pure water, as well as a reduction of the reaction rate when W increases. When they verified the hydrolysis process in the presence of cationic tensioactives, they found a catalytic effect on the reaction rate. These results were confirmed by Zanette et al.[37] on having verified the above mentioned study in *AOT/isooctane/water* microemulsions. Analogous results were found for MG.

Studies on the reactivity of an ionic substrate and a molecular species present a higher complexity. This is the case of the studies on the oxidation reaction of 1,2-benzenediols by hexachloroiridate (IV) that was developed by Minero et al.[35] in microemulsions of $SDS/1$-*butanol/toluene*/H_2O. These researchers proposed a simple model for the microemulsion (which would be formed by three pseudophases), explaining the experimental results by means of a model that considers the distribution of reagents among the pseudophases of the microemulsion, and obtaining values of the rate constant at the interface near those obtained in saturated aqueous solutions of 1-butanol. This implies that its properties must be similar to those of the microemulsion interface.

Rodenas et al.[36] studied the oxidation of alcohols by potassium dichromate in acid medium, in *SDS* reverse micelles in alcohol, finding that the reaction rate decreases with the quantity of alcohol in the inverse micelle. These kinetic studies were explained on the basis of the intermicellar exchange of reagents that depends on the thickness of the interface where the alcohol and the tensioactive are located.

Another example of studies between a neutral substrate and an ionic species is represented by the acid hydrolysis reactions in microemulsions developed by Zanette et al.[37] in o/w and w/o microemulsions of $SDS/1$-*butanol/toluene*/H_2 O, using the formalism of the pseudophase ion-exchange model, and considering that the system is formed by three phases.

Although of great interest, the studies on reactions between molecular species in microemulsions are less frequent, since the confinement of the reagents is more difficult to obtain and the restriction of those reagents to one microemulsion zone solely is improbable. The most interesting works are centred on the processes of enzymatic catalysis with substrates that are distributed among the three phases, standing out the works of Martinek et al.[38] who propose the treatment of kinetic data by means of a model which considers that the enzyme is concentrated in the aqueous phase, so that it would only be active with respect to the substrate molecules dissolved in the aqueous phase. These researchers obtained distribution coefficients of alcohols in different phases from bibliographical results or as a result of the determinations made by means of the flow microcalorimetry technique, achieving a satisfactory interpretation of the experimental results. On having obtained the real values of the Michaelis constant it can be observed that they are very close to the ones obtained in water, thus being concluded that for enzymatic reactions in microemulsions with substrates that are distributed among the three phases, the changes in the Michaelis constant and in the substrate specificity might not be caused by changes in the real catalytic properties of the enzyme but by deficient substrate distribution among the different phases. Moreover, there are interesting applications of these media as reaction medium of lipase-catalysed processes.[39-41]

20.3. The Pseudophase Model

Kinetic studies of reactions in water-in-oil microemulsions can be interpreted in terms of reactivity, only if the local reagent concentrations and the intrinsic rate constants in the various microphases of these organised media can be obtained from the overall, apparent rate data. Our research group has devised a kinetic model based on the pseudophase formalism which can be applied to carry out a quantitative interpretation of the influence of the microemulsion composition on the chemical reactivity.[42-51] In order to apply this formalism to the basic hydrolysis of the NPA we need to consider the microemulsion formed by three strongly differentiated pseudophases: an aqueous pseudophase (w), a continuous medium formed fundamentally by the iC_8 (o) and an interface formed fundamentally by the surfactant (i). The application of the pseudophase model considers that each pseudophase is uniformly distributed in the total volume of the microemulsion. In view of the solubility characteristics of reagents, their distribution among these pseudophases can be described by partition coefficients. Partition coefficients have been expressed in terms of mole ratios in each phase, a choice that makes the kinetic treatment much easier[52] Problems and approximations involved in these definitions have been discussed in the literature.[53]

For the application of the kinetic treatment, it is necessary that the partitioning reagent distribution along the microenvironments of the microemulsions be faster than the reaction rate under study. The kinetics of solubilizate exchange between water droplets of a water-in-oil microemulsion have been widely studied by Robinson et al.[54–55], Fletcher et al.[56] and Pileni et al.[57] Their approach consists of an analysis of a reaction in a water-in-oil microemulsion involving reagent species totally confined within the dispersed water droplets, so that a necessary step prior to their chemical reaction is a transfer of reagents into the same droplet. When the chemical reaction is fast (close to diffusion-controlled), the overall reaction rate is likely to be controlled by the rate of interdroplet transfer or reacting species.

20.4. Nitrosation of Amines by Alkyl Nitrites

In water, the reaction between alkyl nitrites and secondary amines consists of nucleophilic attack by the free (unprotonated) amine upon the nitroso group of the alkyl nitrite in a nonsynchronized concerted mechanism with the rupture of the ONOR bond.[58–60] The same mechanism is assumed for the reactions in microemulsions. This mechanistic hypothesis is supported by the finding that the pseudo-first-order constant kobs depended linearly on total amine concentration in all the reactions studied.[43] The acid or basic hydrolyses of the alkyl nitrites are all negligible under the experimental conditions used.[61–66]

Reaction with Piperazine : The influence of the composition of the microemulsion on the rate of nitrosation of *PIP* by *EEN* was studied.[43] For a given W value, the pseudo-first-order rate constant was almost unaffected by $[AOT]$; for a given $[AOT]$ it increased slightly with W. This behaviour is attributable to the reaction with *EEN* taking place simultaneously in both the *AOT* film and the aqueous pseudophase. Similar results were found when the nitrosation reaction between *PIP* and *BEN* was analyzed.[43] The pseudophase model for the reaction with *EEN* and *BEN*, which treats the microemulsion as a three layer bulk system and ignores its actual micellar structure, is illustrated in Scheme 20.1.

To avoid having to define the volumes of the pseudophases, the partition coefficients assumed to govern the distribution of the reagents among the three pseudophases are defined in terms of their mole per mole concentrations in the pseudophases

$$K_1 = \frac{[PIP]_i}{[PIP]_w} W \quad K_3 = \frac{[RONO]_i}{[RONO]_w} W \quad K_4 = \frac{[RONO]_i}{[RONO]_o} Z, \quad (1)$$

where the subscripts w, o, and i respectively indicate quantities in water, oil, and the surfactant film; square brackets, as usual,[65–74] indicate con-

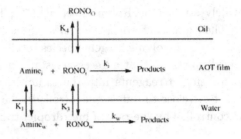

Scheme 20.1.

centrations referred to the total volume of microemulsion; and Z is defined, by analogy with W, as the ratio $[i - C_8]/[AOT]$. With these definitions, the model of Scheme 20.1 implies that the overall pseudo-first-order rate constant is given by

$$k_{obs} = k'_i \frac{1}{1 + \frac{W}{K_3} + \frac{Z}{K_4}} + k'_w \frac{W}{K_3} \frac{1}{1 + \frac{W}{K_3} + \frac{Z}{K_4}}, \qquad (2)$$

where k'_i and k'_w are the pseudo-first-order rate constants in the AOT film and the aqueous pseudophase, respectively; these pseudoconstants are expressed in terms of bimolecular rate constants k_i and k_w as

$$k'_i = k_i \frac{[PIP]_i}{[AOT]} \quad k'_w = k_w \frac{[PIP]_w}{[H_2O]}. \qquad (3)$$

Like the partition coefficients, to avoid having to define the volumes of the pseudophases, k_i and k_w are defined in terms of mole per mole concentrations in the corresponding phases; then eq. 2 becomes

$$k_{obs} = \frac{k_i}{[AOT]} \frac{[PIP]_T K_1}{K_1 + W} \frac{1}{1 + \frac{W}{K_3} + \frac{Z}{K_4}} + \frac{k_W}{[AOT]} \frac{[PIP]_T W}{(K_1 + W) K_3} \frac{1}{\left(1 + \frac{W}{K_3} + \frac{Z}{K_4}\right)}, \qquad (4)$$

$$K_{obs} = \frac{[PIP]_T \left((k_i K_1) + (K_W W/K_3)\right)}{[AOT](K_1 + W)\left(1 + \frac{W}{K_3} + \frac{Z}{K_4}\right)}, \qquad (5)$$

which predicts linear dependence of $[PIP]_T/k_{obs}[AOT](K_1 + W)$ on Z:

$$\frac{[PIP]_T}{k_{obs}[AOT](K_1 + W)} = \frac{1 + (W/K_3)}{K_i K_1 + \left(\frac{K_W W}{K_3} + \frac{Z}{K_4}\right)} + \frac{Z}{K_4 \left(K_i K_1 + \frac{K_W W}{K_3}\right)}, \qquad (6)$$

The validity of the proposed model is therefore supported by the actual observance of this linear dependence of k_{obs} on Z predicted by eq. 6 ($R^2 > 0,9756$, Figure 20.1).

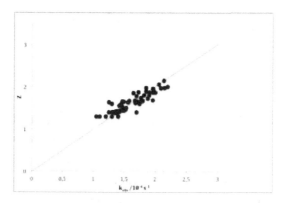

Fig. 20.1. Observed first-order pseudoconstant and calculated first-order pseudoconstant (eq. 5) for the reaction for PIP with EEN.

The parameters k_i, K_3, and K_4 were estimated by fitting with a multidimensional nonlinear regression program based on Marquardt's algorithm;[69] to keep the number of optimized parameters to a minimum, K_1 was assigned the value of 10 previously found for nitrosation of PIP by MNTS,[60] and k_w was calculated as 1.8 s^{-1} by determining the bimolecular rate constant of the nitrosation of PIP by EEN in water. The fit quality is confirmed by the agreement of the values with previously reported k_i values of similar reactions in AOT microemulsions[59] and K_3 values for association of alkyl nitrites to SDS direct micelles.[75-79] The optimized values of k_i, K_3, and K_4 are listed in Table 20.1. Analogous results for the reaction of PIP with BEN are likewise listed in Table 20.1 (see Fig. 20.2).

Reaction with N-Methylbenzylamine : The reactions of EEN and BEN with NMBA in $AOT/i - C_8/water$ microemulsions were studied by means of series of reactions analogous to those described for their reactions with PIP. In this case, however, k_{obs} values increase significantly with $[AOT]$ for given W, and only slightly with W for given $[AOT]$. Because of the poor solubility of NMBA in water, the pseudophase model for this reaction would a priori consider simultaneous reactions in the AOT film and the isooctane pseudophase. However, there are two reasons for ignoring the reaction in the isooctane pseudophase: firstly, the reaction rate in isooctane is several

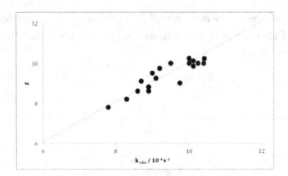

Fig. 20.2. Observed first-order pseudoconstant and calculated first-order pseudoconstant (eq. 5) for the reaction for PIP with BEN.

orders of magnitude less than those observed in this work; secondly, a significant contribution by the reaction in isooctane, in which k_{obs} exhibits a complex dependence on amine concentration,[55] would not be compatible with the linear dependence observed in our case. The pseudophase model adopted for the reaction of NMBA is therefore that shown in Scheme 20.2.

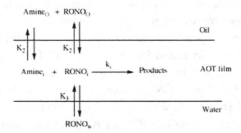

Scheme 20.2.

with K_3 and K_4 defined as above and K_2 by

$$K_2 = \frac{[NMBA]_i}{[NMBA]_o}Z \quad K_3 = \frac{[RONO]_i}{[RONO]_W}W \quad K_4 = \frac{[RONO]_i}{[RONO]_o}Z \quad (7)$$

calculations analogous to those described for piperazine lead to the expression

$$k_{obs} = \frac{k_i}{[AOT]} \frac{[NMBA]_T}{1+\frac{Z}{K_2}} \frac{1}{1+\frac{W}{K_3}+\frac{Z}{K_4}} \quad (8)$$

However, with the values of K_3 and K_4 obtained from the reaction with PIP (Table 20.1), and in the working conditions used, W/K_3 is negligible in comparison with $(1+(Z/K_4))$, so that eq. 8 reduces to

$$k_{obs} = \frac{k_i}{[AOT]} \frac{[NMBA]}{1+\frac{Z}{K_2}} \frac{1}{1+\frac{Z}{K_4}} \quad (9)$$

which can be rewritten in the form

$$\frac{1}{k_{obs}[AOT]} = \frac{(K_2+Z)(K_4+Z)}{K_1 K_2 K_4 [NMBA]_T} \quad (10)$$

Fitting eq. 10 to the experimental data (with $K_4 = 2.82$ for EEN and $K_4 = 2.64$ for BEN fixed as the value obtained in the experiments with PIP) afforded the values of K_2 and k_i listed in Table 20.1. The satisfactory fit obtained for experiments with $W > 12$ supports the validity of the model employed. The anomalous behaviour in systems with $W < 12$, in which the reaction is slower than predicted by the model, is attributed to the fact that under these conditions a large proportion of water molecules are involved in solvation of AOT head groups, which alters the physicochemical properties of the interface medium (dielectric constant, microviscosity, hydrogen bond network, etc.) in ways that hinder reaction. Whereas, this anomalous behaviour observed will be discussed in following chapters.

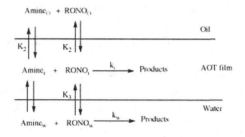

Scheme 20.3.

Reaction with Morpholine: In experiments on the reaction between morpholine (MOR) and BEN, k_{obs} was virtually independent of [AOT] for fixed W, and decreased slightly with increasing W for fixed [AOT]. This behaviour, different from that observed for other amines, should be related to the distribution of reagents among the pseudophases. Thus, since MOR is soluble in both isooctane and water, and since the reaction studied is several orders of magnitude slower in isooctane than was observed in the microemulsions used in this work,[46] the pseudophase model applied is that shown in Scheme 20.3.

Calculations analogous to those described in the previous sections lead to the expression

$$k_{obs} = \frac{k_i}{[AOT]} \frac{[MOR]_T}{1 + \frac{W}{K_1} + \frac{Z}{K_2}} \frac{1}{1 + \frac{W}{K_3} + \frac{Z}{K_4}} + \frac{k_W}{[AOT]} \frac{W[MOR]_T}{K_1 K_3}$$

$$\frac{1}{1 + \frac{W}{K_1} + \frac{W}{K_2}} \frac{1}{1 + \frac{W}{K_3} + \frac{Z}{K_4}} \quad (11)$$

$$k_{obs} = \frac{[MOR]_T \left(k_1 + \frac{k_W W}{K_1 K_3}\right)}{[AOT]\left(1 + \frac{W}{K_1} + \frac{Z}{K_2}\right)\left(1 + \frac{W}{K_3} + \frac{Z}{K_4}\right)} \quad (12)$$

Fitting Eq. 12 to the experimental data for $W > 10$, with K_1, K_3, K_4, and k_w fixed at values obtained as described above (Table 20.1), yielded the values of k_i and K_2 that are likewise listed in Table 20.1 (see Fig. 20.3). The mean discrepancy between the experimental and fitted values of k_{obs} was less than 9%; the greater deviation of the data for media with $W < 10$ is attributed to the same cause as in the case of NMBA, to the fact that under these conditions a large proportion of water molecules are involved in solvation of AOT head groups, which alters the physicochemical properties of the interface medium (dielectric constant, microviscosity, hydrogen bond network, *etc.*) in ways that hinder reaction.

20.5. Nitrosation of Amines by Nitroso Sulfonamides

The reactivity between MNTS and secondary amines in water is well established.[60] The rate equations obtained for the reactions in microemulsions were in all cases similar to those in water, with first-order terms in MNTS and total amine concentration. In our experimental conditions, the protonation of the amines in water is negligible (the percentage of protonated amine in unfavorable conditions is lower than 3%).

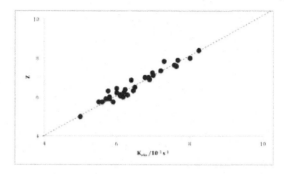

Fig. 20.3. Observed first-order pseudoconstant and calculated first-order pseudoconstant (eq. 12) for the reaction for MOR with BEN.

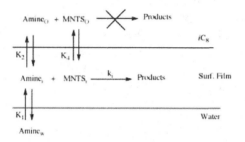

Scheme 20.4.

The competing hydrolysis of MNTS by the small amount of hydroxyl ions liberated by this ionization is insignificant because solutions of MNTS with as much as 2×10^{-2} M of HO^- (referring to the total volume of the solution) are stable for long periods of time (several days). UV-vis spectra at the end of the reaction indicated quantitative nitrosamine formation in every case. Properties of the medium with marked effects on reactivity are known to vary significantly over the range of compositions used. For dimethylamine, the value of k_0 decreases by increasing $[AOT]$ for a given W and also decreases by increasing W for a given $[AOT]$, while for the other amines, k_0 values increase significantly with [AOT] for a given W and only slightly with W for a given $[AOT]$. This behaviour can be attributed to the different distributions of amines between the diverse pseudophases. Dimethylamine will be present in the water phase, in the surfactant film, and also in the oil phase (see Scheme 20.4), but the other five amines will be restricted to both the surfactant film and the oil phase, and their concentrations in the water phase will be negligible (see Scheme 20.5).

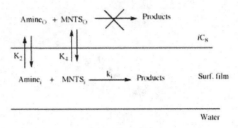

Scheme 20.5.

The pseudophase model for these reactions, which treats the microemulsion as a three-layer bulk system and ignores its actual micellar structure, is illustrated in Schemes 20.4 and 20.5. To avoid having to define the volumes of the pseudophases, the partition coefficients assumed to govern the distribution of the reagents among the three pseudophases are defined in terms of their mole per mole concentrations in the pseudophases (*vide supra*). With these definitions, the model of Scheme 20.4 (dimethylamine) implies that the pseudo-first-order rate constant is given by

$$k_{obs} = \frac{[amine]_i}{[AOT]} \frac{k_i}{\left(1 + \frac{W}{K_1} + \frac{Z}{K_2}\right)\left(1 + \frac{Z}{K_4}\right)} \quad (13)$$

$$k_{obs} = \frac{[amine]_i}{[AOT]} \frac{k_1 K_1 K_2 K_4}{(K_1 K_2 + K_2 W + K_1 Z)(K_4 + Z)} \quad (14)$$

Table 20.1 lists the values of kinetic parameters for each amine. Although in this case (with three parameters to optimize in the equation) there is a significant degree of correlation, the fit is good (discrepancy was less than 10%). This is confirmed by the agreement of the values with previously reported values of similar reactions in AOT microemulsions and K_4 values for association of MNTS to SDS micelles.[61-64] The average value of K_4 is also in good agreement with the value of about 11 estimated for the MNTS-AOT association constant by analysis of changes in the UV spectrum of MNTS at 265-275 nm. It should be borne in mind that precise spectroscopic estimation of this constant is ruled out by the impossibility of having all of the MNTS in the interface. The spectroscopic estimation of K_4 was therefore based on an analysis of the spectroscopic data[38] using a fitting procedure with the absorbance of MNTS when fully bound to the interface as an optimizable parameter. In the case of the other amines (methylethylamine, methylbutylamine, methylhexylamine, methyloctylamine, and methyldodecylamine), the pseudophase model adopted for the reaction is therefore that shown in Scheme 20.5.

$$k_{obs} = \frac{[amine]i}{[AOT]} \frac{k_i}{\left(1+\frac{Z}{K_2}\right)\left(1+\frac{Z}{K_4}\right)} \quad (15)$$

$$k_{obs} = \frac{[amine]_t}{[AOT]} \frac{k_1 K_2 K_4}{(K_2+Z)(K_4+Z)} \quad (16)$$

Calculations analogous to those previously described lead to the expressions Fitting eq. 16 to the experimental data with the same nonlinear regression program as described above provided the values of K_2 and k_i listed in Table 20.1. The satisfactory fit obtained for these experiments supports the validity of the model employed.

20.6. Additional Remarks

The assumption that the mechanism of these reactions in the microemulsions used in this work is the same as the mechanism obtained in bulk water is supported not only by the arguments sketched at the beginning of the Results and Discussion section, but also by the observation of solvent isotope effects k_H/k_D of respectively 2.0 and 1.6 for the reactions of PIP with EEN and NMBA with BEN, both these values being similar to the values of about 1.7 typically found for reactions between alkyl nitrites and amines in water.[60] Furthermore, the more precise hypothesis that the reaction mechanism in the interface is the same as in water is supported by comparison of the reactivities of the alkyl nitrites (relative to MNTS or to each other) in the two media: the reactivities of PIP and NMBA with EEN in the interface are, as in water, similar to those reported for their reactions with MNTS under the same conditions;[51] and the reactivity of BEN in the interface is, as in water, about 1 order of magnitude greater than that of EEN. These similarities between the behaviour of the alkyl nitrites in water and their behaviour in the interface support the notion that the reaction mechanism is the same in both media.

20.6.1. *Nature of the Reaction Mechanism*

The solvent isotope effects reported above are in keeping with the reaction mechanism consisting, in both water and our microemulsions, of nucleophilic attack by the unprotoned amine on the nitroso group of the alkyl nitrite, in nonsynchronized concerted mechanism with the rupture of the ON-OR bond. It is thought that this process requires the accumulation of large negative charge on the alkoxide oxygen atom in the transition state (this would explain why BEN is more reactive than EEN, since electron-withdrawing substituents favour the required charge distribution) so as to ensure strong solvation of the alkoxide group. The existence of strong solvation of the alkoxide group (testified to by the large negative entropy of activation reported for the reaction of PIP with EEN in water, -113.37 J mol^{-1} K^{-1})[74–75] raises the possibility that the existence of the departing alkoxide group may be concerted with its protonation. However, this would imply the loss of an OH^- group from the reaction complex, which in would in turn imply a solvent isotope effect as high as 2.08 according to established theory for reactions with late transition states, the fractionation factor for the triply solvated OH^- ion being 0.48.[78] Hence the solvent isotope effects of around 1.8 that are actually observed suggest that concerted protonation of the alkoxide group is unnecessary for the reaction to proceed.

20.6.2. *Partition Coefficients*

The partition coefficients of the reagents are listed in Table 20.1. Those of the amines, K_1 and K_2, are consistent with those measured in two-phase iC_8/water systems.[43] Those of the alkyl nitrites, K_3 and K_4, show BEN to be more hydrophobic than EEN, as in micellar systems.[57,78,79]

To determine the association constant of MNTS to the AOT Film, we take into account that the presence of micelles or reverse micelles induces changes in the UV-vis spectrum of MNTS. These spectroscopic changes enable us to determine the binding constant[80] of MNTS to the surfactant film, assuming that the spectroscopic changes are due to the association of MNTS from the continuous phase (isooctane) to the interphase of the microemulsion (AOT film). The binding constant can be expressed in terms of mole ratios (17), where the subscripts i and o denote the interphase and isooctane, respectively.

$$K_{MNTS} = X^i_{MNTS}/X^o_{MNTS}. \tag{17}$$

Taking into account the fact that the concentration of MNTS is much lower than the concentration of AOT or than that of the isooctane, eq. 17 can be written as follows:

$$K_{MNTS} = \frac{[MNTS]_t[isooctane]}{[MNTS]_o[AOT]} = \frac{[MNTS]_i}{[MNTS]_o}Z, \tag{18}$$

where all of the concentrations are referring to the total volume of the solution and Z is defined, by analogy with W (vide supra), as the ratio $[iC_8]/[AOT]$. Considering that, for a given total concentration of MNTS, the absorbance to a fixed wavelength is due to the sum of those corresponding to MNTS in the isooctane plus the one associated with the interphase, eq. 18 can be obtained by analogy with the treatment corresponding to normal micelles[42]

$$A = A_0 + \frac{(A_1 - A_0)K_{MNTS}}{K_{MNTS} + Z}, \qquad (19)$$

where A_o and A_i correspond to the absorbance of MNTS in the isooctane and the interphase, respectively. The first one was experimentally determined, while the second one was obtained by a multidimensional optimization process with KMNTS. Fitting experimental data yielded a value of the constant K_{MNTS} of $K_{MNTS} = 11 \pm 2$, compatible with the results obtained by kinetic measurements as reported elsewhere.[42]

Table 20.1. Kinetic and thermodynamic parameters for transnitrosation reaction in AOT based microemulsions, obtained from the pseudophase model formalism.

System	$K_2/M^{-1}s^{-1}$	k_t/s^{-1}	$k_2^i/M^{-1}s^{-1}$	K_1^b	K_2	K_3	K_4
PIP+EEN	3.25 x 10^{-2}	4.39 x 10-3	1.62 x 10^{-3}	10		490	2.82
PIP+BEN	0.36	8.86 x 10^{-2}	3.28 x 10^{-2}	10		740	2.65
NMBA+EEN	3.54 x 10^{-2}	2.64 x 10^{-2}	9.77 x 10^{-4}		18.5	490	2.82
NMBA+BEN	0.38	2.55 x 10^{-2}	9.43 x 10^{-4}		18.5	740	2.64
MOR+BEN	3.62 x 10^{-2}	4.45 x 10^{-3}	1.65 x 10^{-3}	46.9	13.5	740	2.64
DMA+MNTS	3.84 x 10^{-1}	1.60 x 10^{-2}	5.92 x 10^{-3}	83.3	83.3		11.0
MEA+MNTS	8.03 x 10^{-2}	7.48 x 10^{-3}	2.77 x 10^{-3}		36.9		11.0
MBA+MNTS	2.13 x 10^{-1}	1.03 x 10^{-2}	3.81 x 10^{-3}		19.0		11.0
MHA+NNTS	2.12 x 10^{-1}	9.25 x 10^{-3}	3.42 x 10^{-3}		15.3		11.0
MOA+MNTS	1.28 x 10^{-1}	9.14 x 10^{-3}	3.38 x 10^{-3}		15.4		11.0
MDA+MNTS		8.65 x 10^{-3}	3.20 x 10^{-3}		14.5		11.0

20.6.3. Anomalous Behaviour with $W < 12$ and controversies with the droplet model

The reactions with NMBA and MOR were slower than predicted when carried out in media with $W < 12$; similarly anomalous behaviour has previously been reported for the nitrosation of these amines by MNTS in microemulsions with low W.[44] We attribute these anomalies to the scarcity of water molecules in such media, in which a large proportion of the water is engaged in solvating AOT head groups. Numerous studies have shown that the dielectric constant of the interface decreases with decreasing W when W is less than 10-15,[78] and electron solvation experiments,[1,83-87] H-NMR studies,[87] and studies of solvolysis in microemulsions[88] suggest that other kinetically relevant properties of water (solvating power, nucleophilicity, electrophilicity, etc.) also undergo significant alteration in microemulsions with low W. This implies that for media with W less than a certain threshold, the kinetic constant k_i depends on the amount of available water at the interphase and is therefore a function of the composition of the microemulsion.

There is a controversy in the scientific community about the microstructure of water-in-oil microemulsions, mainly at very low water contents. In fact, some authors[89] suggest that for low water content the aggregate microstructure will deviate from the droplet model. Therefore, the pseudophase model considering the existence of three pseudophases is wrong. Correa et al. studied the reduction of ketones[90-91] in $AOT/i-C_8/H_2O$ and in $AOT/toluene/H_2O$ and justify the kinetic behaviour considering that for small W values the water properties are different from those of bulk water, being mostly bounded to the AOT-head group. For this reason, these authors assume that for $W < 10$ ($W = [H_2O]/[AOT]$) in the microemulsion there is no water available to constitute the microdroplet and, hence, only two pseudophases must be considered. They consider that only for $W = 10$ there is free water in the core of the microemulsion. This assumption will imply that a three-pseudophases system for the kinetic model would be not accurate for the study of these ranges of microemulsion compositions.

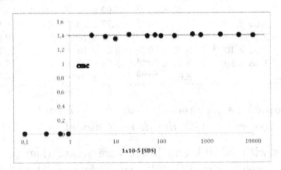

Fig. 20.4. Influence of SDS concentration on the observed rate constant, k_{obs}, for the CV recovery from CV-SO$_3$-. $[CV]=1.00 \times 10^{-5}$ M, $[SO_3^{2-}]=0.5$ M.

Other authors,[92-95] using spectroscopic results (i.e. FT-IR or NMR) have demonstrated that there are at least three kinds of water present in the microemulsion (free-water, trapped-water and bounded-water), they found evidences from the analysis of these spectroscopic data that there is a significant amount of free-water in all W-ranges, even at very small W values.

Other evidence against the existence of two microenvironments at small W values is the possibility of obtaining nanoparticles at these small W values.[96-102] It is well known that the use of microemulsions as reaction media for nanoparticles allows controlling the size of these nanoparticles because there is a relationship between the W value (proportional to the droplets radius) and the particle radius. This relationship between the radii

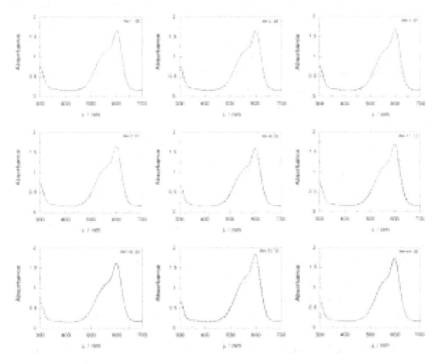

Fig. 20.5. [Solid line] Crystal violet spectra ($[CV] = 1.65 \times 10^{-5} M$) and [dot line] crystal violet recovery from the colourless mixture of crystal violet and sulphite ions ([CV] = $1.65 \times 10^{-5} M$ and $[SO_3^{-2}] = 5.00 \times 10^{-3}$ M) at different W values in $AOT/isooctane/water$ microemulsions.

of the particle and the microdroplet remains satisfactory even at very small values of W. In fact deviations of this relationship appears at very high values of W.

In summary, there are two group of results with contradictory conclusions: (i) at low W values, there is not free water, and hence there is not any water pool and the microemulsion can be considered as a two-phases system, and (ii) the quantity of free water present in the microemulsion is significant in all W values and hence the microemulsion is a three-phases system in all of the range of compositions. As we will show below the fact that the microemulsion can be considered as two-phases system at low W values is an oversimplification and the existence of three well-differentiated microenvironments must be considered.

In order to discriminate these possibilities, Fernndez et al.[103] have studied the reaction between CV and SO_3^{2-} ions in AOT-based microemulsions at very different water contents. They study the influence of microemulsion composition upon the equilibrium between CV (colourer) and CV - SO_3^- (colourless). Previous results, in direct micelles,[104] shown that a compartmentalization of the reagents is necessary for colour recovery in this reaction (Figure 20.4). The presence of SDS micelles implies a complete recovery of the dye because CV - SO_3^- may associate to micelles, and, thus cleaving the CV - SO_3^- molecule and causing sulfite ion to be excluded from the micelle pseudophase by electrostatic repulsion from surfactant heads. As a result, CV would associate to the micelle pseudophase and sulfite ion would be repelled to the aqueous pseudophase.

In all cases, when the AOT microemulsion acts as reaction medium, a significant percentage of recovered CV had been found. This percentage increases on increasing the W value, rising up to a constant value near 100% (at $W > 17$ the percentage is higher than 90%).

When colourless mixture of crystal violet and sulphite ions ($[CV]=$ $1.65 \times 10^{-5} M$ and $[SO_3^{2-}] = 1.00 \times 10^{-3}$ M) is added to an AOT microemulsion (values ranging from $W = 0.56$ to 43.65) the violet colour was completely recovered just after a few seconds, showing a similar behaviour to that observed with direct micelles.[104] Figure 20.5 shows the spectrum of CV in microemulsions ($[CV]=1.65 \times 10^{-5} M$), and the spectrum of recovered CV from the displacement of the equilibrium between the coroled and the colorless product ($[CV]=1.65 \times 10^{-5} M$ and $[SO_3^{2-}] = 5.00 \times 10^{-3} M$) for different composition of microemulsions. The concentration of surfactant, [AOT], was kept constant and equal to 0.5 M, and the molar ratio $W = [H_2O]/[AOT]$ was ranged between 1.92 and 44.09. In all cases, a significant quantity of crystal violet was recovered after the addition of the colourless mixture to the microemulsion. From the ratio between the absorbance maximum at 590 nm in the crystal violet spectra and the crystal violet recovered from the colourless mixture CV/SO_3^{2-}, the percentage of crystal violet recovered can be calculated. Table 20.2 shows the percentage of crystal violet recovered by the displacement of the CV/SO_3^{2-} equilibrium for each initial condition (SO_3^{2-} concentration and microemulsion composition).

Table 20.2. Percentage of recovered crystal violet for each composition of microemulsion and for each initial concentration of sulphite ion in the colourless mixture.

W	%CV recovered	$[SO_3^{2-}]_o$ (M)	W	% CV recovered	$[SO_3^{2-}]_r$ (M)
44.09	–	0.020	8.54	66.52	0.020
	95.18	0.010		84.30	0.010
	96.43	0.005		88.48	0.005
	98.90	0.001		99.35	0.001
33.32	–	0.020	5.05	48.35	0.020
	87.61	0.010		66.39	0.010
	90.66	0.005		79.04	0.005
	92.95	0.001		95.21	0.001
16.9	90.43	0.020	3.97	42.82	0.020
	92.49	0.010		63.14	0.010
	97.61	0.005		71.41	0.005
	99.98	0.001		89.99	0.001
11.12	76.59	0.020	2.92	–	0.020
	82.32	0.010		57.78	0.010
	87.37	0.005		67.38	0.005
	98.67	0.001		89.55	0.001
1.92	–	0.020			
	54.02	0.010			
	61.47	0.005			
	80.73	0.001			

The recovery of the colour, even at very low W values, implies a perfect compartmentalization of the reagents. This compartmentalization of the reagents also remains for very small W values ($W < 10$). Due to this, the fact that microemulsions are formed by three pseudophases must be assumed, independently of the water contents. To sum up, the existence of three well-differentiated zones in the microemulsion is guaranteed.

20.7. Conclusions

In all of the cases considered above, the results can be interpreted by assuming that the reaction occurs at the interface between the water droplet and the isooctane and that the dependence of the observed rate constants on reaction conditions is largely governed by the relative affinities of the amines for the different phases present in the medium. The hydrophobicity of the amine determines its distribution between the three phases and whether the rate constant increases, decreases, or remains essentially constant as the droplet size increases. These results are a particularly satisfactory quantitative explanation in terms of a single theoretical model for all droplet sizes studied. This agreement with the model might seem surprising if we consider that no corrections have been introduced to take into account any increase in the volume of the interface due to the incorporation of amine, but under the conditions used, this influence must be almost negligible. The problem of the volume occupied by the amine at the interface arises only when the concentration of amine at the interface is not negligible with respect to the total concentration of surfactant. Only in experiments at a low concentration of surfactant can the concentration of amine at the interface be considered non-negligible, and even in these cases, the fact that the amine molar volume is much lower than the surfactant molar volume reduces the importance of such possible volume changes. Our results also seem to support our definition of partition coefficients in terms of mole ratios and not of mole fractions (such a difference will only be significant in those experiments in which the concentration of amine incorporated into the surfactant film is not negligible with respect to the total surfactant concentration). All results showed in the present review confirm this choice. In the present study, the pseudophase model allows the thermodynamic problem of partition among the phases to be separated from the kinetic problem and therefore allows one to estimate the reactivity of the different amines in the interfacial region, k_i.

Acknowledgment

G.A. thanks to Xunta de Galicia a "Mara Barbeito" Research Grant. A.C. thanks to Universidad de Vigo a Research Grant. Financial support from Xunta de Galicia (PGIDIT07PXIB383198PR) is gratefully acknowledged.

Nomenclature

AOT	sodium octadecyl sulfosuccinate (Aerosol OT)
BEN	Bromo ethyl nitrite
CV	crystal violet
EEN	Ethoxy ethyl nitrite
k_i	rate constant at the AOT film, defined in terms of mole per mole concentrations and expressed in s-1
K_0	pseudo-first-order rate constant
k_2^i	bimolecular rate constant at the AOT film
K_2	bimolecular rate constant in bulk water
K_1	partition constant of amine between the water phase (isooctane) and the AOT film in $AOT/isooctane/water$ microemulsions
K_2	partition constant of amine between the organic phase (isooctane) and the AOT film in $AOT/isooctane/water$ microemulsions
K_3	partition constant of alkyl nitrite between the water phase (isooctane) and the AOT film in $AOT/isooctane/water$ microemulsions
K_4	partition constant of alkyl nitrite between the organic phase (isooctane) and the AOT film in $AOT/isooctane/water$ microemulsions
$K_{MNTS} = K_4$	partition constant of MNTS between the organic phase (isooctane) and the AOT film in $AOT/isooctane/water$ microemulsions
K_o^w	partition constants of amines between the organic phase (isooctane) and water mixtures
MG	malachite green
$NMBA$	N-Methylbenzylamine
$MNTS$	N-methyl-N-nitroso-p-toluenesulfonamide
MOR	Morpholine
NPA	Nitrophenyl acetate
n_a^o	number of moles of amines in isooctane in the organic phase (isooctane) and water mixtures
N_a^w	number of moles of amines in water in the organic phase (isooctane) and water mixtures
n_o^a	number of moles of isooctane in the organic phase (isooctane) and water mixtures

n_w^a	number of moles of water in the organic phase (isooctane) and water mixtures
o/w	oil-in-water microemulsion
PIP	Piperacine
SDS	Sodium dodecyl sulfate
T_p	temperature of percolation
V	molar volume of AOT in the interface
w/o	water-in-oil microemulsion
W	$[H_2O]/[AOT]$
Z	$[i-C_8]/[AOT]$

References

1. *Reverse Micelles*; P.L. Luisi, B.E. Straub, Eds.; Plenum Press: New York (1984).
2. *Structure and Reactivity in Reverse micelles*; M.P. Pileni, Eds.; Elsevier: Amsterdam, (1989).
3. M. Zulauf, H.F. Eicke, *J. Phys. Chem.*, 83, 480 (1979).
4. J.H. Fendler, *Chem. Rev.* 87, 877 (1987).
5. P. Mukerjee, A.J. Ray, *Phys. Chem.* 70, 2144 (1966).
6. P. Mukerjee, J.R. Cardinal, N.R. Desai, *Micellization, Solubilization and Microemulsions*; K.L. Mittal, Ed.; Plenum Press: New York, (1977).
7. J. Funasaki, *J. Phys. Chem.* 83, 1998 (1979).
8. M.S. Fernandez, P. Fromherz, *J. Phys. Chem.* 83, 1755 (1979).
9. M.F. Ruasse, Y. Blagoeva, P. Gray, Book of Abstrats ESOR IVMMBP; The Royal Society of Chemistry Perkin Division: Newcastle upon Tyne (1993).
10. M. Wong, J.K. Thomas, M.J. Graszel, *Am. Chem. Soc.* 98, 2391 (1976).
11. G. Bakale, G. Beck, J.K. Thomas, *J. Phys. Chem.* 85, 1062 (1981).
12. P.E. Zinsli, *J. Phys. Chem.* 83, 3223 (1979).
13. E. Keh, B. Valeur, *J. Colloid Interface Sci.* 79, 465 (1981).
14. J.K. Thomas, *Chem. ReV.* 80, 281 (1980).
15. W. Marcel, *Proteins: Struct. Funct. Genet.* 1, 4 (1986).
16. E.A. Lissi, D. Engel, *Langmuir* 8, 452 (1992).
17. Y.L. Khmelnitsky, I.N. Neverova, V.I. Polyakov, V.Y. Grinberg, A.V. Levashov, K. Martinek, *Eur. J. Biochem.* 190, 155 (1990).
18. R.M.D. Verhaert, R. Hilhorst, *Recl. TraV. Chm. Pays-Bas* 110, 236 (1991).
19. F.M. Menger, J.A. Donohue, R.F.J. Williams, *Am. Chem. Soc.* 95, 286 (1973).
20. J.M. Blandamer, J. Burges, J. B.J. Clark, *Chem. Soc., Chem. Commun.* 659 (1983).
21. R. Da-Rocha-Pereira, D. Zanette, F. Nome, *J. Phys. Chem.* 94, 356 (1990).
22. S.M. Hubig, M.A.J. Rodgers, *J. Phys. Chem.* 94, 1933 (1990).
23. C.J. O'Coner, E.J. Fendler, J.H. Fendler, *J. Am. Chem. Soc.* 95, 600 (1973).

24. O.A. El-Seoud, M.J. Da Silva, L.P. Barbur, A. Martine, *J. Chem. Soc., Perkin Trans.* 2, 331 (1987).
25. C.A. Martin, P.M. McCrann, M.D. Ward, G.H. Angelos, D.A. Jaeger, *J. Org. Chem.* 49, 4392 (1984).
26. M. Valiente, E. Rodenas, *J. Phys. Chem.* 95, 3368 (1991).
27. C. Izquierdo, J. Casado, *J. Phys. Chem.* 95, 6001 (1991).
28. J.R. Leis, J.C. Mejuto, M.E. Pena, *Langmuir* 9, 889 (1993).
29. C. Gomez-Herrera, M.M. Graciana, E. Munoz, M.L. Moya, F. Sanchez, *J. Colloid Int. Sci.* 141, 454 (1991).
30. M.L. Moya, C. Izquierdo, Casado, *Int. J. Chem. Kin.* 24, 19 (1992).
31. M.L. Moya, C. Izquierdo, Casado, *J. Phys. Chem.*, 95, 6001 (1991).
32. P. Lopez, A. Rodriguez, C. Gomez-Herrera, F. Sanchez, M.L. Moya, *J. Chem. Soc. Faraday Trans.* 88, 2701 (1992).
33. M.J. Blandamer, B. Clark, J. Burgess, J.W.M. Scott, *J. Chem. Soc. Faraday Trans.* 1. 80, 1651 (1984).
34. R.A. Mackay, *Adv. Colloid Interface Sci.* 15, 131 (1981).
35. C. Minero, E. Pramauro, E. Pelizzetti, *Langmuir* 4, 101 (1988).
36. E. Rodenas, E. Perez-Benito, *J. Phys. Chem.* 95(23), 9496 (1991).
37. R. Da Rocha Pereira, D. Zanette, F. Nome, *J. Phys. Chem.* 94, 356 (1990).
38. Y.L. Khmeltnitsky, I.N. Polyakov, V.Y. Grinberg, A.V. Levashov, K. Martinek, *Eur. J. Biochem.* 190, 155 (1986).
39. H. Stamatis, A. Xenakis, U. Menge, F.N. Kolisi, 42, 931 (1993).
40. H. Stamatis, A. Xenakis, E. Dimitriadis, U. Menge, F.N. Kolisi, *Biotechnol. Bioeng.* 45, 33 (1995).
41. F.N. Lolisis, T.P. Valis, A. Xenakis, *Ann. New York Acad. Sci.* 613, 674 (1990).
42. L. Garcia-Rio, J.R. Leis, M.E. Pena, E. Iglesias, *J. Phys. Chem.* 97, 3437 (1993).
43. L. Garcia-Rio, J.R. Leis, J.C. Mejuto *J. Phys. Chem.* 100,10981 (1996).
44. L. Garcia-Rio, J.C. Mejuto, M. Perez-Lorenzo, *J. Chem.* 28, 988 (2004).
45. L. Garcia-Rio, J.C. Mejuto, M. Perez-Lorenzo, *Chem. Eur. J.* 11, 4361 (2005).
46. E. Fernandez, L. Garcia-Rio, J.C. Mejuto, M. Perez-Lorenzo, *New J. Chem.* 29, 1594 (2005).
47. L. Garcia-Ro, J.C. Mejuto, M. Perez-Lorenzo, *Colloid Surface Sci.* A. 115, 270 (2005).
48. L. Garcia-Rio, J.C. Mejuto, M. Perez-Lorenzo, M. *J. Colloid Interface Sci.* 301, 624 (2006).
49. L. Garcia-Rio, J.C. Mejuto, M. Perez-Lorenzo, *J. Phys. Chem.* B 110, 812 (2006).
50. L. Garcia-Rio, J.R. Leis, J.A. Moreira, *J. Am. Chem. Soc.* 122, 10325 (2000).
51. L. Garcia-Rio, J.R. Leis, J.C. Mejuto, *Langmuir* 19, 3190 (2003).
52. L. Garcia-Rio, P. Herves, J.C. Mejuto, J. Perez-Juste, P. Rodriguez-Dafonte, *Ind. Eng. Chem. Res.* 42, 5450 (2003).
53. P.J. Stilbs, *Colloid Interface Sci.* 87, 385 (1982).

54. P.D.I. Fletcher, A.M. Howe, B.H. Robinson, *J. Chem. Soc., Faraday Trans.* 1 83, 985 (1987).
55. B.H. Robinson, D.C. Steytler, R.D. Tack, *J. Chem. Soc., Faraday Trans.* 1 75, 481 (1979).
56. S. Clark, P.D.I. Fletcher, X. Ye, *Langmuir* 6, 1301 (1990).
57. T.K. Jain, G. Cassin, J.P. Badlali, M.P. Pileni, *Langmuir* 12, 2408 (1996).
58. J.R. Leis, J.C. Mejuto, M.E. Pena, *Langmuir* 9, 889 (1993).
59. L. Garcia-Rio, J.R. Leis, E. Iglesias, Submitted for publication.
60. L. Garcia-Rio, E. Iglesias, J.R. Leis, M.E. Pena, A. Ros, *J. Chem. Soc., Perkin Trans.* 2 29 (1993).
61. C. Bravo, P. Hervezs, J.R. Leis, M.E. Pena, *J. Phys. Chem.* 94, 8816 (1990).
62. C. Bravo, J.R. Leis, M.E. Pena, *J. Phys. Chem.* 96, 1957 (1992).
63. L. Garcia-Rio, J.R. Leis, J.C. Mejuto, J. Perez-Juste, *J. Phys. Chem.* B 101, 7383 (1997).
64. P. Hervezs, J.R. Leis, J.C. Mejuto, Perez-Juste, *Langmuir* 13, 6633 (1997).
65. C.A.J. Bunton, *Phys. Org. Chem.* 18, 115 (2005).
66. C.A. Bunton, A. Garreffa, R. Germani, G. Onori, A. Santucci, G. Savelli, *Prog. Colloid Polym. Sci.* 118, 103 (2001).
67. R.A. Mackay, C. Hermansky, *J. Phys. Chem.* 85, 739 (1981).
68. E. Iglesias, J.R. Leis, M.E. Pena, *Langmuir* 10, 662 (1994).
69. W.A. Noyes, *Org. Synth. Coll.* 2, 108 (1943).
70. R.N. Haszecdine, P.J.H. Mattinson, *J. Chem. Soc.* 4172 (1955).
71. D.W. Marquardt, *J. Soc. Ind. Appl. Math.* 11, 431 (1963).
72. L. Garcia-Rio, E. Iglesias, J.R. Leis, M.E. Pena, D.L.H. Williams, *J. Chem. Soc., Perkin Trans.* 2 1673 (1992).
73. M. Kobayashi, *Chem. Lett.* 37 (1972).
74. L.S. Romsted, Surfactants in Solution; B. Lindman, K.L. Mittal, Eds.; Plenum Press: New York, Vol. 2, p 1015 (1984).
75. E. Calle, J. Casado, J.L. Cinos, J.F. Garcia Mateos, M.J. Tostado, *Chem. Soc., Perkin Trans.* 2 987 (1992).
76. S. Oae, N. Asai, K. Fujimori, *Tetrahedrom Lett.* 24, 2103 (1977).
77. J. Casado, A. Castro, F.M. Lorenzo, F. Meijide, *Monatsh. Chem.* 117, 335 (1986).
78. J. Albery, Proton Transfer Reactions; Caldin, E., Gold, V., Eds.; Chapman & Hall: London (1973).
79. L. Garcia-Rio, E. Igleisas, J.R. M.E. Pena, *Langmuir* 9, 1263 (1993).
80. L. Garcia-Rio, P. Herves, J.C. Mejuto, M. Parajo, Perez-Juste, *J. Chem. Res.* 716 (1998).
81. L. Garcia-Rio, E. Igleisas, A. Fernandez, J.R. Leis, *Langmuir* 11, 1917 (1995).
82. M. Belete'te, G.J. Durocher, *Colloid Interface Sci.* 134, 289 (1990).
83. P.M. Pileni, B. Hickel, C. Ferradini, Pucheault, *J. Chem. Phys. Lett.* 92, 308 (1982).
84. M. Wing, M. Graszel, J.K. Thomas, *Chem. Phys. Lett.* 92, 329 (1975).
85. V. Calvo-Perez, G.S. Beddard, J.H. Fendler, *J. Phys. Chem.* 85, 3216 (1981).
86. G. Bakale, G. Beck, J.K. Thomas, *J. Phys. Chem.* 85, 1062 (1981).

87. M. Wong, J.K. Thomas, T. Nowak, *J. Am. Chem. Soc.* 99, 4730 (1977).
88. L. Garcia-Rio, E. Iglesias, J.R. Leis, *J. Phys. Chem.* 99, 12318 (1995).
89. J.J. Silber, M.A. Biasutti, E. Abuin, E. Lissi, *Adv. Colloid Interface Sci.* 82, 189 (1999).
90. N.M. Correa, D.H. Zorzan, L. D'Anteo, E. Lasta, M. Chiarini, G. Cerichelli, *J. Org. Chem.* 69, 8224 (2004).
91. N.M. Correa, D.H. Zorzan, L. D'Anteo, E. Lasta, M. Chiarini, G. Cerichelli, *Org. Chem.* 69, 8231 (2004).
92. G. Onori, A. Santucci, *J. Phys. Chem.* 97, 5430 (1993).
93. M.B. Temsamani, M. Maeck, I. El Hassani, H.D. Hurwith, *J. Phys. Chem.* B 102, 3335 (1998).
94. D. Fioreto, M. Freda, S. Mannaioli, G. Onori, A. Santucci, *J. Phys. Chem.* B 103, 2631 (1999).
95. T.K. De, A. Maitra, *Adv. Colloid Interface Sci.* 59, 95 (1995).
96. K. Kurihara, J. Kizling, P. Stenius, J.H. Fendler, *J. Am. Chem. Soc.* 105, 2574 (1983).
97. C. Petit, P. Lixon, M.P. Pileni, *J. Phys. Chem.* 97, 12974 (1993).
98. A.R. Kortan, R. Hull, R.L. Opila, M.G. Bawendi, M.L. Steigerwald, P.J. Carroll, L.E. Brus, *J. Am. Chem. Soc.* 112, 1327 (1990).
99. S.K. Haram, A.R. Mahadeshwar, S.G. Dixit, *J. Phys. Chem.* 100, 5868 (1996).
100. P. Ayyub, A.N. Maitra, D.O. Shah, *Phys.* C 168, 571 (1990).
101. E. Joselevich, I. Willner, *J. Phys. Chem.* 98, 7628 (1994).
102. C.L. Chang, H.S. Fogler, *Langmuir* 13, 3295 (1997).
103. E. Fernandez, L. Garcia-Rio, J.C. Mejuto, M. Perez-Lorenzo, *Coloids Surf.* A 295, 248 (2007).
104. E. Fernandez, L. Garcia-Rio, J.C. Mejuto, M. Perez-Lorenzo, *J. Chem. Res.* 1, 52 (2006)

Chapter 21

Electric Field Induced Gel Formation and Fracture in Layers of Laponite

Suparna Sinha and Sujata Tarafdar

Condensed Matter Physics Research Centre,
Department of Physics, Jadavpur University,
Kolkata 700032, India

The synthetic clay Laponite forms a gel, when a dilute solution with concentration of about 2% is left to age. The structure of the gel and interactions between particles leading to its formation are yet unknown. We observe that an electric current flowing through the sol leads to very rapid gel formation at the anode. The gels formed with and without field may not be identical in structure and properties. It is also observed that the gel forms radial cracks, when it dries in an electric field, in absence of the field, the cracks have a typical disordered pattern, without radial symmetry

21.1. Introduction

The synthetic clay Laponite[1,2] is used in a variety of consumer products, such as paints, cosmetics, ointments as well as in industry e.g. oil drilling. This has led to wide interest resulting in a large number of publications on different properties of Laponite.[3-5] Laponite forms a gel in water at very low concentrations, hence its usefulness as a thickening agent, its other attractive features are : it is nontoxic and the gel is transparent. Laponite consists of flat disc-like nano-sized particles. It has the chemical formula $Si_8Mg_{5.45}Li_{0.4}H4O_{24}Na_{0.7}$ and the discs have approximately 20-30 nm diameter and $\sim$ 1 nm thickness.[1-2] Compared to natural smectite clays, the particles are reasonably mono-disperse and much smaller in size. Hence, Laponite serves as a convenient model for experiments on clay, without the troublesome features of heterogeneity and polydispersity of natural clay.

The particles have a surface charge in aqueous solution, the large flat surfaces of the discs being negatively charged and the thin rims charged positively or negatively depending on the pH and ionic strength. There is a net excess of negative charge. The distribution has zero dipole moment but a nonzero quadrupole moment.[6]

The most important and useful property of Laponite is its gel formation ability, so it is necessary to understand the structure and formation kinetics of the gel. There is an ongoing debate about the interactions between the flat clay particles and how they aggregate, in fact whether the aggregate should be called a gel at all or a glass.[7-8] According to Mourchid and Levitz[9] at high pH about 10, a repulsive glass is formed in preference to a 'house of cards' with edge-to-face attraction at low pH.

The present work shows that gelation in Laponite can happen either by allowing a solution (of concentration 0.0225 gm /ml and pH 9.5) to stand for at least 4 hours (at 70% humidity) or by placing the sol in an electric field, when a gel forms around the anode in about 20 minutes. We study gelation in the presence of an electric field and explore the possibility that this gel has a structure different from that formed without the field.

Another striking effect of an electric field on Laponite gels is the pattern of crack formation. On desiccation, layers of Laponite gel crack, forming typical patterns depending on the substrate and drying conditions.[10] The crack patterns are different when a non-polar solvent, like methanol dries.[11] When the gel is allowed to dry in an electric field, the crack patterns follow the line of force of the field, producing a radial pattern in a field with cylindrical symmetry.

21.2. The Experimental Setup

21.2.1. *Gelation in Electric Field*

Laponite RD from Rockwood Additives (UK) was mixed in distilled water at concentration 2.25% w/w and stirred in a magnetic stirrer for 30 minutes. At higher concentration an in-homogenous lumpy gel forms within 15 minutes of stirring, so we avoid concentrations higher than 2.25%. The transparent sol prepared, which had a pH 9.5, was placed in a Petri-dish of diameter 10 cm. The anode, a graphite rod was placed at the center and the cathode was made in a cylindrical form with aluminum foil. The cathode was fitted outside the periphery of the Petri-dish and was electrically connected to the sol. The desired voltage was applied through a power supply for a period of 20 min. The field was switched off and the gel formed at the anode was collected and weighed. The mass of gel formed in

20 minutes under supply voltage varying from 60V to 240V DC was noted. We find that 20 minutes is the optimum time to get a measurable mass of gel for the lowest voltage.

21.2.2. Fracture of Desiccating Gel

Laponite sol was prepared with the same concentration as in 2.1 and was left standing in a Petri-dish for 4-5 hrs. until gelation. Then electrodes were placed as before and the voltage was applied. After 8-9 hours cracks started to appear. They proceeded in a radial direction from the positive to negative terminal. A similar setup without the field resulted in cracks without the radial pattern. These experiments are discussed elaborately in Mal et al.[11,12] Figure 21.1 shows the radial crack patterns formed in an electric field and Figure 21.2 in a similar set up without the field.

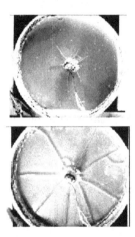

Fig. 21.1. Formation of radial cracks with centre terminal positive, after 24 hours (left) and 72 hours (right).

Figure 21.3. shows results when the center terminal is negative and the periphery positive. Here the cracks proceed from the periphery towards the center. Another setup with a different geometry of the electric field was also studied. Here the electrodes were in the form of two semicircular arcs, placed along the periphery of the Petri-dish. The crack patterns here follow the lines of force of the electric field as shown in Figure 21.4. However, the

Fig. 21.2. A disordered pattern formed in the same set up as in Fig. 21.1, but without the electric field.

layer of Laponite must not be more than 1 mm in thickness to obtain the curved cracks shown.

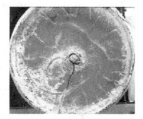

Fig. 21.3. The cracks formed with the center terminal negative, here cracks start at the periphery.

21.3. Results and Discussion

21.3.1. *Gelation Results*

The results of section 21.1 can be summarized as follows — the mass of gel formed in equal intervals of time increases with the voltage. The variation in gelation rate with the effective voltage read in the voltmeter is shown in Figure 21.5. These results can be fitted to a quadratic relation as shown. This may be compared with previous work, where a variation in gelation rate proportional to the square of voltage is obtained.[13]

The gel formation under the action of an electric field is similar to electrophoresis in clays. A number of such studies have been reported.[13-15] We have done the same experiment with an Al anode. Here the gel formed in a very short time. Since Al ions appeared to enter into the electrolyte, it

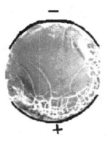

Fig. 21.4. Curved cracks following the lines of force, formed in an electric field, with semicircular electrodes.

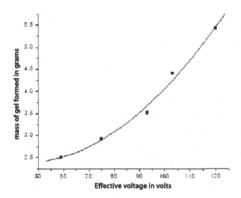

Fig. 21.5. Graph showing variation of gelation with voltage. The mass of gel formed at the anode in 20 min. is plotted for different values of the effective voltage.

was not possible to ascertain whether coagulation was due to these ions or the electric field. To confirm the role of the electric current, fine Al powder was suspended in the sol to look for any gelation in absence of field, but no gelation was observed in this case. With carbon as the anode material this question does not arise.

Considered in conjunction with the question of gel structure in Laponite, it is quite likely that the gel formed in an electric field may be different in structure from the gel formed simply on standing.

The debate regarding the exact structure of the semi-solid laponite 'gel' or 'glass' still continues. It is clear, however, that Laponite shows a rich phase behavior, depending on pH as well as ionic strength (due to addition of salt), a phase diagram is reported by Levitz et al.[16] It is generally agreed[17] agreed that for varying pH in the range 7.5 − 10, configuration of adjacent platelets changes from edge-face to face-face as pH is increased. The flat circular surfaces of the discs always have negative charge, but the rim has positive or negative charge at low and high pH respectively.

Earlier literature proposed the 'house of cards' structure for Laponite gel. It was suggested[18] that the flat platelets with negative charge on the surface and positive charge on the rims, attract to form a 'T' type configuration, these again aggregate to form the 'house of cards', as shown in references.[1,13] So the house of cards, if observed should be at lower pH, i.e. at lower clay concentration. However, the argument against this structure is that at the low concentration, where gel forms, platelets are too far apart for the 'house of cards' arrangement suggested.[16] SEM studies indicate that the particles do not touch.[11] In such case, the mechanism for gel formation is suggested to be through repulsive interaction between the particles due to their similar surface charge.[13] Bonn et al.[15] propose that a 'repulsive glass' is the more appropriate description for the semi-solid Laponite.

In view of so much conflicting evidence, we can only offer a conjecture regarding the gel structure, based on our observations. As the pH measured for our sol $\sim$ 9.5, is on the higher side, a repulsive glass structure is likely to be expected. We suggest here that the gels formed on standing belong to this category, while the gel formed in the field may have a 'house of cards' structure. The presence of the external field drives the particles together overcoming their mutual repulsion. In close proximity, the plates may thus prefer to take the energetically favorable 'T' configuration relative to each other. Since the interaction potential between laponite platelets is attractive at short range, but repulsive at longer range, as suggested by Li et al.[19] A Monte Carlo simulation by Dijkstra et al.[20] shows aggregation of T-shaped units.

Earlier SEM[10] photos reveal linear arrangements of nano-discs when there is no external field. It is suggested to be due to scratches or linear defects on the substrate. However an alternative explanation for the observed linear patterns may be that this ordering is due to repulsion between adjacent discs. The SEM micrographs for the samples near the anode in a radial electric field,[10] showed on the other hand, wormlike aggregates. These nearly closed loop-like structures may be the signature of edge-face attracting configurations. Further structural studies using different probes are necessary to establish this conjecture.

21.3.2. Crack formation

The radial crack patterns for different field geometries are shown in Figures 21.1, 21.3 and 21.4. The cracks clearly proceed from the positive electrode towards the negative. With the positive electrode at the periphery, the field is weaker here and the gel dries out before the cracks reach the cathode (Fig 21.3).

On connecting the power supply, the temperature shows a rise of several degrees at the anode just after application of the field, however the temperature returns to normal after about 30 min. The initial appearance of cracks (Fig. 21.1) occurs much after that. So we presume that crack formation is due to the electric field distribution rather than Joule heating.

The formation of hierarchical cracks can be understood from theory and simulations,[21] but the role of the electric field in producing directional cracks is not yet clear. Probably the ions in the electrolyte look for low energy paths from anode to cathode and open up existing weak spots or micro-cracks in the right direction. Once a path is opened up, it becomes easier for other ions to follow, rather than start new cracks.

Acknowledgments

Authors sincerely thank Professor S P Moulik for valuable suggestions. Discussion and comments from participants at the International Symposium on Recent Trends in Surface and Colloid Science is of great help in improving this study. Authors are grateful to DST, Govt. of India for providing a research grant.

References

1. *Laponite Technical Information* (Rockwood Additives Ltd.).
2. *Southern Clay Products*, Inc. Product brochure.
3. M. Kroon, W. L. Vos, G. H. Wegdam, *Phys. Rev.* E, **57**, 1962 (1998).
4. R. D. Leonardo, F. Ianni, G. Ruocco, *Phys. Rev* E, **71**, 011505 (2005).
5. D. Bonn, S. Tanase, B. Abou, H. Tanaka, J. Meunier, *Phys. Rev. Lett.*, **89**, 015701 (2002).
6. M. Dijkstra, J.P. Hansen, P. A. Madden, *Phys. Rev.* E., **55**, 3044 (1997).
7. D. Bonn, H. Kellay, H. Tanaka, G. Wegdam, J. Meunier, *Langmuir*, **15**, 7534 (1999).
8. Xi Angrong, W. Lianze, W. Cheng, H. Fenglei, *Aerosol Science*, **37**, 1370 (2006).
9. A. Mourchid, P.Levitz, *Phys.Rev.* E., **57**, 5, 4887 (1998).
10. D. Mal, S. Sinha, T. R. Middya, S. Tarafdar, Appl. Clay Sci. in press doi:1016/j.clay.2007.05.005.

11. D. Mal, S. Sinha, T. Dutta, S. Mitra, S. Tarafdar, *J. Phys. Soc. Jpn*, **76**, 014801 (2007).
12. D. Mal, S. Sinha, T. Dutta, S. Mitra, S. Tarafdar, *Fractals* **14** 283, (2007).
13. W. D. Ristenpart, I. A. Aksay, D. A. Saville, *Phys. Rev. E.*, **69**, 021405 (2004).
14. F. F. Reuss, *Mem. Soc. Imp. Natur. Moscow*, **2**, 327 (1809).
15. J. Janca, F. Checot, N. Gospodinova, S. Touzain, M. Spirkova, *J. Colloid Interface Sci.*, **229**, 423 (2000).
16. A. Mourchid, A. Delville, J. Lambard, E. Lecolier, P. Levitz, *Langmuir*, **11**, 1942 (1995).
17. J M. Saunders, J. W. Goodwin, R. M. Richardson, B. Vincent, *J. Phys. Chem. B.*, **103** 9211, (1999).
18. S. Kutter, J.P. Hansen, M. Sprik, E. Boek, *J. Chem. Phys.*, **112**, 311 (2000).
19. Li Li, L. Harnau, S. Rosenfeldt, M. Ballauff, *Phys. Rev. E.*, **72**, 051504 (2005).
20. M Dijkstra, J.P. Hansen, P.A. Madden *Phys. Rev. Lett.*, **75** 2236 (1995).
21. S. Sadhukhan, S. Roy Majumder, D. Mal, T. Dutta, S. Tarafdar, *J. Phys. Cond. Mat.* **19** 356206 (2007).

Chapter 22

Temperature Dependent Structural Insignia of Cinnamic Acid

B. Nandi Ganguly, Nagendra Nath Mondal

Saha Institute of Nuclear Physics,
1/AF Bidhannagar, India

S.K. Bandopadhyay, Pintu Sen

Variable Energy Cyclotron Centre,
1/AF Bidhannagar, India

Oil of cinnamon contains cinnamic acid (phenyl acrylic acid) as the essential constituent, which is known for its medicinal values. Main structural feature of cinnamic acid is strong hydrogen bonding between the carboxylic groups. The dimers are also interconnected by $CH\ldots O$ intermolecular hydrogen bonds, with $C-H$ unit mostly originating from aromatic ring. These bonds keep the dimers and higher oligomers together and thus the compound exists in a crystalline form. Since the hydrogen bonded structure is susceptible to rupture at temperature changes above the ambient, a preliminary observation has been made using positron annihilation spectroscopy, varying temperature as a parameter to investigate the changes of molecular structure, in the crystalline phase. These subtle changes were investigated independently through the kinetic results of differential scanning calorimetry, which sheds light on the activation energy of the molecular state at different temperature regimes.

22.1. Introduction

Cinnamic acid (phenyl acrylic acid) is an essential constituent of cinnamon extract (oil), well known for its medicinal values, and most importantly is a crucial pathway for useful pharmaceuticals (for high blood pressure and stroke prevention) known as coumarin or oxy-cinnamic acid (a derivative of cinnamic acid). As a flavonoid, cinnamic acid also exerts antitumor activity against human colon cancer cells. Its main structural feature[1] is the strong hydrogen bonding between the carboxyl groups, the dimmers are interconnected by $C-H\ldots O$ intermolecular hydrogen bonds. In most cases the C atom of the $C-H$ unit is a member of the aromatic ring. Intra molecular (olefinic) $C-H\ldots O$ are frequent, fixing the synperiplannar and antiperiplannar $C=C-C=O$ conformations in the same abundance. These bonds keep the dimmers and higher oligomers packed together and thus the compound exists in a crystalline form (which is evidenced from the crystallographic data). Since the hydrogen bonded structure is susceptible to changes with variation in temperature, the present work has been aimed to study the perturbation effect due to weaker hydrogen bond interaction that helps to maintain the closed packed crystalline structure. The peculiarities as observed through various experimental probes such as: (i) temperature dependent positron annihilation spectroscopic (PAS) investigations both with Doppler broadening (DB) and life time (LT) experiments from (265-340 K) depicting the subtleties of hydrogen bonding interaction within the molecular conformation and ii) kinetic results of differential scanning calorimetry (DSC), which sheds light on the activation energy of the molecular sate at different temperature regimes (at slightly higher than the ambient) has been be presented. The experimental results obtained so far, primarily points to the crucial role of hydrogen bonding interaction, and stability in the molecular system which has been manifested through temperature dependent investigation in the experimental sections mentioned above.

22.2. Experiments

22.2.1. *Positron Annihilation Spectroscopy*

A carrier free positron source (^{22}Na in the form of NaCl) of strength ~ 5 (μ Ci in a Ni foil was embedded in the cinnamic acid powder in a glass vial that was further degassed and the temperature was controlled through a Lakeshore temperature controller. The positron annihilation life times were measured with a slow fast coincidence system using BaF_2 scintillators coupled to XP2020Q photo multipliers; the details of this kind of measurement is found elsewhere.[2] The Doppler Broadening of the annihilation radiation (DB) around 511 keV was measured using HPGe detector whose intrinsic resolution is around 1.1 keV for 569 keV gamma line, for ^{207}Bi and ~ 1.8 keV for ^{60}Co 1.33 MeV line.

22.2.2. Differential Scanning Calorimetry (DSC)

The kinetic measurements were performed at different heating rate ($K\,min^{-1}$) for the sample (Ozawa method)[3] using Seiko Instruments Inc. DSC 6200. Activation energy of the system was calculated over the different temperature ranges between 298 − 343 K which actually showed a varied nature.

22.3. Results and Discussions

The initial results of the DB analysis with respect to the 'S' parameter[2] exhibited almost a periodic variation as shown in Fig. 22.1, which can be ascribed to perturbation of the molecular structure on the basis of the changes in electronic momentum distribution. This is also in concurrence with the observed changes in the full width at half maximum (fwhm) of the Gaussian distribution of the DB annihilation data at those temperatures.

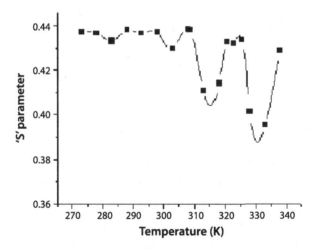

Fig. 22.1. S parameter results of the DB annihilation radiation around 511 keV line for the temperature variation in the system. The experimental error is of the point size, typically ±0.0002.

In the analysis of the life time spectroscopy results we observed two components, τ_1 and τ_2 with the respective associated intensities I_1 and I_2. The faster annihilation rate $\lambda_1 = 1/\tau_1$ is ascribed to bulk annihilation with the corresponding intensity I_1 (the bulk rate changes only slightly towards higher temperature with corresponding increase in intensity parameter). The other component (τ_2, I_2) senses the different electronic density regions in the molecular medium, as given below.

Furthermore, the results from life time spectroscopy consistently depict definite repetitive changes more or less around the same temperature regimes as compared from the Figs. 22.2 and 22.3. This necessitates an explanation from the structural point of view, as the life time results sense the electronic micro environment of the molecular structure relative to the temperature in the system. Results shown in Fig. 22.2, exhibit a periodic change (appearance of the maxima and minima) after an interval of temperature $\sim$ 15K.

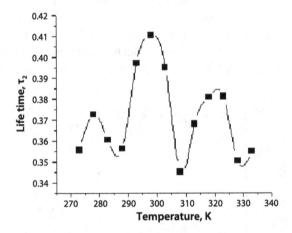

Fig. 22.2. The life time τ_2 results (error point size, typically, $\pm.0005$ns) on the temperature variation of cinnamic acid

From the relevant literature studies at least two types of molecular hydrogen bonding are pertinent in cinnamic acid, viz., i) strong hydrogen bonding between the carboxylic groups ($\sim 25 - 30$ kJ mole^{-1}) which holds the dimmer and is responsible for the Centro-symmetric structure of the assembly, the other is ii) $C - H \ldots O$ bonding of lower strength (~ 5 to 10 kJ mole^{-1}) that occur in multimers within solid state and these were confirmed through IR spectroscopy.[4] Configurational energy change in the molecular system here is relevant, which is a likely manifestation of the hydrogen bond dissociation. But repetitions of pattern in the experimental curves namely Figs. 22.2 and 22.3, would also suggest multiple traps for positron at definite temperature intervals upon its propagation into the molecular medium (with in the minute grains). These trapping sites could originate from a layered arrangement[1,4] of the molecular structure with the strongly electronegative polar group existing in a layer followed by the non polar zone (as in the lamellar structure of the amphiphiles). This structural arrangement has been reported in the spectroscopic data.[4]

In Fig. 22.3, we see the maxima in intensity $I_2\%$, when the annihilation rate is faster (annihilation rate, $(\lambda = 1/\tau)$, if one compares Fig. 22.2 with Fig. 22.3. and *vice-versa*. One could be inclined to interpret that rapture of hydrogen bonds sets the surrounding electron density free for the positrons to annihilate (if it is assumed that no other kind of interactions exists) as the temperature increases, but this alone cannot explain all the changes observed here.

But, from the positron annihilation spectroscopy results, one can assess the local rearrangement in the molecular (oligomer) configuration, which would mean a higher annihilation rate from the higher electro negative site (i.e. the head groups in the layered structure) and correspondingly a relaxed annihilation rate and lower intensity percentage from the non polar zone as positron propagates in the medium through diffusion. One can also notice in Fig. 22.3., that the $I_2\%$ falls at a steady rate as the temperature of the sample increase. This is due to the increase in the annihilation rate of the bulk component with the increase in temperature of the sample.

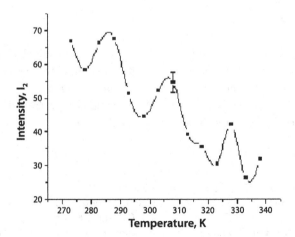

Fig. 22.3. Relative intensity I_2 percentage associated with the life time component.

From our astute interest in the above temperature induced molecular process, we have performed a DSC kinetic study of the system at various heating rates and calculated the activation energy changes at different temperature regimes (Ozawa method). The results are given in Table 22.1.

Table 22.1. Activation energy of the molecular system of cinnamic acid at different temperature regimes.

Temperature range	303-313 K	323-333 K
Activation Energy (-E) kJ mole^{-1}	-53.87	-115.15

In a stacked arrangement of hydrogen bonded system, some energy is expended in breaking the network of these bonds, which would result in different transient states (activated state)[5] and $(-E)$ should increase with temperature. The activation energies in the range 303–313 K and 323–333 K correspond to the changes related to $C - H \ldots O$ bond of lower strength of hydrogen bonding (5 – 10 kj mole^{-1}, either involved in olefinic or aromatic hydrogen) which could be corroborated with vibrational energy = $3/2\ kT$ at the temperature range stated (298 – 343K), k is the Boltzmann constant. Thus, this observation confirms the notion of changes in molecular electronic environment, through heat content of the system, but no gross critical transition leading to structural change has been observed.

22.4. Conclusion

From our preliminary study on the temperature dependent characteristics of cinnamic acid, it has been found that although there could be some definite changes in electronic environment via hydrogen bonding network in the molecular arrangement, however there is no critical transition observed so far in the temperature range $(265 - 340K)$ studied.

References

1. I. Palinko; *Acta Cryst*, **B55**, 216 (1999).
2. Dhanadeep Dutta, *Positron Annihilation Study on Molecular Substances with Special Reference to Porous Materials*. (Ph. D. 2005)
3. Manual and Hand book for *DSC measurement Seiko Instruments Inc.*
4. I Palinko and J.T. Kiss, *Microkim. Acta. Supp.* **1**, (14), 253 (1997).
5. S Glasstone, *Text Book of Physical Chemistry* (Tata Macmillan, New York, p. 499 (1977).